Algorithmic Algebraic Combinatorics
and Gröbner Bases

Mikhail Klin · Gareth A. Jones · Aleksandar Jurišić ·
Mikhail Muzychuk · Ilia Ponomarenko
Editors

Algorithmic Algebraic Combinatorics and Gröbner Bases

Springer RISC

Mikhail Klin
Ben-Gurion University of the Negev
Department of Mathematics
POB 653
84633 Beer Sheva
Israel
klin@cs.bgu.ac.il

Gareth A. Jones
School of Mathematics
University of Southampton
Southampton SO17 1BJ
United Kingdom
G.A.Jones@soton.ac.uk

Aleksandar Jurišić
Laboratory for Cryptography and Computer Security
Faculty of Computer Science
Tržaška cesta 25
1000 Ljubljana
Slovenia
ajurisic@valjhun.fmf.uni-lj.si

Mikhail Muzychuk
Department of Mathematics and Computer Science
Netanya Academic College
University st. 1
42 365 Netanya
Israel
muzy@netanya.ac.il

Ilia Ponomarenko
Laboratory of Representation Theory and Computational Mathematics
V.A. Steklov Mathematical Institute
Fontanka 27
191023 St. Petersburg
Russia
inp@pdmi.ras.ru

ISBN 978-3-642-42438-0 ISBN 978-3-642-01960-9 (eBook)
DOI 10.1007/978-3-642-01960-9
Springer Heidelberg Dordrecht London New York

Mathematics Subject Classification (2000): 05E30, 13P10, 20B25, 68-04

Cover design: WMX Design GmbH, Heidelberg

Printed on acid-free paper

springer.com

Preface

In 2006 a special semester on Gröbner bases and related methods was organized by RICAM and RISC, directed by Bruno Buchberger and Heinz Engl. The main focus of the semester were the development of the formal theory of Gröbner bases (briefly GB), the efficient implementation of all algorithms related to this theory, and the promotion of recent and new applications of GB.

The workshop D1 "Gröbner bases in cryptography, coding theory and algebraic combinatorics", Linz, May 1–6, 2006 (chairmen M. Klin, L. Perret, M. Sala) was one of the main ingredients of the semester. The last two days of this workshop, devoted to combinatorics, made it possible to bring together experts in algorithmic problems related to coherent configurations and association schemes with a community of people working in the area of GB. Each side was interested in understanding the computational problems and current algorithmic possibilities of the other, with a particular objective of introducing the practical use of GB in algebraic combinatorics.

Materials (mainly slides of lectures and posters) available from the site http://www.ricam.oeaw.ac.at/specsem/srs/groeb/schedule_D1.html provide a helpful and vivid picture of the successful exchange of scientific information during the workshop D1.

As a follow-up to the special semester, 10 volumes of proceedings are being published by different publishers. The current collection of papers reflects diverse investigations in the area of algebraic combinatorics (with or without explicit use of GB), but with a definite emphasis on algorithmic approaches.

Ever since its initial inception in 1965 by Buchberger, the algorithmic theory of GB has been computer-oriented, and nowadays this method is available in all major software systems. The theory of GB has fundamental computational significance in algebraic geometry, and it is also applied in many other areas, including coding theory, cryptography, statistics and optimization. In the last few decades computational methods have proved to be highly effective in algebraic combinatorics, particularly in the search for new distance regular and strongly regular graphs and partial geometries, the determination of automorphism groups of association schemes, and identification of graphs. It is now very timely to combine these efforts in the two areas of GB and algebraic combinatorics, with a special emphasis on algorithmic aspects.

A common feature of many problems in combinatorics and geometry is the existence of huge search spaces. Elements of these spaces are parametrized by integer vectors, subject to suitable feasibility conditions, usually expressed in terms of certain equations and inequalities. Typical research tasks for a prescribed feasible set of parameters consist of proving the existence of a desired object, classifying all objects up to isomorphism, and investigating the symmetry properties of discovered objects. Techniques developed within GB provide attractive possibilities for all these types of investigation in combinatorics and geometry. This collection aims to promote the interplay between algorithmic aspects of GB and its existing and possible new combinatorial applications, starting with the immediate efficient use of GB and ending with the consideration of algorithmic approaches which are currently based on alternatives to GB.

Part A of this book consists of five tutorials prepared by participants of the workshop D1. Three of these tutorials, written by members of the same scientific group, introduce the reader to the methodology of computer-aided investigation of coherent configurations, fusion association schemes and related combinatorial structures, with a special emphasis on the use of the computer algebra system GAP.

Sections 1 and 2 of the tutorial T2 by Klin, Reichard and Woldar may serve as a brief introduction to the whole scope of those concepts from algebraic combinatorics which are considered in many other contributions in this book. The main body of T2 describes a new class of Siamese combinatorial objects which appear through the interplay of color graphs, association schemes, Steiner designs and generalized quadrangles. Tutorial T1 by Heinze and Klin aims to present natural links between diverse objects from algebra, geometry and graph theory such as loops, Latin squares, nets, association schemes, strongly regular graphs and partial difference sets. Tutorial T4 by Pech and Reichard deals with the problem of enumeration of orbits of a permutation group acting on subsets. It describes an efficient implementation of an algorithm for the solution of this problem, based on a depth-first approach using the techniques of dynamic programming.

Tutorial T3 by Leonard introduces the reader to innovative techniques applying GB in the area of association schemes. We are pleased to mention that this approach was conceived at Linz, in the course of the workshop D1. Together with the above-mentioned texts, this paper provides a vivid picture of computational and theoretic aspects of the theory of association schemes. Special themes of the tutorial T3 are flag algebras of generalized quadrangles and Steiner designs, their finitely-presented form, and the explicit enumeration of all fusion association schemes via the use of the Buchberger algorithm. Codes in MAGMA, as well as some computational results, greatly strengthen the role of T3 as a striking pattern for future new applications of the suggested techniques.

Tutorial T5 by Peretz is concerned with the Jacobian conjecture, a famous and ambitious problem which has remained open since 1939. Posed originally in the language of classical analysis and algebraic geometry, it nowadays has very deep links with algebra. A computational approach to the 2-dimensional version of this conjecture based on the use of GB is presented in the framework of diverse reformulations of the problem and its interesting links with recent activities of many mathematicians. It brings the reader to the forefront of innovative attempts to understand the origins of the difficulty of this conjecture.

The individual tutorials, and indeed the whole of Part A, can play a significant training role for researchers: they provide a relatively self-contained introduction to several flourishing areas of modern combinatorics in the framework of a constructive algorithmic approach, with emphasis on the practical use of modern software. Moreover each tutorial contains new scientific results, generally obtained via the use of the algorithmic tools described here.

Part B consists of seven research papers, many of which also serve an instructive purpose by maintaining the ingredients of a tutorial style.

Paper R1 by Felszeghy and Rónyai deals with several problems from algebra and combinatorics, which are formulated in terms of a set of points in an n-dimensional vector space over a field F. Using techniques of GB and the concept of a Hilbert function of an F-algebra, the authors present the Lex Game method and a few of its striking applications, such as new proofs of Garsia's generalization of the fundamental theorem on symmetric polynomials, and of the famous rank formula of R. M. Wilson, and an investigation of set families which do not shatter large sets.

Paper R6 by Moorhouse also covers a wide range of problems, applying methods of algebraic geometry to diverse problems in finite geometry. Techniques of GB appear here in a hidden form, for instance via a computational example using the software Macaulay 2. The study of p-ranks of incidence matrices of geometric structures creates a nice interplay with the methodology of paper R1.

Paper R4 by Kohnert is devoted to innovative computational efforts in the overlap between coding theory and finite geometry. Starting with the classical correspondence between two-weight codes and sets of type (d_1, d_2) in projective spaces, the projective Hjelmslev planes over Galois rings are considered. An extensive computer search leads to new geometric structures and to codes corresponding to them.

Techniques of GB may also be applied in the area of mathematical chemistry – this is the main message communicated in R2 by Gugisch. Following the ideas of A. Dreiding & A. Dress, and of N. S. Zefirov & S. S. Tratch, techniques of oriented matroids are revisited in order to provide adequate tools to model conformations of organic compounds. This requires the efficient enumeration of diverse geometrical objects, in particular partial chirotopes. Corresponding algorithmic implementations are described, which are tested on a number of attractive examples.

Paper R7 by Ziv-Av describes another possible area for the application of GB. The notion of a total graph coherent configuration is introduced and investigated for two classical infinite series of strongly regular graphs – triangular graphs and lattice square graphs. In both cases all fusion association schemes are completely classified. The methodology used has an obvious overlap with tutorial T3, providing a challenge to apply GB in future for the solution of wider classes of similar problems.

The remaining two papers acquaint the reader with different algorithmic approaches to the solution of combinatorial problems.

Paper R3 by Jørgensen deals with non-symmetric association schemes with three classes. A list of feasible parameters for such primitive schemes of order ≤ 100 is calculated. It contains 24 parameter sets, three of which are eliminated in R3 with the aid of computer search techniques. There remain ten open cases, providing a striking challenge for future researchers. Imprimitive schemes are also carefully investigated, and new theoretical results are presented in conjunction with the discovery of four examples on 36 points which are equivalent to 'skew' Bush-type Hadamard matrices of order 36 (examples of such matrices were not known before).

Finally, paper R5 by Miyamoto introduces a computational approach to doubly transitive permutation groups via the concept of a superscheme. This approach is investigated for a particular case: 3-designs on $q+2$ points are constructed, which are invariant with respect to the action of the group $PSL(2,q)$ on the projective line, extended by an extra point. The resulting designs are studied for some small values of q with the aid of the computer package GAP.

We are pleased to thank Bruno Buchberger and Peter Paule for their essential support of this project at all stages of its development. The kind patience and attention to detail of Ruth Allewelt (Springer) were crucially helpful. Last but not least we wish to thank the two dozen referees who helped us to ensure hopefully high scientific standards for the entire collection.

Mikhail Klin
Gareth A. Jones
Aleksandar Jurišić
Mikhail Muzychuk
Ilia Ponomarenko
March 2009

Contents

Contributors

Bálint Felszeghy Computer and Automation Institute, Hungarian Academy of Science, Budapest, Hungary; Institute of Mathematics, Budapest University of Technology and Economics, Budapest, Hungary, fbalint@math.bme.hu

Ralf Gugisch Mathematisches Institut, University of Bayreuth, 95440 Bayreuth, Germany, ralf.gugisch@uni-bayreuth.de

Aiso Heinze Department of Mathematics, Leibniz Institute for Science Education, Olshausenstraße 62, 24098 Kiel, Germany, heinze@ipn.uni-kiel.de

Leif K. Jørgensen Department of Mathematical Sciences, Aalborg University, Fr. Bajers Vej 7, 9220 Aalborg, Denmark, leif@math.aau.dk

Mikhail Klin Ben-Gurion University of the Negev, Beer Sheva 84105, Israel, klin@cs.bgu.ac.il

Axel Kohnert Mathematisches Institut, University of Bayreuth, 95440 Bayreuth, Germany, axel.kohnert@uni-bayreuth.de

Douglas A. Leonard Department of Mathematics and Statistics, Auburn University, Auburn, AL, USA, leonada@auburn.edu

Izumi Miyamoto Department of Computer Science and Media Engineering, University of Yamanashi, Kofu 400-8511, Japan, imiyamoto@yamanashi.ac.jp

G. Eric Moorhouse Department of Mathematics, University of Wyoming, Laramie, WY 82071, USA, moorhous@uwyo.edu

Christian Pech Department of Mathematics, Ben-Gurion University of the Negev, Beer Sheva, Israel, pech@cs.bgu.ac.il

Ronen Peretz Department of Mathematics, Ben-Gurion University, Beer Sheva, Israel, ronenp@math.bgu.ac.il

Lajos Rónyai Computer and Automation Institute, Hungarian Academy of Science, Budapest, Hungary; Institute of Mathematics, Budapest University of Technology and Economics, Budapest, Hungary, lajos@ilab.sztaki.hu

Sven Reichard University of Western Australia, Crawley 6009, Western Australia, reichard@maths.uwa.edu.au

Andrew Woldar Villanova University, Villanova, PA 19085, USA, andrew.woldar@villanova.edu

Matan Ziv-Av Department of Mathematics, Ben-Gurion University of the Negev, Beer Sheva, Israel, matan@svgalib.org

Part A

Tutorials

Loops, Latin Squares and Strongly Regular Graphs: An Algorithmic Approach via Algebraic Combinatorics

Aiso Heinze[1] and Mikhail Klin[2]

[1] Department of Mathematics, Leibniz Institute for Science Education, Olshausenstraße 62, 24098 Kiel, Germany. heinze@ipn.uni-kiel.de
[2] Ben-Gurion University of the Negev, Beer Sheva, 84105, Israel. klin@cs.bgu.ac.il

Summary. Using in conjunction computer packages GAP and COCO we establish an efficient algorithmic approach for the investigation of automorphism groups of geometric Latin square graphs. With the aid of this approach an infinite series of proper loops is presented which have a sharply transitive group of collineations. The interest in such loops was expressed by A. Barlotti and K. Strambach.

Key words: Latin square graph, Loop, Net, Transversal design, Regular subgroup, Computer algebra, Partial difference set, Association scheme

1 Introduction

The goal of this tutorial paper is to introduce the reader to links between loops, Latin squares, nets, association schemes, strongly regular graphs and partial difference sets via an algorithmic approach based on the use of computer algebra packages.

Simultaneously we are pursuing a serious scientific objective, introducing an infinite series of proper loops Q_{2p} of order $2p$, p a prime, $p \equiv 3 \pmod 4$, for which the group $G = \mathrm{Aut}(\Gamma)$ contains a regular subgroup of order $4p^2$. Here $\Gamma = SRG(Q)$ is the Latin square graph naturally associated with the loop Q. The problem of the existence of such loops goes back to A. Barlotti and K. Strambach [14].

Originally, we first found Q_6, examining Edward Spence's catalogue [95] of strongly regular graphs that have a relatively small number of vertices. By creating a computer free description of all necessary features of Q_6 and related combinatorial structures, we came to the conclusion that in fact Q_6 is just the first member of an infinite series.

The methodology presented by us is based (for every value of p) on a careful inspection of a suitable auxiliary structure say $\mathcal{S}$. We describe the complete

M. Klin et al. (eds.), Algorithmic Algebraic Combinatorics and Gröbner Bases, DOI 10.1007/978-3-642-01960-9_1, © Springer-Verlag Berlin Heidelberg 2009

automorphism group of $\mathcal{S}$, which turns out to be isomorphic to the desired group G. After that, starting from the structure $\mathcal{S}$, we justify all requested properties of the graph Γ and loop Q.

The topics considered here are the subject of various investigations in diverse areas of mathematics. According to the experience accumulated by us, in some extent these areas still remain isolated. A researcher acting in one area may not even be aware of the existence of a "parallel world" in another area.

A genetical style of exposition, utilized in this paper, is designed to help the reader to overcome quickly artificial terminological barriers and to exploit all advantages resulted from simultaneous operation with adequate algebraic, geometric and combinatorial objects.

The article consists of ten sections. In Sect. 2 the most important preliminaries are introduced in order to make this text self-contained as much as it is possible. Classical and folklore results about the links between the considered structures are presented in Sect. 3. These and some other links are illustrated with the aid of examples in Sect. 4. In Sects. 5 and 6 we explain the origins of our interest and discuss briefly how various computer tools were used on the initial stages of the project.

Section 7 is the main part of the article. In fact, by examining Q_6, we are able to illustrate all of the important properties of the members of the whole infinite series. The next Sect. 8 serves as a bridge from Q_6 to the general case considered in Sect. 9.

Section 10 contains miscellaneous information which was postponed to the end of the presentation in order to not interrupt the main, we believe quite friendly, line of exposition. In this section, in particular, an alternative approach is outlined which makes it possible to generalize our results for all prime values of the parameter p. We are trying to provide also all important credits, in particular to the crucial inputs of A. Sprague, R. Wilson and K. Kunen. Our comprehensive bibliography may serve as a reasonable complement to the one included in [69].

2 Preliminaries

2.1 Main Notions

2.1.1 We start from a classical definition of a Latin square (see e.g., [65]). A *Latin square of order n* is a quadruple $(R, C, S; L)$ where R, C, S are sets of cardinality n, called rows, columns and symbols respectively and L is a mapping $L : R \times C \to S$ such that for any $i \in R$ and any $x \in S$ the equation $L(i, j) = x$ has a unique solution $j \in C$, and for any $j \in C, x \in S$ the same equation has a unique solution $i \in R$. In a more naive way the above rigorous definition may be interpreted as an $n \times n$ array with n different entries, $n \geq 2$, such that each entry occurs exactly once in any row and in any column of the array. As a rule, we will set $R = C = S = [1, n]$, where $[1, n] = \{x \in N | 1 \leq x \leq n\}$ for $n \in N$.

A Latin square is said to be *reduced* or to be in *standard form*, if in the first row and column its elements $1, 2, \ldots, n$ occur in natural order.

A *quasigroup* is a set Q with a binary operation "$\cdot$" such that for all $a, b \in Q$ the equations $a \cdot x = b$ and $y \cdot a = b$ have a unique solution in Q. It is easy to see that every Latin square may be interpreted as a multiplication table of a quasigroup, and for each quasigroup its Cayley table provides a Latin square.

A *loop* L is a quasigroup with an identity element $e \in L$ with the property $ex = xe = x$ for every $x \in L$. Usually, the identity element e is identified with $1 \in [1, n]$. Then we may say that each loop naturally defines a reduced Latin square and each reduced Latin square may be interpreted as a Cayley table of a loop. An associative loop is a group. We refer to [34] and [79] as to classical sources in the theory of Latin squares and quasigroups respectively.

2.1.2 There are a few alternative representations of Latin squares, as well as some related combinatorial structures, which are proved to be very efficient in the classification of Latin squares.

Let V be a set of cardinality n and let $L \subseteq V^3$ be a ternary relation over V (that is a collection of ordered triples whose components belong to V) such that $|L| = n^2$. Then the relation L is called a Latin square of order n if and only if each of the following three sets has n^2 distinct elements:

$$L_1 = \{(i, j) | (i, j, k) \in L\}, \qquad L_2 = \{(i, k) | (i, j, k) \in L\},$$
$$L_3 = \{(j, k) | (i, j, k) \in L\}.$$

We refer to [6] for a detailed algorithmic analysis of this *triple representation* for the goals of a constructive enumeration.

In the case when a Latin square is originally represented as the Cayley table of a group H, it is convenient to identify its triple representation with a set of n^2 ordered triples $\{(i, j, k) | i, j, k \in H, ijk = 1\}$, where 1 is the identity element of H.

An *orthogonal array* $OA(n, 3)$ *of order* n *and depth 3* is a $3 \times n^2$ array with entries from $[1, n]$, such that for any two rows of the array the n^2 vertical pairs occurring in these rows are different.

2.1.3 A *3-net of order* n is an incidence structure $\mathcal{S} = (\mathcal{P}, \mathcal{L})$ which consists of an n^2-element set $\mathcal{P}$ of points and a $3n$-element set $\mathcal{L}$ of lines. The set $\mathcal{L}$ is partitioned into three disjoint families $\mathcal{L}_1, \mathcal{L}_2, \mathcal{L}_3$ of (parallel) lines, for which the following conditions hold:

(i) every point is incident with exactly one line of each family $\mathcal{L}_i$ $(i = 1, 2, 3)$;
(ii) two lines of different families have exactly one point in common;
(iii) two lines in the same family do not have a common point;

(iv) there exist three lines belonging to three different families which are not incident with the same point.

The families $\mathcal{L}_1, \mathcal{L}_2, \mathcal{L}_3$ sometimes are called the *directions or parallel classes of* $\mathcal{S}$. It is easy to understand that each Latin square L of order n naturally produces a 3-net. Points of this net are formed by the cells of L, while its directions correspond to horizontal lines, vertical lines and the lines occupied in L by the same element.

Thus, a 3-net $\mathcal{S}$ is a uniform and regular incidence structure with the parameters $v = n^2, b = 3n, k = n, r = 3$ (see e.g., [52] for all definitions). Note also that a 3-net $\mathcal{S}$ is a particular case of a partial geometry (see e.g., [65]). It is also clear that we can get a Latin square from a 3-net, identifying one direction with rows, a second with columns and the last one with the symbols. Note that a 3-net is a particular case of a k-net ($k \geq 3$). In what follows we will call a 3-net simply a net, provided that there is no risk of a confusion.

2.1.4 Now we may consider a dual structure $\mathcal{S}^T = (\mathcal{L}, \mathcal{P})$ which has $\mathcal{L}$ as points and $\mathcal{P}$ as lines, and with the incidence relation transposed. (Recall that in our case the original incidence relation coincides with the set-theoretical inclusion.) Then $\mathcal{S}^T$ has three families of points each of cardinality n, which are called *groups*[3], and n^2 blocks (lines). The parameters of $\mathcal{S}^T$ are $v = 3n$, $b = n^2, k = 3, r = n$. By definition two distinct points from the same group are not collinear, while there is exactly one line through two distinct points from distinct groups.

A structure $\mathcal{S}^T$ will be called a *transversal design* $TD(3, n)$. Note that a transversal design is a particular case of a partial linear space (or configuration in other terms), see e.g., [21].

2.1.5 The reader is referred to [13, 21, 40] for various aspects of the theory of association schemes. Our presentation here follows mainly [45].

Let L be a Latin square represented in a traditional manner as an $n \times n$ array and let Ω be the n^2-element set of cells of L. Let $R_0 = \{(x, x) | x \in \Omega\}$ be the complete diagonal relation over Ω. Define R_1 as follows: $(x, y) \in R_1$ for $x, y \in \Omega$ if and only if x, y are in the same row of L. Similarly R_2 is defined with respect to the columns of L, while for two cells x, y we have $(x, y) \in R_3$ if and only if x, y are occupied in L by the same symbol. Finally, let $R_4 = \Omega^2 \setminus \{R_0 \cup R_1 \cup R_2 \cup R_3\}$.

Then it is easy to see, that the relational structure

$$\mathcal{M} := \mathcal{M}(L) = (\Omega, \{R_0, R_1, R_2, R_3, R_4\})$$

is an association scheme with four classes. (This is a particular case of amorphic association schemes introduced in [45].) Merging of any classes in $\mathcal{M}$ provides again an association scheme. In particular, $(\Omega, \{R_0, R_1 \cup R_2, R_3, R_4\})$ is

[3] Note that the term "group" in the definition of $TD(3, n)$ is very unfortunate causing confusion with the term "group" in algebra.

an association scheme of L with three classes, and $(\Omega, \{R_0, R_1 \cup R_2 \cup R_3, R_4\})$ is an association scheme with two classes. Each of its classes corresponds to a strongly regular graph. A strongly regular graph $\Gamma = SRG(L) = (\Omega, R_1 \cup R_2 \cup R_3)$ is commonly called a *Latin square graph*, it has parameters $v = n^2, k = 3(n-1), \lambda = n, \mu = 6$.

2.1.6 Two permutations are discordant if no symbol has the same image under both permutations. A set of n pairwise discordant permutations of degree n is called a complete set of discordant permutations (an alternative name is a sharply transitive set of permutations).

There are three important ways to establish a one-to-one correspondence between Latin squares of order n and complete sets of discordant permutations of degree n. For the first correspondence we regard permutations as various rows of a Latin square L. The second correspondence (dual to the first) is linked to columns of L. Usually this correspondence is not distinguished from the first one. Another correspondence attributes to each permutation a permutation matrix, thus showing positions in L where a certain element i appears. In our consideration these correspondences will sometimes be viewed implicitly. We refer to [35, 36] for a detailed consideration of these ways.

Very close to permutation representations are graphical representations of Latin squares, using certain factorizations of the complete n-vertex undirected graph K_n, complete directed graph K_n^* and complete directed pseudograph $\overline{K}_n^*$ with a loop in each vertex.

2.2 Classification

Here we briefly discuss the most significant ways to classify Latin squares.

2.2.1 Let L_n denote the total number of Latin squares of order n considered as $n \times n$-arrays over $[1, n]$. Let R_n denote the total number of reduced Latin squares of order n. It is easy to see that $L_n = n!(n-1)!R_n$.

2.2.2 In what follows we denote by $\mathcal{L}_n$ a set of all Latin squares of order n in their ordered triple representation, i.e., for $L \in \mathcal{L}_n$ the triple, $(i, j, k) \in L$ means that the cell on the intersection of row i and column j is filled by the symbol k. Thus, there exists a natural bijection between the sets of triple representations and the traditional $n \times n$-array representations.

We denote by S_n a symmetric group of degree n and order $n!$ in its natural action on $[1, n]$. Following [6], let us denote by $\mathcal{G}_n$ the exponentiation $S_n \uparrow S_3$ of the symmetric group of degree n with the symmetric group of degree 3. (Sometimes the notation $[S_n]^{S_3}$ is used instead of $S_n \uparrow S_3$.) Recall that $S_n \uparrow S_3 = \{[g, h(x)] | g \in S_3, h(x) \in S_n \text{ for } x \in [1, 3]\}$ consists of formal tables $[g, h(x)] = [g; h_1, h_2, h_3]$, where $g \in S_3, h_1, h_2, h_3 \in S_n$. In order to get the image $(\alpha_1, \alpha_2, \alpha_3)^{[g,h(x)]}$ for a triple $(\alpha_1, \alpha_2, \alpha_3) \in [1, n]^3$ we first apply h_i to α_i, $i = 1, 2, 3$, and after that permute the components of the resulted triple according to the permutation g.

Thus, $S_n \uparrow S_3$ is a permutation group of degree n^3 and of order $6(n!)^3$. This group acts transitively on the set $[1, n]^3$. As an abstract group the group $S_n \uparrow S_3$ is isomorphic to the wreath product of the groups S_n and S_3 acting on a $3n$-element set. We refer to [59] and [40] for a more detailed discussion of the operation of the exponentiation and its combinatorial applications.

Proposition 1. *The group $\mathcal{G}_n = S_n \uparrow S_3$ acts on the set $\mathcal{L}_n$.*

2.2.3 Let $(Q_1, \circ)$ and $(Q_2, \diamond)$ be two quasigroups. We introduce an *isomorphism of quasigroups* in a traditional algebraic manner, that is, if there exists a bijection f from Q_1 to Q_2 such that for $a, b, c \in Q_1$ we have

$$(a \circ b = c) \quad \Longleftrightarrow \quad (a^f \diamond b^f = c^f).$$

The set $\mathrm{Aut}(Q)$ of all isomorphisms of Q onto itself (in other words the set of automorphisms of Q) evidently forms a group, which can be regarded as a permutation group of degree $n = |Q|$. Establishing isomorphism classes of quasigroups is an important ingredient in the classification of Latin squares. A similar definition and problem may be formulated for loops. Here of course, any isomorphism of loops L_1 and L_2 should send the identity of L_1 to the identity of L_2.

2.2.4 Let $(S_n)^3 = S_n \uparrow E_3$ be a subgroup of $S_n \uparrow S_3$. Here E_3 is the trivial subgroup in S_3, which is generated by the identity $e \in S_3$. In other words, $(S_n)^3 = \{[e; h_1, h_2, h_3] | h_1, h_2, h_3 \in S_n\}$.

Two Latin squares $L_1, L_2 \in \mathcal{L}_n$ are called *isotopic*, if they belong to the same orbit under the action of $(S_n)^3$ on $\mathcal{L}_n$, e.g., $L_2 = L_1^{[e;h_1,h_2,h_3]}$. Intuitively, we may explain that two Latin squares L_1, L_2 represented as $n \times n$-arrays are isotopic, if we can find three independent permutations $h_1, h_2, h_3 \in S_n$, such that the action of h_1 on rows, h_2 on columns, h_3 on symbols of L_1 brings L_1 to L_2. The orbits (equivalence classes) of this relation are called *isotopy classes of Latin squares.* The stabilizer $I_s(L) = \{\sigma \in (S_n)^3 | L^\sigma = L\}$ is called the *autotopy group of* L, the elements of $I_s(L)$ are *autotopisms of* L.

Note, that in these terms isomorphism (automorphism) of a quasigroup (loop) may be formulated as a particular case of isotopy. Namely, $\sigma = [e; h_1, h_2, h_3] \in (S_n)^3$ is an isomorphism of L_1 and L_2 if and only if it is an isotopism between L_1 and L_2, for which $h_1 = h_2 = h_3$.

2.2.5 Let us now pay special attention to the group S_3 which participated in the definition of $\mathcal{G}_n = S_n \uparrow S_3$. This group S_3 is acting on the set $[1, 3] = \{1, 2, 3\}$. Let $t = (1, 2)(3) \in S_3$ be one of the transpositions from S_3.

Then $\langle t \rangle$ is evidently a subgroup of order 2 in S_3 and we may also consider a permutation group $S_n \uparrow \langle t \rangle$ of order $2 \cdot (n!)^3$, where

$$S_n \uparrow \langle t \rangle = \{[g; h_1, h_2, h_3] | g \in \langle t \rangle, h_1, h_2, h_3 \in S_n\}.$$

We say that the Latin squares $L_1$and L_2 are of the same *type* if $L_2 = L_1^\sigma$ for $\sigma \in S_n \uparrow \langle t \rangle$.

Type equivalence can be explained intuitively in terms of $n \times n$-arrays, if in addition to isotopisms we allow the operation of the transposition of arrays.

2.2.6 Let us consider the orbits of a natural action $(\mathcal{G}_n, \mathcal{L}_n)$; they will be called the *main classes of Latin squares of order* n. Two squares from the same main class will be called *paratopic.* (Recall that we postpone the discussion of the origins of all the terminology used by us, as well as other alternatives to Sect. 10.)

The stabilizer $MC(L) = \{\sigma \in \mathcal{G}_n | L^\sigma = L\}$ is called the *autoparatopy group of* L, and its elements are the *autoparatopisms of* L.

2.2.7 Another traditional equivalence class of Latin squares is related to the consideration of the exponentiation $E_n \uparrow S_3$, where E_n is the identity permutation group of degree n, i.e.,

$$E_n \uparrow S_3 = \{[g; e, e, e] | g \in S_3, e \text{ is the identity in } S_n\}.$$

Evidently $E_n \uparrow S_3$ is a permutation group of order 6, acting on the set $\mathcal{L}_n$. The orbits of this action are called the *conjugate classes of Latin squares of order* n. Two Latin squares belonging to the same conjugacy class are called *conjugates.* We can also introduce a stabilizer of a Latin square in the group $E_n \uparrow S_3$. It is playing a significant role in the quasigroup theory, with its use one may introduce a few important classes of quasigroups.

Note that the action of a group $E_n \uparrow S_3$ on $\mathcal{L}_n$ has a very natural intuitive interpretation. This means that we allow permutations of the sets of rows, columns and symbols, and conjugate squares are those, which appear one from another via such a permutation.

2.2.8 Let L be a Latin square of order n, let $\mathcal{N}(L)$ be the 3-net, defined by L. Then we say that two Latin squares L_1 and L_2 are *net-equivalent* if and only if $\mathcal{N}(L_1) \cong \mathcal{N}(L_2)$. Here, by an isomorphism of 3-nets, we mean a traditional isomorphism of incidence structures.

It is quite evident, that the introduced net-equivalence is coinciding with the *AS-equivalence of Latin squares.* By AS-equivalence of L_1 and L_2 we mean that the 4-class (amorphic) association schemes $\mathcal{M}(L_1)$ and $\mathcal{M}(L_2)$ are isomorphic. The automorphism group $\mathrm{Aut}(\mathcal{M}(L))$ of the association scheme $\mathcal{M} = (\Omega, \{R_0, R_1, R_2, R_3, R_4\})$ is the traditional automorphism group of the association scheme, that is $\mathrm{Aut}(\mathcal{M}) = \{g \in S(\Omega) | R_i^g = R_i,\ 0 \le i \le 4\}$, where $S(\Omega)$ is the symmetric group of the set Ω.

It is also important to consider the group of *color (or weak) automorphisms of* $\mathcal{M}$, namely $\mathrm{CAut}(\mathcal{M}) = \{g \in S(\Omega) | R_i^g = R_j,\ 0 \le i, j \le 4\}$. Permutations from $\mathrm{CAut}(\mathcal{M})$ may permute classes of $\mathcal{M}$, while permutations from $\mathrm{Aut}(\mathcal{M})$ preserve each class of $\mathcal{M}$. These two natural groups also have other names, provided that we consider 3-nets instead of association schemes.

The *collineation group* Σ *of a quasigroup* Q is the (full) collineation group of the 3-net $\mathcal{N}(Q)$. Collineation is defined to be a permutation of points of $\mathcal{N}(Q)$, which maps a line onto a line. The group Σ has a normal subgroup $\mathcal{T}$ of index ≤ 6, which maps every class of (parallel) lines onto itself. This group may be called the group of *direction preserving collineations of* $\mathcal{N}(Q)$. If it is necessary to attribute $\mathcal{T}$ and Σ to Q, we write $\mathcal{T}(Q)$ and $\Sigma(Q)$.

2.2.9 Let $L_1, L_2 \in \mathcal{L}_n$ and let $\Gamma_1 = SRG(L_1), \Gamma_2 = SRG(L_2)$ be strongly regular graphs defined by L_1 and L_2 respectively. We say that L_1 and L_2 are *SRG-equivalent* if Γ_1 and Γ_2 are isomorphic graphs. If L is a Latin square and $\Gamma = SRG(L)$, then we also consider $\mathrm{Aut}(\Gamma)$, the classical automorphism group of the graph Γ. This group will play the most significant role in this paper, providing a natural graph-theoretical way for measuring the symmetry of L.

2.3 Regular Subgroups

In this paper we are especially interested in regular subgroups of the collineation group Σ of a quasigroup (loop). Recall that a finite permutation group (H, Ω) is called *regular*, if it is transitive and the order $|H|$ is equal to the degree $|\Omega|$. Each regular group H is similar (as a permutation group) to a right action of H on its elements. Therefore, we always can regard H as acting on itself and thus denote it by (H, H).

Following [14], we will call a regular subgroup of $\Sigma(Q)$, where Q is a quasigroup, a *sharply transitive group of collineations of* Q. We can also consider a regular subgroup H of the group $\mathcal{T}(Q)$, where Q is a quasigroup. It is clear that in this case elements of H preserve each of the three directions of the net $\mathcal{N}(Q)$. Following A. Sprague [96] such a subgroup H will be called a *translation group of* $\mathcal{N}(Q)$, while a net $\mathcal{N}(Q)$, for which a translation group exists, will be called a *translation net*.

In Sect. 3 we will show that if Q is a group, then $\Sigma(Q)$ contains a sharply transitive collineation group H, moreover H is also a translation group.

A loop Q will be called a *proper loop* if its main class does not contain a group. This paper was originally motivated by the following remark by Barlotti and Strambach, see [14, p. 79]: "We were not able to decide whether there exists a proper finite loop having a sharply point transitive group of collineations."

A question about the existence of sharply transitive collineation groups can be reformulated in terms of graphs. Let H be a group (in additive notation) and let X be a subset of H. Then a (directed) graph $\Gamma = Cay(H, X) = (H, R)$, where $R = \{(h, x+h) | h \in H, x \in X\}$ is called a *Cayley graph over H with a connection set X*. The graph Γ does not have loops, if the identity element $0 \in H$ does not belong to X. We can identify Γ with a simple (undirected) graph, if $-X = X$, where $-X = \{-x | x \in X\}$.

Finally, if $\Gamma = Cay(H, X)$ is a strongly regular Cayley graph (over a group H) then its connection set X is called a *partial difference set in H*. The investigation of the partial difference sets in groups of small order in [50] was, in fact, the starting point of this project.

2.4 Principal Loop-Isotopes of Quasigroups

There is one more extremely important class of groups, which may be associated to a given quasigroup Q. We start from a subgroup $\mathcal{P}$ of $\mathcal{G}_n$, which consists of tables $[e; g_1, g_2, \epsilon]$, where e is the identity of S_3, $g_1, g_2 \in S(Q)$ and ϵ is the identity of $S(Q)$. This subgroup $\mathcal{P}$ has order $(n!)^2$. Two quasigroups are called *principally isotopic*, if their isotopism can be realized by an element $\theta \in \mathcal{P}$. Those loops L, which are principally isotopic to a quasigroup Q, are called its *principal L-isotopes.*

For any $a, b \in Q$ we can define in a sense a canonical principal L-isotope of Q, which is denoted by $L(a, b)$, see [26] for a precise definition. In terms of these n^2 objects R. Bryant and H. Schneider define in [26] a certain group, which is associated with a given quasigroup Q. Following D. A. Robinson [82], we will call this group the *Bryant-Schneider group of* Q, or briefly $BS(Q)$.

It turns out that the Bryant-Schneider group plays a significant role in the enumeration of loops which are isotopic to Q.

3 Classics and Folklore

In this section we collect the most important facts about the links between Latin squares and various kinds of algebraic and combinatorial objects. We will give only brief formulations without any attempts to the consideration of proofs. Some of the results are of a definite folklore nature. The most important references will be mentioned immediately, the other ones will be discussed in the last section of the article. However, a complete characterization of all bibliographical sources is beyond the scope of this paper. Some part of the material here is also considered in [32].

3.1 General Links Between Groups

Let L be a Latin square of order n, and let $\mathcal{N}(L)$ be a 3-net which is defined by L. Let $\mathcal{M} = \mathcal{M}(L)$ be an association scheme with four classes, naturally attributed to L. Three of the classes of $\mathcal{M}$ define imprimitive strongly regular graphs, each isomorphic to $n \circ K_n$. Each such graph consists of n copies of the complete graph on n vertices and thus, corresponds to one parallel class of the net $\mathcal{N}(L)$. Conversely, if we have a net $\mathcal{N}(L)$, we easily determine an association scheme with four classes, which is denoted here as $\mathcal{M}(L)$.

From these observations it is easy to get that $\mathrm{Aut}(\mathcal{M}(L)) = \mathcal{T}(L)$ as it was defined in Sect. 2.2.8. One may also consider a group $\mathrm{CAut}(\mathcal{M}(L))$ of weak (or color) automorphisms of the association scheme $\mathcal{M}(L)$. A combinatorial definition of this group was also given in Sect. 2.2.8. We see that the group $\mathrm{CAut}(\mathcal{M}(L))$ coincides with the group $\Sigma(L)$ of all collineations of L. Let us now start to discuss the link between $\mathrm{Aut}(\mathcal{N}(L))$ and $\mathrm{Aut}(SRG(L))$, where $\Gamma = SRG(L)$ is a strongly regular graph defined by L.

Here we have the particular case of a more general situation. Namely, $\mathcal{N}(L)$ is a special case of a partial geometry, while Γ is the point graph of this geometry. Clearly Γ is defined by $\mathcal{N}(L)$, however $\mathcal{N}(L)$ is not necessarily uniquely reconstructed from Γ. For more details we refer to the classical papers [22, 24, 20] and to textbook [65].

Therefore, in general we may only say that $\mathrm{Aut}(\mathcal{N}(L)) \leq \mathrm{Aut}(SRG(L))$, and moreover, it may happen that the first group is a proper subgroup of the second.

3.2 Geometrical and Nongeometrical SRG's

Let Γ be a strongly regular graph with the parameters $v = n^2, k = 3(n-1)$, $\lambda = n, \mu = 6$. Such a graph is usually called a *pseudo-Latin square graph.* The name refers to the fact that for an arbitrary Latin square L of order n the graph $SRG(L)$ is indeed a strongly regular graph with the same parameters. In this case $SRG(L)$ is called simply a *Latin square graph.*

Note again that a similar situation appears for an arbitrary partial geometry with the parameters K, R, T. In our case we have $K = n, R = 3, T = 2$.

The question, when a pseudo-geometrical strongly regular graph is indeed a geometrical graph, was considered by Bruck and Bose in [22, 24, 20]. We formulate their general answer for a particular case of 3-nets.

Proposition 2. *A strongly regular graph Γ with the parameters $v = n^2$, $k = 3(n-1), \lambda = n$, $\mu = 6$ is a Latin square graph, provided that $n > 23$ (see extra comments in Sect.* 10*).*

We will discuss the small cases $n = 3, 4, 5$ in the next section.

3.3 Factorization of Latin Square Graphs

Now we are interested in the following question. Suppose that Γ is a $SRG(L)$ for a Latin square graph L of order n. Then the edge set of Γ is attained via the merging of edge sets of three copies of graphs $n \circ K_n$, which correspond to the three parallel classes of the net $\mathcal{N}(L)$. Therefore, there exists a special factorization of the graph Γ. In what follows for a Latin square graph Γ we will call an arbitrary collection of n disjoint cliques of size n in Γ a *spread of* Γ. Then a 3-net is nothing but a factorization of Γ into three spreads. In general, such a factorization is not unique (see examples in Sect. 4).

The following lemma is also of a folklore nature.

Lemma 1. *Let L be a Latin square and $\Gamma = SRG(L)$. If $n \geq 5$, then the cliques of Γ necessarily correspond to lines of an associated 3-net $\mathcal{N}(L)$.*

Lemma 2 (cf. [7]). *For $n \geq 5$ we can reconstruct the 3-net $\mathcal{N}(L)$ uniquely from the graph $\Gamma = SRG(L)$.*

This lemma immediately implies the following important graph theoretical reformulation.

Proposition 3. *For $n \geq 5$ we have* $\mathrm{Aut}(SRG(L)) = \mathrm{Aut}(\mathcal{N}(L)) = \Sigma(L)$.

3.4 The Group Case

In this section we pay special attention to a case when a Latin square L is a Cayley table of a group H. Such a case of a Latin square will be called a *group case*. Sometimes we may write $L = L(H)$ to stress that L is coming from a group H.

Lemma 3 (cf. [32]). *Let H be a finite group and $L := \{(x,y,z) \in H^3 | \; xyz = 1\}$ an associated Latin square considered in a special triple representation. Define mappings $L \to L$ by*

(a) $\iota : (x,y,z) \to (z^{-1}, y^{-1}, x^{-1})$,
(b) $\sigma : (x,y,z) \to (y,z,x)$,
(c) $\tau_{a,b,c} : (x,y,z) \to (a^{-1}xb, b^{-1}yc, c^{-1}za)$, *where* $(a,b,c) \in H^3$,
(d) $\alpha_f : (x,y,z) \to (x^f, y^f, z^f)$ *where* $f \in \mathrm{Aut}(H)$.

Then each of the mappings above is an automorphism of L and

$$\mathrm{Aut}(L) = \langle \iota, \sigma, \tau_{a,b,c}, \alpha_f \rangle_{(a,b,c) \in H^3, f \in \mathrm{Aut}(H)}.$$

Corollary 1. *Let $L = L(H)$ be a group Latin square. Then*

$$\mathrm{Aut}(\mathcal{N}(L)) \cong (H^2 : \mathrm{Aut}(H)).S_3.$$

Theorem 1 (cf. [32]). *Let $L(H)$ be a group Latin square with $|H| \geq 5$. Then*

$$\mathrm{Aut}(SRG(H)) \cong (H^2 : \mathrm{Aut}(H)).S_3.$$

Clearly, the proof is based on the combination of Proposition 3 with Corollary 1. In the next section we will discuss some special examples.

Corollary 2. *Let H be a group of order n and $\Gamma = SRG(H)$. Then*

(a) $\mathrm{Aut}(\Gamma)$ *is a transitive group of degree* n^2,
(b) $\mathrm{Aut}(\Gamma)$ *contains a regular subgroup* H^2 *of order (and degree)* n^2.

3.5 Main Class of a Group Case

We refer to [72], see also [44], Exercise 10.23 as strict formulations of the following facts of folklore nature.

Proposition 4. *Let H be a group of order n and let Q be a loop of order n. Then $H \cong Q$ if and only if the corresponding 3-nets $\mathcal{N}(H)$ and $\mathcal{N}(Q)$ are isomorphic.*

Corollary 3.

(a) If H_1 and H_2 are nonisomorphic groups of order n, then $SRG(H_1) \ncong SRG(H_2)$.
(b) If a Latin square L does not appear in a main class of any group, then $SRG(L)$ is not isomorphic to any Latin square graph over a group.

This culminating corollary will play an essential role in all further considerations.

4 Garden of Small Examples

4.1 Main Results

The goal of this section is to provide the reader an opportunity to digest many of the introduced notions on a level of simple small examples. We restrict ourselves to a consideration of examples of order $n \leq 6$.

Table 1 provides general information about the size of data which is relevant to our consideration.

Clearly, $n = 1$ is an absolutely trivial case. Thus, we start to consider our examples from $n = 2$. Though this case still should be regarded as trivial, it is quite convenient on this level to illustrate some concepts.

Table 1. Data for $n \leq 6$

n	Main classes	Types	Isotopy classes	Loops	Reduced squares R_n	Quasigroups	Total amount of squares $\lvert\mathcal{L}_n\rvert$
1	1	1	1	1	1	1	1
2	1	1	1	1	1	1	2
3	1	1	1	1	1	5	12
4	2	2	2	2	4	35	576
5	2	2	2	6	56	1411	161280
6	12	17	22	109	9408	1130531	812851200

4.2 The Case $n = 2$

It is easy to check, that $|\mathcal{L}_2| = 2$. The two Latin squares of order 2 are depicted below.

1	2
2	1

2	1
1	2

The group $\mathcal{G}_2$ has order $3! \cdot (2!)^3 = 48$. If we regard $\mathcal{G}_2$ as the exponentiation $S_2 \uparrow S_3$ then the natural action of the latter group is on the set $[1,2]^3$ of all binary sequences of length 3 (here it is more convenient for us to substitute the alphabet $\{0,1\}$ by $\{1,2\}$). In this notation $\mathcal{G}_2$ is nothing else but the automorphism group of a very classical object: the 3-dimensional cube Q_3. Following [13], for this graph we will also use the notation $H(3,2)$ and call it the Hamming graph (with prescribed parameters). Its vertices are all binary strings of length 3, where two vertices are adjacent if and only if corresponding strings differ in exactly one position.

Let us now consider cocliques of $H(3,2)$ of the size $2^2 = 4$, i.e., induced subgraphs of Q_3 which are isomorphic to the 4-vertex empty graph E_4. Clearly, Q_3 contains just two such cocliques, which are depicted in Fig. 1. The reader will easily attribute these cocliques to the above mentioned Latin squares of order 2. Indeed, we interpret a string (α, β, γ) as a cell of the square on the intersection of a row α and a column β, which is filled with the element γ.

Clearly, the group $\mathcal{G}_2$ has just one orbit on such cocliques. This justifies (an evident) fact, that the number of main classes for $n = 2$ is equal to 1.

4.3 The Case $n = 3$

Here $|\mathcal{G}_3| = (3!)^4 = 1296$ and $|[1,3]^3| = 27$. Thus $S_3 \uparrow S_3$ acts naturally on the 27-element set. Again we regard $\mathcal{G}_3$ as the automorphism group of the suitable Hamming graph $H(3,3)$. This is a distance regular graph of diameter 3 and valency 6. Easy reasonings show that there are exactly 12 different Latin squares of order 3, which are presented in Table 2.

Again we would like to interpret each Latin square as a certain subgraph of $H(3,3)$. Starting from a definition of a Latin square as an array of order 3, we get its interpretation as a coclique of $H(3,3)$ of size 9. A prescribed vertex of $H(3,3)$, say $(1,1,1)$, appears in exactly $\frac{12 \cdot 9}{27} = 4$ such cocliques. One of these

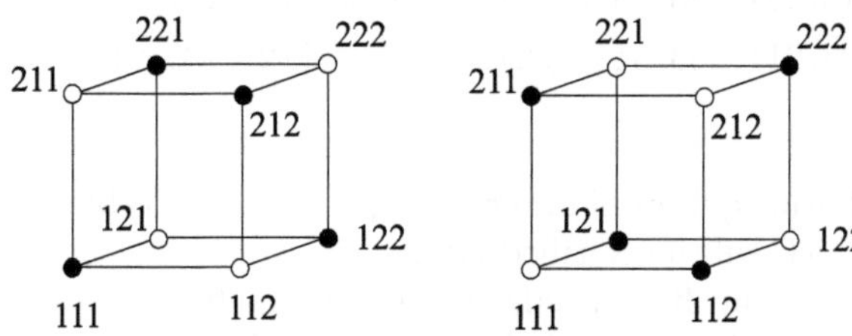

Fig. 1. Cocliques of $H(3,2)$

Table 2. All Latin squares of order 3

1 2 3 2 3 1 3 1 2	1 2 3 3 1 2 2 3 1	1 3 2 2 1 3 3 2 1	1 3 2 3 2 1 2 1 3
$\mathbb{Z}_{3,1}$	LU_1	RU_1	I
2 1 3 1 3 2 3 2 1	2 1 3 3 2 1 1 3 2	2 3 1 1 2 3 3 1 2	2 3 1 3 1 2 1 2 3
SQ_1	RU_2	LU_2	$\mathbb{Z}_{3,2}$
3 1 2 1 2 3 2 3 1	3 1 2 2 3 1 1 2 3	3 2 1 1 3 2 2 1 3	3 2 1 2 1 3 1 3 2
$\mathbb{Z}_{3,3}$	LU_3	RU_3	SQ_2

cocliques is depicted in the diagram of $H(3,3)$ in Fig. 2, it corresponds to the Latin square $\mathbb{Z}_{3,1}$ in Table 2. (Each line in the picture designates a clique of size 3 in $H(3,3)$.) Considering the orbits of this coclique with respect to suitable subgroups of $\mathcal{G}_3$, we may find a confirmation of the first four entries in the row for $n = 3$ in Table 1.

The only non-trivial result, which is relevant to $n = 3$, is concerned with the number of isomorphism classes of quasigroups.

For this purpose, we have to make visible to the reader the action of the diagonal subgroup $\mathcal{D}_3 = \{[e; g, g, g] | g \in S_3\}$ of order 6 in $\mathcal{G}_3$. An easy task is to describe orbits of $\mathcal{D}_3$ on $[1,3]^3$. There are five such orbits ($\alpha\beta\gamma$ is a short notation for (α, β, γ)):

$$A = \{111, 222, 333\},$$
$$B = \{112, 113, 221, 223, 331, 332\}, \qquad C = \{121, 131, 212, 232, 313, 323\},$$
$$D = \{122, 133, 211, 233, 311, 322\}, \qquad E = \{123, 132, 213, 231, 312, 321\}.$$

All twelve Latin squares of order 3 are presented in Table 2, together with their names, which will be clarified later.

Now we are taking five specific Latin squares from the table and count for them a number of numerical characteristics. Namely, for each of the five orbits of $\mathcal{D}_3$ we find the size of the intersection of it with a subset of $[1,3]^3$

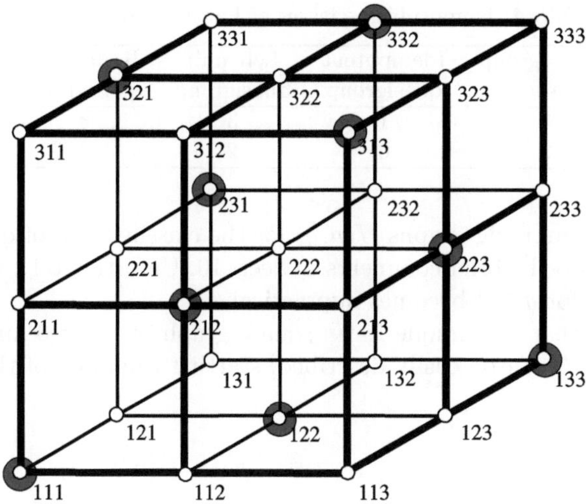

Fig. 2. Hamming graph $H(3,3)$

corresponding to a square. Besides this, for two invariant subsets of $\mathcal{D}_3$ we describe the isomorphism type of a colored subgraph of $H(3,3)$, induced by the square on the subset (here color means the distance in $H(3,3)$, the sign [x] shows the induced subgraph). All these data are presented in Table 3.

The reader is able to check that the selected squares may be distinguished with the aid of the columns A, $[A \cup B]$ and $[C]$. This means that we have at least five different orbits of quasigroups of order 3.

Finally, a simple routine inspection shows that we indeed have exactly five such orbits.

Information about these isomorphism classes together with their names (traditionally attributed in quasigroup theory) is presented in Table 4. We believe that the case $n = 3$ provides a nice illustration of the opportunities

Table 3. Characteristics for five Latin squares

Square	A	B	C	D	E	$[A \cup B]$	$[C]$
$\mathbb{Z}_{3,1}$	1	2	2	2	2	triangle 111, 223, 332; edges 3, 3, 3	
I	3	0	0	0	6		
LU_1	1	2	2	2	2	triangle 111, 221, 331; edges 2, 2, 2	232 –3– 323
RU_1	1	2	2	2	2	triangle 111, 221, 331; edges 2, 2, 2	212 –2– 313
SQ_1	0	3	3	3	0		

Table 4. Isomorphism classes of Latin squares of order 3

Name	Cyclic group of order 3	Idempotent quasigroup	Left unit quasigroup	Right unit quasigroup	Commutative quasigroup
Amount	3	1	3	3	2
$\lvert \mathrm{Aut}(Q) \rvert$	2	6	2	2	3

of the use of Hamming graphs $H(n, q)$ for the classification of quasigroups of order q, see also the brief comments in Sect. 10. Unfortunately, the direct use of the graphs for $q \geq 4$ becomes impractical.

Note also that the unique Latin square graph of order 3 and the unique 3-net of order 3 can be easily described, starting from any of the considered Latin squares.

4.4 The Case $n = 4$

There are exactly four reduced squares which are presented below.

1	2	3	4
2	1	4	3
3	4	1	2
4	3	2	1

1	2	3	4
2	1	4	3
3	4	2	1
4	3	1	2

1	2	3	4
2	3	4	1
3	4	1	2
4	1	2	3

1	2	3	4
2	4	1	3
3	1	4	2
4	3	2	1

Thus, the total amount of $|\mathcal{L}_4|$ is equal to $4! \cdot 3! \cdot 4 = 576$.

In principle, the classification of quasigroups may still be arranged by hand computations in the same spirit as for $n = 3$. However, the limits of our exposition do not allow to present the necessary computations. Therefore, we leave the resulting number 35 without justification. Clearly, for larger values of n this part of the problem requires definitively the use of a computer.

Let us now turn to the left end of the row in Table 1 corresponding to $n = 4$. There exist two groups of order 4: the cyclic group $\mathbb{Z}_4$ and the elementary abelian group E_4. Thus, according to Corollary 3, there exist at least two main classes. To prove that these are all main classes, we may employ a classical theorem by Shrikhande [91], which claims that a strongly regular graph with the parameters $v = n^2, k = 2(n-1), \lambda = n-2, \mu = 2$ is for $n \neq 4$ unique up to isomorphism and is isomorphic to the lattice square graph $L_2(n)$. The vertices of this graph are elements of $[1, n]^2$ and two pairs (a, b) and (c, d) are adjacent if and only if they coincide in exactly one coordinate. It is easy to prove that $\mathrm{Aut}(L_2(n)) \cong S_n \uparrow S_2$ is a group of order $2 \cdot (n!)^2$.

The same theorem claims that for $n = 4$ there is one more exceptional strongly regular graph, which is called *Shrikhande graph* and is denoted by Sh.

Now we turn to the Latin square graphs for $n = 4$. Clearly, these graphs have parameters $v = 16, k = 9, \lambda = 4, \mu = 6$. It turns out that this parameter set corresponds to the complements of the strongly regular graphs with parameters of $L_2(4)$. Thus, taking into account that there exist exactly two non-isomorphic groups of order 4, we obtain the desired result: there are exactly two main classes of Latin squares of order 4.

Let us first associate $L_2(4)$ to one of the two main classes. Consider the graph $\Gamma_1 = \overline{L_2(4)}$: $\mathrm{Aut}(\Gamma_1)$ has order $2 \cdot (4!)^2 = 1152$ and acts transitively on the vertices. We want to describe all cliques of size 4 in Γ_1. Let us fix one vertex, say $(1,1)$ and consider the neighbor subgraph of it in Γ_1. Clearly, it is isomorphic to $\overline{L_2(3)}$. An easy exercise is to prove that $L_2(3) \cong \overline{L_2(3)}$. It is evident that $L_2(3)$ (and thus $\overline{L_2(3)}$) has exactly six different 3-cliques. Therefore, altogether there are $\frac{6 \cdot 16}{4} = 24$ 4-cliques in Γ_1. In order to get a 3-net of order 4 from the graph Γ_1 we need just twelve cliques. Note also that the graph $L_2(4)$ itself has exactly eight cliques, which come from two parallel classes.

At this point we consider the graph $SRG(\mathbb{Z}_4)$ for the group $\mathbb{Z}_4$, which corresponds to the third Latin square above. Consider $\overline{SRG(\mathbb{Z}_4)}$ and detect that the neighbor graph of a vertex in it, say $(1,1)$ is isomorphic to the hexagon C_6. Clearly, C_6 does not have 3-cliques, and as a consequence $\overline{SRG(\mathbb{Z}_4)}$ does not contain 4-cliques. This implies finally, that $L_2(4)$ is isomorphic to $\overline{SRG(E_4)}$, where E_4 is represented by the first Latin square above.

Recall that $\mathrm{Aut}(E_4) \cong S_3$, and using Corollary 1, we get that $\mathrm{Aut}(\mathcal{N}(E_4))$ is a group of order $4^2 \cdot 3! \cdot 3! = 576$. In other words $\mathrm{Aut}(\mathcal{N}(E_4))$ is a subgroup of index 2 in $\mathrm{Aut}(SRG(E_4))$. We are obtaining a desired example, which shows that the assumption $|H| \geq 5$ in Theorem 1 is important: For groups of order 4 this theorem is not correct. Moreover, the two independent portions of information, which are presented above, now converge. Indeed, 24 cliques in $\overline{L_2(4)}$ provide two distinct 3-nets, which are isomorphic. The point graphs of both nets are identical. An "extra" automorphism of $L_2(4)$ interchanges the two nets. Let us keep the notation $\Gamma_1 \cong \overline{L_2(4)} \cong SRG(E_4)$. Thus, naturally, we consider $\Gamma_2 \cong SRG(\mathbb{Z}_4)$. We already know that $\overline{\Gamma}_2 \cong Sh$, and $\overline{\Gamma}_2$ is not geometrical, that is, it is not the point graph of a 2-net.

We now want to describe the order of the group $\mathrm{Aut}(Sh)$. The Shrikhande graph is a Cayley graph over $\mathbb{Z}_4^2$ with a connection set $X = \{11, 12, 21, 23, 32, 33\}$ (here $\mathbb{Z}_4 = \{0,1,2,3\}$). We invite the reader to check that Sh can be depicted with the aid of Fig. 3 (which should be considered as drawn on the torus).

The group $\mathrm{Aut}(Sh)$ also acts transitively on the vertices of Sh. An easy inspection shows that the stabilizer of a point acts on its neighbor set as the dihedral group D_6 of order 12, and this action is faithful. Thus, we have $|\mathrm{Aut}(Sh)| = 16 \cdot 12 = 192$. On the other hand, $|\mathrm{Aut}(\mathbb{Z}_4)| = 2$, therefore, $|\mathrm{Aut}(\mathcal{N}(\mathbb{Z}_4))| = 4^2 \cdot 2 \cdot 3! = 192$.

Thus, we obtain that for the second main class Γ_2 the groups of the 3-net and of the point graph coincide.

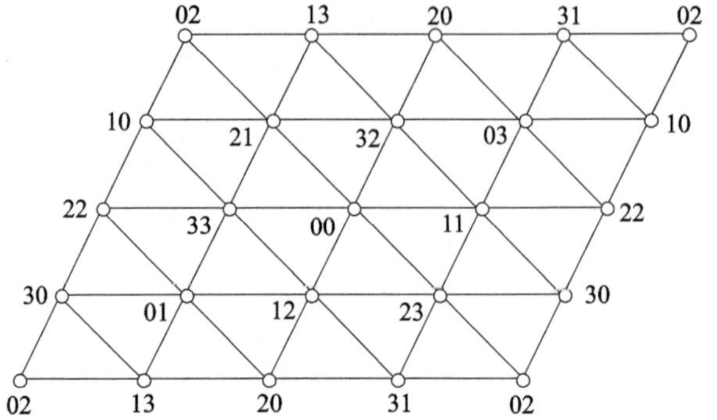

Fig. 3. The Shrikhande graph Sh

Finally, inspecting suitable subgroups of $\mathrm{Aut}(\Gamma_1)$ and $\mathrm{Aut}(\Gamma_2)$ we may prove that for $n = 4$ the results of the classification of types, isotopy classes and loops do not provide surprises, that is, the number 2 appears in the corresponding row of Table 1 three more times.

4.5 The Case $n = 5$, Part a

Counting the reduced squares in this case is still quite an easy job. It was Cayley, who found in [28] that $R_n = 56$. Here, proper loops appear for the first time.

There are exactly 15 strongly regular graphs with the parameters $(v, k, \lambda, \mu) = (25, 12, 5, 6)$. Only two of them are geometrical, in other words, there are exactly two main classes of Latin squares of order 5.

One of these main classes, clearly, corresponds to the cyclic group of order 5. Denoting the corresponding Latin square graph as $\Gamma_1 = SRG(\mathbb{Z}_5)$, we get that $|\mathrm{Aut}(\Gamma_1)| = 5^2 \cdot 4 \cdot 6 = 600$. A corresponding partial difference set over $(\mathbb{Z}_5)^2$ consists of a union of non-identity elements in three (arbitrary) subgroups of order 5 in $(\mathbb{Z}_5)^2$. This is a rank 3 graph.

There is, however, another classical way to get a rank 3 graph with the same parameters. Indeed, let us consider the Paley graph $P(25)$. Its vertices are the elements of $GF(25)$. Two vertices x, y are adjacent if and only if $y - x$ is a square in $GF(25)$. The graph $P(25)$ is invariant with respect to the Frobenius group $E_{25} \rtimes \mathbb{Z}_{12}$, that is to the subgroup of index 2 in $AGL(1, 25)$. This observation immediately implies that $P(25)$ is a rank 3 graph. In fact, $\mathrm{Aut}(P(25))$ has the order twice as large as $E_{25} \rtimes \mathbb{Z}_{12}$ (add to it the automorphism of $P(25)$ generated by the non-trivial automorphism of the field $GF(25)$). Finally, elementary considerations show that $\mathbb{Z}_{12}$ acts transitively on non-identity elements from three (of six) proper cyclic subgroups of the additive group of $GF(25)$. Thus, we conclude that Γ_1 and $P(25)$ may be regarded as the same Cayley graphs over E_{25}, that is, they are isomorphic.

The main class corresponding to Γ_1 produces just a single type, a single isotopy class and a single loop (which is a group). At this point we are faced with the necessity to consider another geometrical Γ_2. One can take any proper loop and obtain from it Γ_2. We, however, will prefer a more sophisticated procedure: to construct Γ_2 in such a manner that will introduce its group $\mathrm{Aut}(\Gamma_2)$.

Note that Γ_1 is a self-complementary graph, whereas Γ_2 is not self-complementary. Indeed, otherwise, a pair $(\Gamma_2, \overline{\Gamma}_2)$ will provide another affine plane of order 5 (different from the classical one), contradicting the well-known fact that the projective plane of order 5 is unique (up to isomorphism).

4.6 The Case $n = 5$, Part b

Let us first consider the direct sum $S_4 + S_3$ and the direct product $S_4 \times S_3$ of the symmetric groups S_4 and S_3, acting as permutation groups of degree 7 and 12 respectively (see [59] for the discussion of these operations for permutation groups). Both groups are isomorphic as abstract groups and have order 144. For a short while it is more convenient for us to work with $S_4 + S_3$. An easy exercise is to show that $S_4 + S_3$ has three different subgroups of index 2, namely, $A_4 + S_3, S_4 + A_3$, and $(S_4 + S_3)^{\mathrm{pos}}$.

Here, for a permutation group (H, Ω), the group $(H^{\mathrm{pos}}, \Omega)$ denotes the intersection of H with the alternating group of Ω. In other words, H^{pos} contains only even permutations from H.

Consider now the following geometry presented in Fig. 4(a). Its set of lines contains three 4-element horizontal lines and four 3-element vertical lines. We will denote it by $L_{3,4}$ (lattice of size 3×4) and we will call it briefly a *lattice*. Its point graph $\Gamma(L_{3,4})$ is depicted in Fig. 4(b). It is evident that $\mathrm{Aut}(L_{3,4}) = \mathrm{Aut}(\Gamma(L_{3,4})) = S_4 \times S_3$. Thus, we may use the more simple diagram in Part (a), even if we think of the lattice in terms of its point graph.

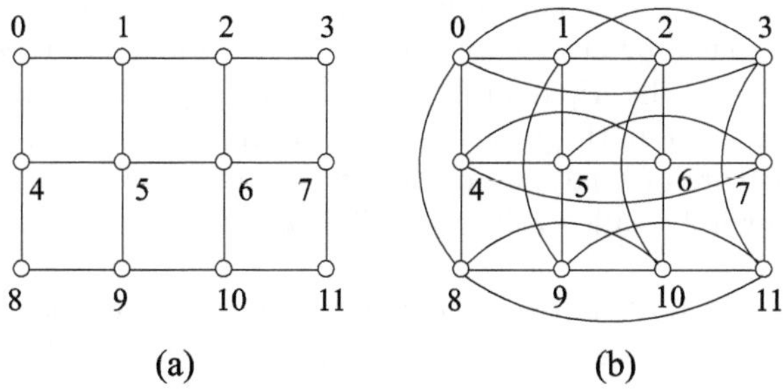

Fig. 4. The lattice $L_{3,4}$ and its point graph

Now we consider the group $G \cong (S_4 + S_3)^{\rm pos}$ of order 72 as a subgroup of $\mathrm{Aut}(L_{3,4})$. We may use the following presentation for G:

$$G = \left\langle \begin{array}{l} (0,1,2)(4,5,6)(8,9,10), (1,2,3)(5,6,7)(9,10,11), \\ (0,4,8)(1,5,9)(2,6,10)(3,7,11), (0,5)(1,4)(8,9)(2,6)(3,7) \end{array} \right\rangle.$$

Here the fourth generator corresponds to a simultaneous transposition of the first two rows and the first two columns of the lattice.

Note that G in its action of degree 12 contains odd permutations. We now associate more geometrical objects to $L_{3,4}$ and G.

The graph $\Gamma(L_{3,4})$ has anticliques of size 3, that is induced empty subgraphs of order 3. It is easy to see that altogether there are 24 such anticliques, which we will call *triples.*

Let $\{0,5,10\}$ be one of these triples. Then the stabilizer $G_{\{0,5,10\}}$ is a subgroup of order 6 in G:

$$G_{\{0,5,10\}} = \langle (0,5,10)(1,6,8)(2,4,9)(3,7,11), (0,5)(1,4)(8,9)(2,6)(3,7) \rangle.$$

This implies that the set of triples is split into two orbits of length 12 under the action of G.

Triples from the orbit $\{0,5,10\}^G$ will be called *right triples*, while ones from $\{2,5,8\}^G$ will be called *left triples.* For the reader's convenience we list the right triples T_r and the left triples T_l below:

$$\begin{aligned} T_r = \{&\{0,5,10\},\{0,6,11\},\{0,7,9\},\{1,4,11\},\{1,6,8\},\{1,7,10\}, \\ &\{2,4,9\},\{2,5,11\},\{2,7,8\},\{3,4,10\},\{3,5,8\},\{3,6,9\}\}, \\ T_l = \{&\{0,5,11\},\{0,6,9\},\{0,7,10\},\{1,4,10\},\{1,6,11\},\{1,7,8\}, \\ &\{2,4,11\},\{2,5,8\},\{2,7,9\},\{3,4,9\},\{3,5,10\},\{3,6,8\}\}. \end{aligned}$$

It is easy to observe that the incidence structures with the point set $[0,11]$ and the line sets T_r, T_l are each partial linear spaces; that is, in each geometry any two points are joined by at most one line. Let us associate to these geometries a graph: vertices are the triples and two triples are adjacent if they are disjoint.

It turns out that both graphs are isomorphic to $\Gamma(L_{3,4})$. Moreover, we also get a structure of a lattice on the sets T_r and T_l. The diagrams of both lattices, which will be called *right* and *left dual lattices* respectively, are shown in Fig. 5.

An analysis of both lattices shows that the group G may be described as the intersection of the automorphism groups of two incidence structures, namely $G = \mathrm{Aut}(([0,11],T_r)) \cap \mathrm{Aut}(([0,11],T_l))$. This representation turns out to be a crucial observation in the coming description of our model for the graph Γ_2, which was implicitly introduced above.

Recall that for $n \geq 5$ the following concepts are equivalent: the main class of a Latin square Q of order n, its graph $SRG(Q)$, the 3-net $\mathcal{N}(Q)$, and the

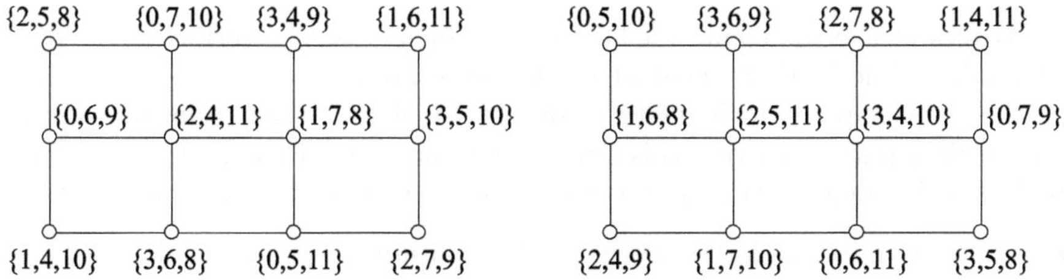

Fig. 5. Left and right dual lattices

transversal design $TD(Q)$. This means in particular, that each of these objects is uniquely reconstructible from the others.

Here we prefer to describe firstly the incidence structure $\gamma_2 = TD(3,5)$ which corresponds to Γ_2.

First we have to decide which of the two dual lattices will play a "special role". (Note that both selections provide different, though isomorphic structures.) Thus, let us decide, e.g., the right one. Then the points of our structure are formed by the union $\mathcal{P}_1 \cup \mathcal{P}_2$ of two sets, where $\mathcal{P}_1$ is the set of rows of the right dual lattice, while $\mathcal{P}_2$ is formed by 12 vertical pairs in the right dual lattice. Here, a vertical pair is a subset consisting of two triples in the same column. The points are divided into three disjoint groups, with each group containing a row from the right lattice and four vertical pairs which appear in the two remaining rows.

Below is a list of three groups (as it was produced by a computer package COCO), which, in principle, may be obtained by simple routine enumeration.

0. $\{\{0,5,10\},\{1,4,11\},\{2,7,8\},\{3,6,9\}\}$
1. $\{\{0,7,9\},\{1,6,8\},\{2,5,11\},\{3,4,10\}\}$
2. $\{\{0,6,11\},\{1,7,10\},\{2,4,9\},\{3,5,8\}\}$
3. $\{\{0,5,10\},\{1,6,8\}\}$
4. $\{\{1,6,8\},\{2,4,9\}\}$
5. $\{\{0,6,11\},\{2,7,8\}\}$
6. $\{\{0,5,10\},\{2,4,9\}\}$
7. $\{\{2,7,8\},\{3,4,10\}\}$
8. $\{\{0,7,9\},\{1,4,11\}\}$
9. $\{\{0,7,9\},\{3,5,8\}\}$
10. $\{\{0,6,11\},\{3,4,10\}\}$
11. $\{\{2,5,11\},\{3,6,9\}\}$
12. $\{\{1,4,11\},\{3,5,8\}\}$
13. $\{\{1,7,10\},\{3,6,9\}\}$
14. $\{\{1,7,10\},\{2,5,11\}\}$

In what follows we will keep the numbers from [0,14] for the points from the set $\mathcal{P} = \mathcal{P}_1 \cup \mathcal{P}_2$. Thus, in such a notation the groups are

$$\{0,4,9,10,14\},\{1,5,6,12,13\},\{2,3,7,8,11\}.$$

The set of lines is $\mathcal{L}_1 \cup \mathcal{L}_2 \cup \mathcal{L}_3$. Here $\mathcal{L}_1$ includes just one line, which corresponds to the whole right lattice. $\mathcal{L}_2$ and $\mathcal{L}_3$ correspond respectively to the triples in the right and left lattice. The incidence is described as follows: The unique line in $\mathcal{L}_1$ is incident to the three points in $\mathcal{P}_1$. The line defined by the right triple is incident to a row, in which this triple appears, and to two vertical pairs involving this triple. The line defined by the left triple say $\{a,b,c\}$ is incident to the three vertical pairs which are obtained as follows:

- find in the right lattice three triples containing $\{a,b\},\{a,c\}$ and $\{b,c\}$ respectively;
- each of these triples defines an "additional" vertical pair, that is two remaining triples in the same column of the right lattice;
- these additional pairs are incident to $\{a,b,c\}$.

For the reader's convenience we show (in a new numeration of the elements of γ_2) three lines; a line attributed to the whole lattice is $\{0,1,2\}$, a line for $\{0,5,10\}$ is $\{0,3,6\}$ and a line for $\{2,5,8\}$ is $\{8,10,13\}$.

We again provide a list of all lines obtained in such a manner, each line is represented as a 3-element set of incident points.

0. $\{0,1,2\}$	9. $\{1,11,14\}$	18. $\{3,5,14\}$
1. $\{0,3,6\}$	10. $\{2,9,12\}$	19. $\{6,10,11\}$
2. $\{1,3,4\}$	11. $\{0,11,13\}$	20. $\{3,10,12\}$
3. $\{2,5,10\}$	12. $\{2,13,14\}$	21. $\{3,9,13\}$
4. $\{2,4,6\}$	13. $\{8,10,13\}$	22. $\{4,7,13\}$
5. $\{0,5,7\}$	14. $\{7,12,14\}$	23. $\{4,5,8\}$
6. $\{0,8,12\}$	15. $\{6,8,14\}$	24. $\{6,7,9\}$
7. $\{1,8,9\}$	16. $\{5,9,11\}$	
8. $\{1,7,10\}$	17. $\{4,11,12\}$	

Now we have to check that the constructed incidence system γ_2 is indeed a $TD(3,5)$. An inspection may be simplified, if we observe that our group G has three orbits on pairs of points from $\mathcal{P}$, belonging to different groups, namely: $\{0,1\}^G$ of length 3, $\{0,3\}^G$ of length 36 and $\{8,10\}^G$ of length 36. Clearly, γ_2 contains one line through each of the representatives. Counting the total number of lines, we multiply it by 3 and deduce that indeed γ_2 is a partial linear space. Now it remains to describe $\mathrm{Aut}(\gamma_2)$. By construction, $G \leq \mathrm{Aut}(\gamma_2)$. Denote $\overline{G} = \mathrm{Aut}(\gamma_2)$. Then sequentially using a well-known orbit-counting lemma, we may obtain the order of $\overline{G}$. For this purpose we first have to show that 0 and 3 belong to different orbits of $\overline{G}$. This seems to be a slightly cumbersome job. For example, we may get the Latin square graph Γ_2 corresponding to γ_2, construct its subgraphs $\Gamma_2(0)$ and $\Gamma_2(1)$ induced by the neighbors of 0 and 1 respectively (recall that the vertices of Γ_2 are lines of γ_2). It turns that these subgraphs are non-isomorphic and may be distinguished by,

Table 5. Parallel classes of lines of $\tilde{\gamma}_2$

1. $\{0,1,5,6,11\}$	1. $\{0,2,7,8,9\}$	1. $\{0,3,4,10,12\}$
2. $\{2,4,17,22,23\}$	2. $\{1,4,15,19,24\}$	2. $\{1,2,18,20,21\}$
3. $\{3,8,13,19,20\}$	3. $\{3,5,16,18,23\}$	3. $\{5,8,14,22,24\}$
4. $\{7,10,16,21,24\}$	4. $\{6,10,14,17,20\}$	4. $\{6,7,13,15,23\}$
5. $\{9,12,14,15,18\}$	5. $\{11,12,13,21,22\}$	5. $\{9,11,16,17,19\}$

e.g., a number of maximal 3-cliques, indeed $\Gamma_2(0)$ has four maximal 3-cliques while $\Gamma_2(1)$ has only one maximal 3-clique, cf. also [92].

Thus, we now get that the lines 0 and 1 are in different orbits of $\overline{G}$, and thus points 0 and 3 are also in different orbits. Finally, we obtain (using each time suitable information about G):

$$\begin{aligned}|\overline{G}| &= |\overline{G}_3| \cdot |3^{|\overline{G}|}| = 12 \cdot |\overline{G}_3| = 12 \cdot |8^{\overline{G}_3}| \cdot |\overline{G}_{3,8}| = 36 \cdot |\overline{G}_{3,8}| \\ &= 36 \cdot |0^{\overline{G}_{3,8}}| \cdot |\overline{G}_{0,3,8}| = 72 \cdot |\overline{G}_{0,3,8}|.\end{aligned}$$

Now an easy brute force inspection shows that the automorphism, which fixes 0, 3 and 8, leaves all points of γ_2 in place.

Thus, $|\overline{G}| = 72 \cdot 1 = 72$, and therefore $\overline{G} = G$. The introduced group G is indeed the full automorphism group of γ_2. We note that, like for Γ_1, there corresponds just one type and one isotopy class to the graph Γ_2. It turns out, however, that the graph Γ_2 corresponds to the smallest case, where one main class contains a few non-isomorphic loops.

Let us first try to extract at least one loop from Γ_2. We consider the incidence structure $\tilde{\gamma}_2$, which is dual to γ_2. This is a 3-net with 25 points. It has three parallel classes of lines and each line contains five points. Let us (arbitrarily) distinguish the first, second and third parallel class. We now have freedom to manipulate the order of the members of each class. Thus, finally, we label five blocks in each of the three classes by elements from $[1,5]$. A suitable ordering is presented in Table 5. Now, the first class defines the rows, the second the tables and the third the contents of the cell of the tables. We pick up a row i, a column j, find the intersection $\{x\}$ of the corresponding blocks, detect the unique block in the third class containing x and put its label on the intersection of row i and column j.

Clearly, we obtain a Cayley table of a quasigroup. However, our lucky ordering provides us more: a loop. Its Cayley table is the following

1	2	3	4	5
2	1	4	5	3
3	5	1	2	4
4	3	5	1	2
5	4	2	3	1

L_1

It turns out that using the same graph Γ_2 we may also obtain the following Cayley tables of loops:

1	2	3	4	5
2	1	5	3	4
3	4	2	5	1
4	5	1	2	3
5	3	4	1	2

L_2

1	2	3	4	5
2	1	4	5	3
3	4	5	1	2
4	5	2	3	1
5	3	1	2	4

L_3

1	2	3	4	5
2	1	4	5	3
3	4	5	2	1
4	5	1	3	2
5	3	2	1	4

L_4

1	2	3	4	5
2	3	4	5	1
3	5	2	1	4
4	1	5	3	2
5	4	1	2	3

L_5

Considering multisets of elements on a diagonal, we may claim that the loops L_1, L_2, and L_5 each form a simple isomorphism class.

It remains to distinguish combinatorially the loops L_3 and L_4. For this purpose we will associate to each loop a certain set of permutations (see comments in Sect. 10). It is convenient to consider sets of permutations defined by columns. Then we associate to L_3 one permutation of type x_1^5 and four permutations of type x_2x_3, while to L_4 one of type x_1^5, one of type x_2x_3 and three of type x_5. Thus, the loops L_3 and L_4 are also not isomorphic.

4.7 The Case $n = 6$

There are 12 main classes of squares of order 6, three of them have a transitive automorphism group. In particular, $\mathrm{Aut}(SRG(\mathbb{Z}_6))$ has order $6^2 \cdot 2 \cdot 6 = 432$ and $\mathrm{Aut}(SRG(S_3))$ has order $6^2 \cdot 6 \cdot 6 = 1296$. The third main class with a transitive group will be presented in the next sections, its careful investigation is the central point in this article.

Here we will consider one more main class, which has an intransitive automorphism group, though quite large and clear. Our consideration is influenced by [16], though we act in a different fashion (see Sect. 10).

For current purposes it will be convenient for us to use the language of association schemes. First we will present a scheme on a set Ω with four classes of valency 5, 5, 5, 20, which corresponds naturally to a 3-net. Three schemes with three classes, each obtained by merging of two classes of valency 5, will correspond to the different types. Finally, a merging of three classes of valency 5 leads to the Latin square graph and its complement.

Let us start with the group A_5 of order 60 in its natural action on five points. The set of cardinality 36 comprises two orbits Ω_1 and Ω_2 defined by the induced action of $(A_5, [0,4])$. Let us consider a pentagon, that is a regular connected undirected graph of valency 2. Its automorphism group D_5 consists of even permutations only. Therefore, an orbit of a pentagon under the action of A_5 has cardinality 6. This is the set Ω_1, it is represented in Fig. 6.

Let us consider an ordered pair of two disjoint 2-element subsets of $[0,4]$. Clearly, all such pairs form a single orbit under the action of A_5. This is set Ω_2 of cardinality 30.

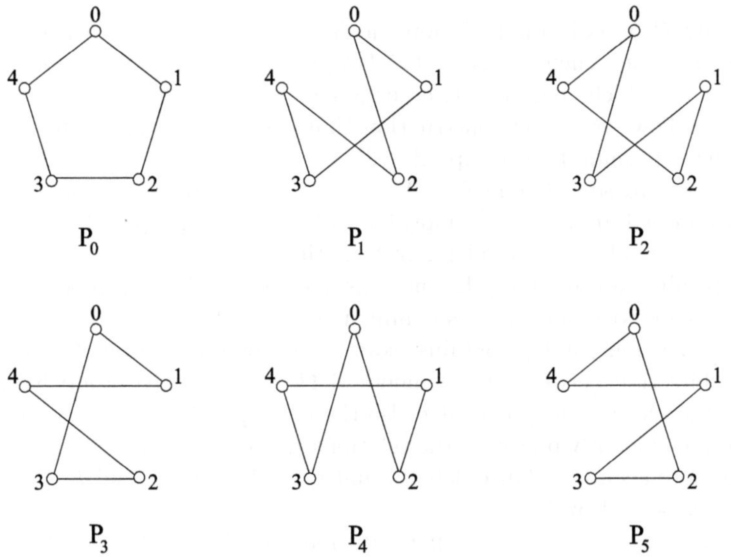

Fig. 6. The orbit Ω_1 of A_5

Now we need to define classes of valency 5 in the association scheme. Each such class corresponds to a graph of the form $6 \circ K_6$. Therefore, in order to represent such a specific binary relation, it is enough to define a partition of Ω into six sets of equal cardinality.

The partition corresponding to R_1, contains the set Ω_1 as one cell, five other cells are defined by elements of $[0,4]$. For example, a cell defined by 4 contains ordered pairs

$$(\{0,1\},\{2,3\}),\ (\{0,2\},\{1,3\}),\ (\{0,3\},\{1,2\}),\ (\{1,2\},\{0,3\}),\ (\{1,3\},\{0,2\}),\ (\{2,3\},\{0,1\}).$$

The partition corresponding to R_2 contains six cells, each is defined by an element of Ω_1 like

$$\Big\{P_0,(\{0,1\},\{2,4\}),(\{1,2\},\{0,3\}),(\{2,3\},\{1,4\}),(\{3,4\},\{0,2\}),(\{0,4\},\{1,3\})\Big\}.$$

Similarly, a partition corresponding to R_3 is defined by a typical cell

$$\Big\{P_0,(\{2,4\},\{0,1\}),(\{0,3\},\{1,2\}),(\{1,4\},\{2,3\}),(\{0,2\},\{3,4\}),(\{1,3\},\{0,4\})\Big\}.$$

It follows immediately from the definition that the partitions R_1, R_2, R_3 are orthogonal, in a sense, that corresponding graphs are edge-disjoint.

Finally, R_0 is a diagonal relation and $R_4 = \Omega^2 \setminus (\bigcup_{i=0}^{3} R_i)$. We get an amorphic association scheme $\mathcal{M}_0 = (\Omega, \{R_0, R_1, R_2, R_3, R_4\})$ with four classes (cf. [45]), to which corresponds a certain main class of Latin squares of order 6. It follows from the construction that $\mathcal{M}_0$ is invariant with respect to the induced permutation group (A_5, Ω).

It is easy to see, that in fact $\mathcal{M}_0$ is invariant with respect to the larger group of order 120, which is isomorphic to S_5. For this purpose let us consider A_5 above as a subgroup of index 2 of S_5 acting on the same set $[0, 4]$. Then S_5 acts naturally on the set Ω_2. Let us consider the set Ω_1' of all pairs consisting of a pentagon together with its complement. Clearly, $|\Omega_1'| = |\Omega_1|$; moreover, because each pair of Ω_1' contains exactly one pentagon from Ω_1, we get a natural bijection between the elements of Ω_1 and Ω_1'. The group S_5 acts on Ω_1' and therefore we may consider also the corresponding action of S_5 on Ω_1. This action evidently preserves the relations R_1 and R_4 and the union $R_2 \cup R_3$, while a more accurate glance detects that the relations R_2 and R_3 separately are also preserved by S_5.

Let us now define one more permutation τ on Ω, which fixes all elements of Ω_1, while each ordered pair of the form $(\{a, b\}, \{c, d\})$ is transposed with $(\{c, d\}, \{a, b\})$. Again very easy reasonings show that τ preserves the relations R_1 and R_4 and transposes the relations R_2 and R_3. Moreover, τ commutes with all elements of S_5, therefore we may define the group $S_5 \times \langle \tau \rangle$ which is isomorphic to $S_5 \times S_2$.

Now we consider the association schemes

$$\mathcal{M}_1 = (\Omega; \{R_0, R_1, R_2 \cup R_3, R_4\}), \qquad \mathcal{M}_2 = (\Omega; \{R_0, R_1 \cup R_3, R_2, R_4\}),$$
$$\mathcal{M}_3 = (\Omega; \{R_0, R_1 \cup R_2, R_3, R_4\}), \qquad \mathcal{M}_4 = (\Omega; \{R_0, R_1 \cup R_2 \cup R_3, R_4\}).$$

It is again evident from the construction that $\mathrm{Aut}(\mathcal{M}_1)$ (and thus also $\mathrm{Aut}(\mathcal{M}_4)$) contains a subgroup isomorphic to $S_5 \times S_2$ of order 240. Simple combinatorial arguments may be used to show that we already know all the automorphism groups, namely $\mathrm{Aut}(\mathcal{M}_1) = \mathrm{Aut}(\mathcal{M}_4) = S_5 \times S_2$ and $\mathrm{Aut}(\mathcal{M}_0) = \mathrm{Aut}(\mathcal{M}_2) = \mathrm{Aut}(\mathcal{M}_3) = S_5$.

Moreover, each permutation from $(S_5 \times S_2) \setminus S_5$ establishes an isomorphism between $\mathcal{M}_2$ and $\mathcal{M}_3$, while $\mathcal{M}_1$ is not isomorphic to $\mathcal{M}_2$ and $\mathcal{M}_3$. Thus, we have obtained a desired main class of Latin squares, which is split into two types, one corresponds to $\mathcal{M}_1$ and the other to $\mathcal{M}_2$ and $\mathcal{M}_3$.

One more attractive step would be a description of the isotopy classes in our main class, as well as the evident detection of at least one loop from this class. We, however, stop here, referring again to [16] for a nice description of a loop in terms of the icosahedron. (Note the well-known fact that the automorphism group of the icosahedron graph is exactly $A_5 \times S_2$.) Here we just present a multiplication table of a loop from the considered main class:

1	2	3	4	5	6
2	1	6	3	4	5
3	6	1	5	2	4
4	3	5	1	6	2
5	4	2	6	1	3
6	5	4	2	3	1

5 The Remark of Barlotti and Strambach

Now we are coming back to the main line of our presentation: the consideration of "group-like" quasigroups.

Proposition 5. *Consider the following Latin square Q_6 (No 3.1.1 in [34]):*

1	2	3	4	5	6
2	3	1	5	6	4
3	1	2	6	4	5
4	6	5	2	1	3
5	4	6	3	2	1
6	5	4	1	3	2

Then:

(a) The main class of Q_6 does not contain a group;
(b) $G = \mathrm{Aut}(\Gamma(Q_6))$ is a transitive permutation group of degree 36 and order 648;
(c) G has a regular subgroup.

Note that the quasigroup Q_6 is a well-known object (see more details in Sect. 10). In particular, parts (a) and (b) of our claim can be extracted from many sources in the literature.

Part (c) may be extracted from [96]. Nevertheless, it seems that, as an entity, the whole proposition appeared for the first time in [50], where it was proved with the aid of a computer.

6 Computer Aided Answer

The computer aided proof of part (c) in Proposition 5 was in a sense, a non-expected by-product of a general project, which was undertaken by A. Heinze in his Ph.D. Thesis [50].[4]

[4] Supervisors M. Klin and U. Knauer.

6.1 Computer-based Analysis of the Example with GAP

The use of the computer system GAP (see [43]) allows us to obtain some set of generators for our group G of order 648.

Recall that originally we started from a catalogue of strongly regular graphs [95], inspected a concrete graph with its automorphism group and checked whether this graph Γ is geometric, i.e., if $\Gamma = SRG(Q)$.

However, as soon as we agree that Q is our source of information we just create a description of $SRG(Q)$, put it into GAP (more exactly into its share package GRAPE [94]) and then determine a desired set of generators for the group G. Moreover, using some of the simple routines, which are described in [50] (cf. also [53]), we easily observe that G has two (up to isomorphism) regular subgroups H_1 and H_2 (of order 36). One of these groups, say H_2, can be easily identified by GAP as $H_2 = S_3 \times S_3$, that is the direct product of two symmetric groups of degree 3. For the whole group G it was not so easy to find its structure. Thus, using ad hoc arguments we formulated the following question:

> Is it true that $G \cong (S_3 \wr S_3)^{\mathrm{pos}}$. In other words does G consist of all even permutations in the standard action of the wreath product of S_3 with S_3 on nine points?

The immediate answer of GAP was "yes".

In a similar manner we guessed that for any point $x \in \Omega$ (here (G, Ω) is a transitive permutation group of degree 36) $G_x \cong D_9$, where G_x is the stabilizer of x in G and D_9 is a dihedral group of degree 9 and order 18. Again, we obtained from GAP the confirmation of our guess. At this moment we may complete to use GAP. We know the whole group G (as an abstract group), we know the subgroup D_9 of G, which returns the stabilizer. Therefore, the action (G, Ω) is permutationally isomorphic to the action of G on the cosets modulo D_9.

6.2 Further Computer-based Analysis with COCO

Now we are prepared to use another computer package COCO (COherent COnfigurations, cf. [39]). We would like to obtain more structural information about our permutation group (G, Ω). First we need to construct this action. For this purpose we have to create for COCO a description of G and a description of its subgroup D_9 as an input.

We describe G as a group of degree 9 acting on the set [0,8] with the aid of generators. Thus $G = \langle g_1, g_2, g_3, g_4 \rangle$, where $g_1 = (0, 1, 2, 3, 4, 5, 6, 7, 8)$, $g_2 = (0, 3, 6), g_3 = (0, 3)(1, 4)$, $g_4 = (0, 1)(3, 4, 6, 7)$. We present extra comments on this set below.

Instead of a description of the subgroup D_9, COCO requires (and this is in fact one of its main paradigms) a relational structure S on the same set [0,8], preferably as simple as possible, such that $\mathrm{Aut}(S) \cap G = D_9$. It is clear how to

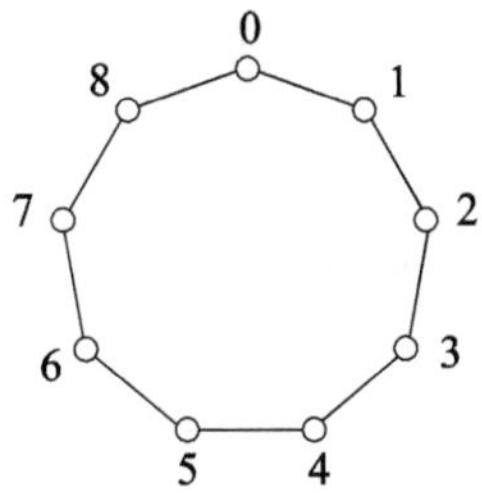

Fig. 7. Cycle C_9 of length 9

manage a fulfillment of such a requirement. Consider an unordered cycle C_9 of length 9 and let us set $S := C_9$ (we mean here by a cycle a regular connected graph of valency 2). It is convenient to select a canonical labeling of vertices in our copy of C_9 like in Fig. 7. All automorphisms from $D_9 = \mathrm{Aut}(C_9)$ are even permutations. Thus, we can select in advance (knowing our copy of C_9) such a representation of G that our general requirement above, that is $\mathrm{Aut}(C_9) \cap G = D_9$, will be substituted by a simpler condition: $D_9 < G$. This provides the way to obtain a description of G. Consider the disconnected graph $3 \circ K_3$ as shown in Fig. 8.

Evidently $D_9 < \mathrm{Aut}(3 \circ K_3)$ and $G \cong (\mathrm{Aut}(3 \circ K_3))^{\mathrm{pos}} = (S_3 \wr S_3)^{\mathrm{pos}}$. Finally, we just need to check that $g_1, g_2, g_3, g_4 \in (S_3 \wr S_3)^{\mathrm{pos}}$ and moreover that $(S_3 \wr S_3)^{\mathrm{pos}} = \langle g_1, g_2, g_3, g_4 \rangle$. Thus, we provide the promised justification of our generators.

Now we are ready to discuss the output of COCO. It consists of the following pieces of information:

- the generators $\widetilde{g_1}, \widetilde{g_2}, \widetilde{g_3}, \widetilde{g_4}$ of the induced action of G on the set Ω of all different copies of C_9 with respect to permutations from G; this action has degree $\frac{648}{18} = 36$;
- the 2-orbits of this action and the corresponding subdegrees: $1, 1, 1, 6, 9, 9, 9$;
- the intersection numbers of a corresponding association scheme $\mathcal{M} = (\Omega, \text{2-orb}(G, \Omega))$ with six classes;
- all mergings of classes of this association scheme.

In particular, we obtain that there exist three non-Schurian mergings with two classes with valencies $1, 15, 20$. It turns out that these three mergings correspond to three (isomorphic) copies of the strongly regular graph

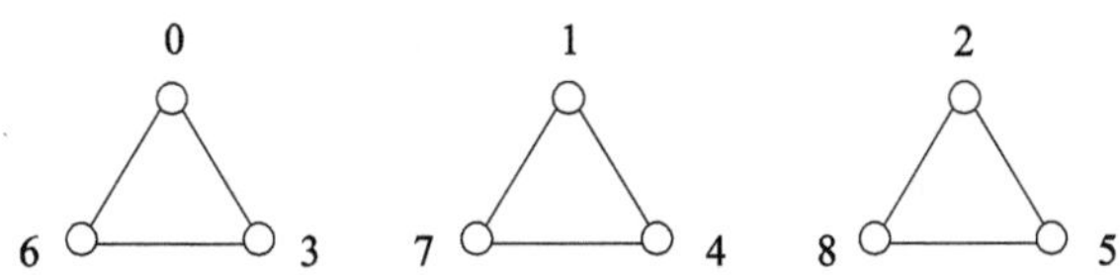

Fig. 8. Graph $3 \circ K_3$

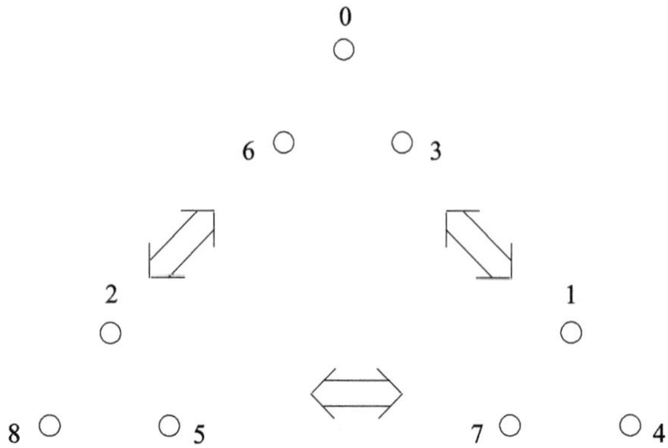

Fig. 9. The auxiliary graph $\Delta = \overline{3 \circ K_3}$

$\Gamma = SRG(Q_6)$ with parameters $v = 36, k = 15, \lambda = 6, \mu = 6$. (One may again use GAP to confirm this information.) Thus, we have taken a first step towards understanding our graph Γ. More exactly we approach a computer free explanation of some results obtained by the conjunction of GAP and COCO.

Consider the auxiliary graph $\Delta = \overline{3 \circ K_3}$ as in Fig. 9.

Here the sign $\Longleftrightarrow$ in Fig. 9 means a set of nine edges of a complete bipartite graph $K_{3,3}$. An easy combinatorial exercise shows that Δ has exactly $3 \cdot 3 \cdot 2 \cdot 2 \cdot 2 = 72$ spanning subgraphs which are Hamiltonian cycles in Δ. Our initial canonical copy of C_9 is shown once more in Fig. 10.

There are two orbits of length 36 in the action of G on these 72 cycles. Recall that G is a "half" of $\mathrm{Aut}(\Delta)$, while all permutations in D_9 are even. Thus, we have the typical Buridanian Donkey problem: How do we select one of these two isomorphic orbits? The criterion which we are using is to take the orbit which includes our canonical cycle C_9. Now, analyzing the output of COCO, we are obtaining representatives of neighbors of a canonical C_9 in all six 2-orbits of (G, Ω) (see Fig. 11).

The huge numbers in Fig. 11 show the corresponding subdegrees of the 2-orbits. Now we have three possibilities to merge a relation of valency 6 with

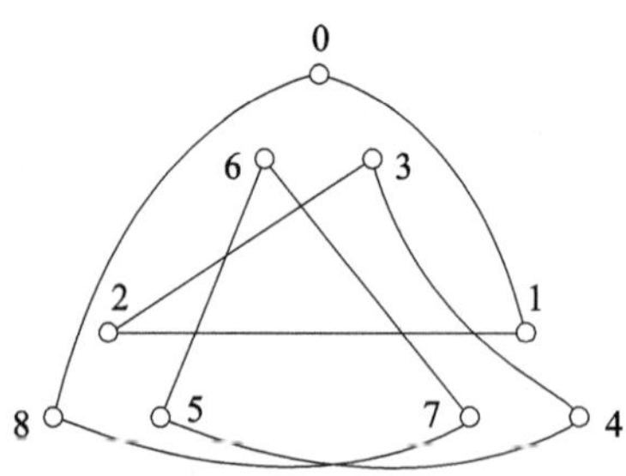

Fig. 10. Copy of C_9 as Hamiltonian cycle in Δ

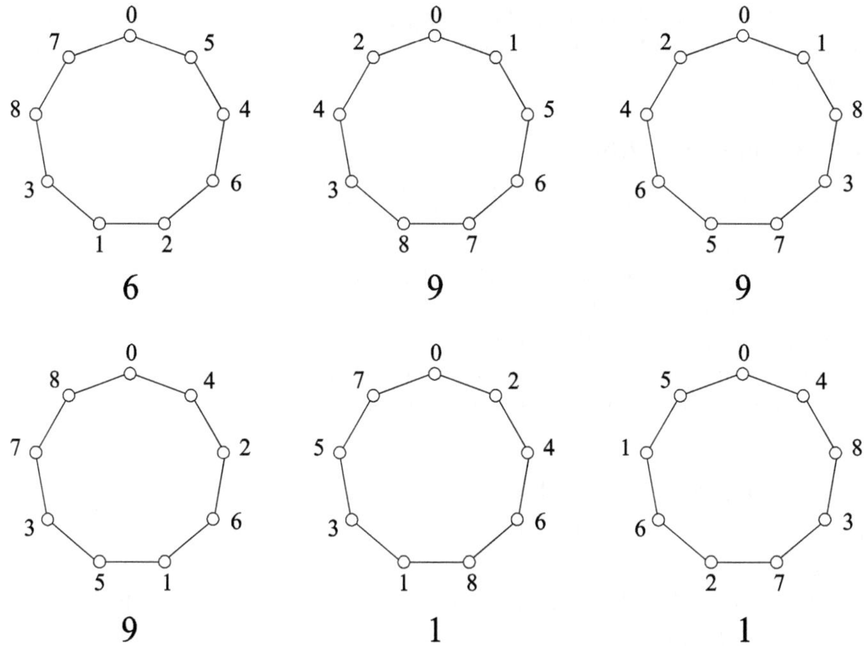

Fig. 11. Representatives of neighbors of a canonical C_9 in all six 2-orbits of (G, Ω)

a relation of valency 9. Each time we reach a desired strongly regular graph of valency 15. We call the obtained description an *explanation* of the graph Γ. We mean that we, in principle, may now explain *post factum* vertices and edges of Γ without the use of a computer.

In the next section we present a real computer free *interpretation*: such way of reasonings where everything, including proofs, are independent of routine computations.

7 Computer Free Interpretation

This section is the central one in our presentation. We are trying to make it self-contained as much as it is possible.

7.1 General Idea

Starting from a given Latin square Q (of order 6) we are getting a corresponding 3-net $\mathcal{N}(Q)$ with 36 points and 18 lines. This is a partial geometry. Recall that the dual incidence structure is also a partial geometry namely a transversal design $TD(3,6)$.

It turns out that in our case of Q_6 it is much easier to work with $TD(3,6)$ than with $\mathcal{N}(Q_6)$. The reason is that we have a very transparent explicit description of the points of $TD(3,6)$ and still above an implicit description of lines (= points of the net). This description will be enough to check that all axioms of $TD(3,6)$ are satisfied and that we may justify all requested properties.

7.2 Axioms of $TD(3,6)$

We need an incidence structure $\mathcal{S} = (\mathcal{P}, L)$ such that:

(i) $|\mathcal{P}| = 18, |\mathcal{L}| = 36$.
(ii) $\mathcal{P} = P_1 \cup P_2 \cup P_3$ is a partition of $\mathcal{P}$ into three classes of size 6, which are called *groups of* $\mathcal{S}$.
(iii) Each block has cardinality 3.
(iv) Every unordered pair of distinct points either is contained in exactly one group, or is contained in exactly one block, but not both.

7.3 Description of a Model

Our starting basic object will be again the graph $\Delta = \overline{3 \circ K_3}$, as it was depicted in Fig. 9.

By a *partial 1-factor of* Δ we mean a set of three edges of Δ which form a 1-factor of a graph $K_{3,3}$ (a subgraph of Δ which is depicted by the sign "$\Longleftrightarrow$" in Fig. 9). An example of such a 1-factor is provided in Fig. 12.

Clearly we have $3 \cdot 3! = 18$ partial 1-factors. The set $\mathcal{P}$ of our incidence structure $\mathcal{S}$ consists of all such objects. The blocks are exactly 36 selected copies of C_9 from our orbit Ω as they were defined in Sect. 6.2. It is clear from our construction that each Hamiltonian cycle can be split into three partial 1-factors. This provides us with a natural incidence relation between points and blocks.

7.4 Fulfillment of Axioms

It is immediately clear from our definition that the Axioms (i) and (iii) are satisfied. The edge set of Δ is naturally split into edge sets of three copies of $K_{3,3}$ (i.e., three signs $\Longleftrightarrow$ in Fig. 9). Each copy of $K_{3,3}$ evidently contains six partial 1-factors. Thus, we have a partition of $\mathcal{P}$ which is requested in (ii). It remains to check if Axiom (iv) is satisfied. More exactly, we have to take two

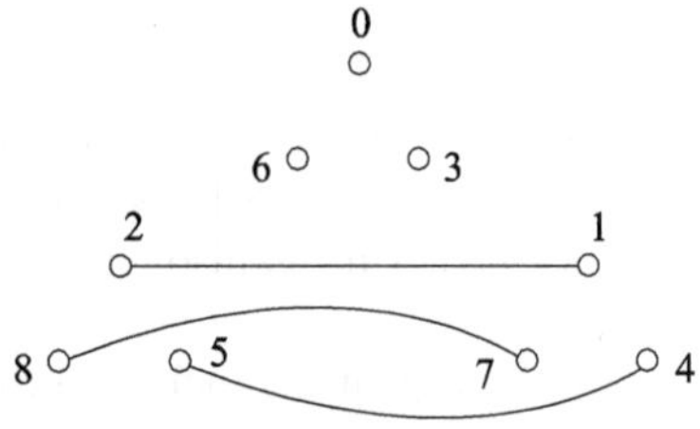

Fig. 12. Example of a 1-factor of Δ

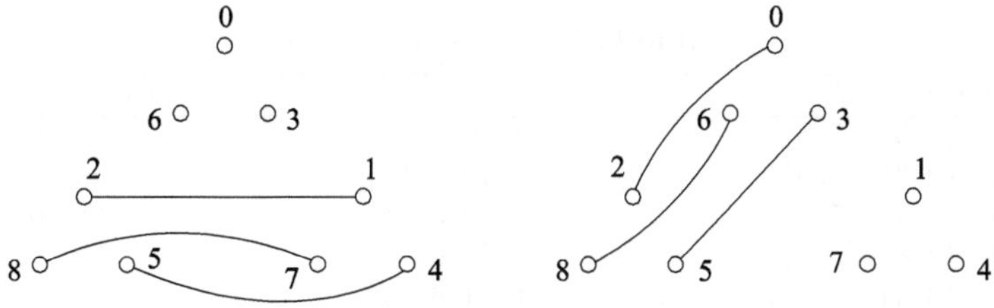

Fig. 13. Two partial 1-factors of Δ

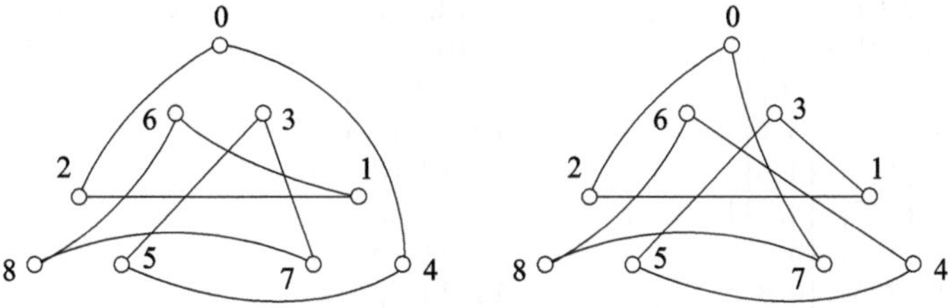

Fig. 14. Hamiltonian cycles C', C'' of Δ

partial 1-factors in distinct groups and to show that they appear in exactly one cycle in Ω. Note that the group G acts transitively on such pairs. This implies that we can present a purely pictorial proof. Two partial 1-factors are depicted in Fig. 13.

It is easy to see that together they appear in exactly two Hamiltonian cycles C', C'' of Δ, see Fig. 14.

Let us now find an arbitrary permutation g which sends C' to C'', for example $g = (4,7)(5,8)(3,6)$. The set of all such permutations is a coset $D_9 \cdot g$ in S_9, which evidently consists of odd permutations only. Therefore, exactly one of C' and C'' belongs to our orbit Ω. We are done (and we do not care to know, which of the two is actually in Ω). Therefore, indeed, $\mathcal{S}$ is a model of a transversal design $TD(3,6)$.

7.5 The Group Aut($\mathcal{S}$)

It is clear from the construction that $\overline{G} = \mathrm{Aut}(\mathcal{S}) \geq G$. We need to prove that, in fact, $|\mathrm{Aut}(\mathcal{S})| = |G|$. This evidently will imply that $\mathrm{Aut}(\mathcal{S}) = G$. For this part of the proof we prefer to use a list of points and lines of $\mathcal{S}$. We obtained it using COCO, though any alternative hand or computer tools may be exploited by the reader.

Below we provide a list of elements of $\mathcal{P}$ labeled from 0 to 17, each label corresponds to a partial 1-factor of Δ (see Table 6). Note that we have the following groups (partition of $\mathcal{P}$):

$$\{0,2,6,10,15,16\},\ \{1,3,5,7,9,14\},\ \{4,8,11,12,13,17\}.$$

Table 6. Elements of $\mathcal{P}$ and $\mathcal{L}$

Elements of $\mathcal{P}$	Elements of $\mathcal{L}$	
0. $\{\{1,2\},\{4,5\},\{7,8\}\}$	0. $\{0,1,4\}$	18. $\{6,7,17\}$
1. $\{\{0,8\},\{2,3\},\{5,6\}\}$	1. $\{0,5,11\}$	19. $\{1,11,15\}$
2. $\{\{1,5\},\{2,4\},\{7,8\}\}$	2. $\{2,3,4\}$	20. $\{0,3,17\}$
3. $\{\{0,2\},\{3,8\},\{5,6\}\}$	3. $\{3,6,12\}$	21. $\{7,12,15\}$
4. $\{\{0,1\},\{3,4\},\{6,7\}\}$	4. $\{1,6,13\}$	22. $\{5,16,17\}$
5. $\{\{0,5\},\{2,6\},\{3,8\}\}$	5. $\{0,13,14\}$	23. $\{2,5,8\}$
6. $\{\{1,8\},\{2,4\},\{5,7\}\}$	6. $\{2,7,13\}$	24. $\{9,15,17\}$
7. $\{\{0,8\},\{2,6\},\{3,5\}\}$	7. $\{3,8,15\}$	25. $\{1,8,16\}$
8. $\{\{0,4\},\{1,3\},\{6,7\}\}$	8. $\{0,7,8\}$	26. $\{9,10,13\}$
9. $\{\{0,5\},\{2,3\},\{6,8\}\}$	9. $\{2,9,11\}$	27. $\{1,2,12\}$
10. $\{\{1,2\},\{4,8\},\{5,7\}\}$	10. $\{5,10,12\}$	28. $\{3,13,16\}$
11. $\{\{0,7\},\{1,3\},\{4,6\}\}$	11. $\{8,10,14\}$	29. $\{2,14,17\}$
12. $\{\{0,1\},\{3,7\},\{4,6\}\}$	12. $\{6,8,9\}$	30. $\{4,9,16\}$
13. $\{\{0,4\},\{1,6\},\{3,7\}\}$	13. $\{1,10,17\}$	31. $\{6,11,14\}$
14. $\{\{0,2\},\{3,5\},\{6,8\}\}$	14. $\{4,7,10\}$	32. $\{0,9,12\}$
15. $\{\{1,5\},\{2,7\},\{4,8\}\}$	15. $\{4,14,15\}$	33. $\{12,14,16\}$
16. $\{\{1,8\},\{2,7\},\{4,5\}\}$	16. $\{4,5,6\}$	34. $\{7,11,16\}$
17. $\{\{0,7\},\{1,6\},\{3,4\}\}$	17. $\{3,10,11\}$	35. $\{5,13,15\}$

In Table 6 we also provide a list of elements of $\mathcal{L}$ where each line in $\mathcal{L}$ is identified with a 3-element subset of $\mathcal{P}$ as it is labeled from the left.

Now we would like to find a pointwise stabilizer $H = \overline{G}_{0,1}$ of the points $0, 1$ in $\mathrm{Aut}(\mathcal{S})$. Clearly, H leaves also point 4 in place. In what follows we will associate the gray color to the first group, the black color to the second and the white color to the third group. Besides line $\{0,1,4\}$ there are exactly 15 lines each of which contains exactly one of the elements $0, 1, 4$. Depending on the element, we will also attribute a color to such a line. Thus, we come to an auxiliary structure with colored vertices and edges as follows (see Fig. 15, white lines are thin).

The automorphism group H^* of this structure is generated by two permutations:

$$h_1 = (3,7,9)(2,10,16)(8,12,17) \quad \text{and} \quad h_2 = (5,14)(6,15)(11,13),$$

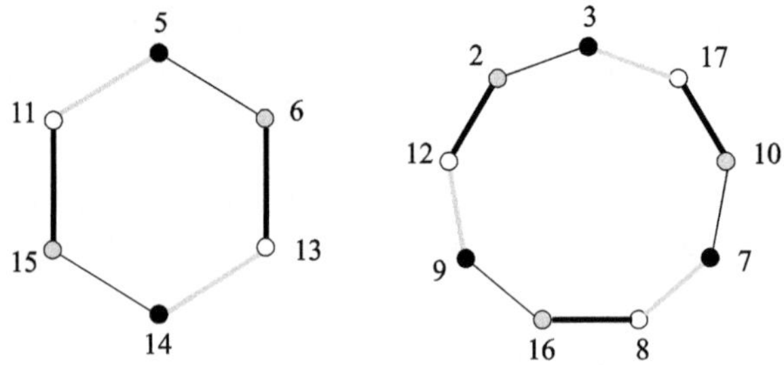

Fig. 15. Auxiliary structure with colored vertices and edges

that is H^* has order 6. However, for $\{3,6,12\} \in \mathcal{L}$ we have $\{3,6,12\}^{h_2} = \{3,15,12\} \notin \mathcal{L}$. Therefore $h_2 \notin H$ and $|H| \leq 3$. The reader can easily check that, in fact, $h_1 \in G$ (and thus in H), it corresponds to an automorphism $(1,4,7)(2,5,8)(0,3,6)$ of Δ. Therefore, $|H| = 3$. Now we obtain

$$|\overline{G}| = |\overline{G}_0| \cdot |0^{\overline{G}}| = 18 \cdot |\overline{G}_0| = 18 \cdot |\overline{G}_{0,1}| \cdot |1^{\overline{G}_0}| = 18 \cdot 12 \cdot |\overline{G}_{0,1}| = 18 \cdot 12 \cdot 3 = 648.$$

Therefore, $\overline{G} = \mathrm{Aut}(\mathcal{S}) = G$.

7.6 Transversal Designs of Groups of Order 6

Recall now that according to Theorem 1

$$|\mathrm{Aut}(SRG(\mathbb{Z}_6))| = 6^2 \cdot 2 \cdot 6 = 432 \quad \text{and} \quad |\mathrm{Aut}(SRG(S_3))| = 6^2 \cdot 6 \cdot 6 = 1296.$$

Evidently, there are exactly two distinct groups of order 6. Therefore, because the order of G is not equal to 432 or 1296, we conclude with the aid of Corollary 3, that our transversal design $\mathcal{S}$ is not coming from a group. In other words, any loop corresponding to this design is indeed a proper loop.

7.7 Regular Subgroup for the Loop Case

Now we have to find a regular subgroup in the action $(G, \mathcal{P})$. For this purpose we may use the action of G on the graph Δ. It is sufficient to find a subgroup H of G such that all $h \in H$, $h \neq e$ do not preserve any of the 36 copies of the cycle C_9, which form the set $\mathcal{L}$. First we will present one copy of such a subgroup.

Let $K_1 := \langle (0,3,6), (0,3)(2,5) \rangle$, $K_2 := \langle (1,4,7), (1,4)(2,5) \rangle$ and define $H_2 := K_1 \times K_2$. Then:

(a) $K_1 \cong K_2 \cong S_3$, moreover $H_2 \cong S_3 \times S_3$;
(b) $|H_2| = 36$;
(c) $H_2 \leq G$;
(d) there is no copy of C_9 in $\mathcal{L}$ which is preserved by H_2.

This means that we proved (more or less without essential use of a computer) that we have an example which provides a positive answer on the question of Barlotti-Strambach.

In fact, our group G contains one more copy of a regular subgroup (up to isomorphism). Let now $a = (0,3,6), b = (1,4,7), c = (0,1,3,4)(6,7)$ and define $H_1 := \langle a, b, c \rangle$. It is easy to check that $c^{-1}ac = b$ and $c^{-1}bc = a^2$. Therefore, $\langle a, b \rangle \cong (\mathbb{Z}_3)^2$ is a normal subgroup of H_1.

Thus, $H_1 \cong (\mathbb{Z}_3)^2 \rtimes \mathbb{Z}_4$ is a group of order 36. This group is non-isomorphic to H_2, because H_2 does not have elements of order 4.

Clearly $H_1 \leq G$. Each element of H_1 fixes the vertices $2, 5, 8$ of Δ. Taking into account that each non-identical permutation in D_9 has at most one fixed point we have shown that H_1 acts regularly on the set $\mathcal{L}$. Thus, G indeed has at least two regular subgroups. In fact, H_1, H_2 are the only regular subgroups in G (up to isomorphism). This fact was established, using GAP (see details in [50]).

8 Exceptional Quasigroup Re-visited

Recall our methodology. We started from the exceptional quasigroup Q_6, constructed its graph Γ, found the group $G = \mathrm{Aut}(\Gamma)$ and after that restarted our consideration again. Thus we have assumed to pretense of not remembering Q_6. All we know is the group G, the action of G on future points $\mathcal{P}$ and lines $\mathcal{L}$ and the incidence in $\mathcal{S}$. We check that indeed $\mathcal{S}$ is a $TD(3,6)$ and $G = \mathrm{Aut}(\mathcal{S})$.

Now we would like to come back to Q_6. Clearly, there are many options to establish a quasigroup belonging to $\mathcal{S}$; but we will reconstruct Q_6 exactly.

In principle, the methodology of our behavior is clear. As soon as we have $\mathcal{S}$, first we should attribute to it three groups that will be the rows, columns and elements of a future Latin square (we have here 3! options). After that we label the elements of each group by the numbers from the set $\{1,2,3,4,5,6\}$ – altogether $(6!)^3$ options. Finally, we read our list of lines in L and fill, step by step each of the 36 cells of a square. We, of course, may use for this purpose labels of elements from $\mathcal{P}$ and $\mathcal{L}$.

It turns out that it is a bit tricky to find a possible labeling of the elements from $\mathcal{P}$ and $\mathcal{L}$. However, by a simple trial-and-error method we found a labeling which is suitable. In the following Table 7 we did not label the rows, columns and elements of the Latin square by numbers from the set $\{1,2,3,4,5,6\}$ but by permutations of the symmetric group S_3. As we will see later, there is a correspondence between our quasigroup and the group S_3.

Table 7 gives a one-to-one correspondence between elements of each of the three groups forming $\mathcal{P}$ (in the labeling of Table 6) and the permutations of S_3. To create the quasigroup Q_6 we have to define the products of the rows and the columns. For example, take the second row – the permutation (a,b,c) – and the fourth column – the permutation (a,b). The corresponding numbers

Table 7. Labeling of points in $\mathcal{P}$

Rows		Columns		Symbols	
Permutation	No. in $\mathcal{P}$	Permutation	No. in $\mathcal{P}$	Permutation	No. in $\mathcal{P}$
e	4	e	15	e	14
(a,b,c)	13	(a,b,c)	6	(a,b,c)	5
(a,c,b)	11	(a,c,b)	0	(a,c,b)	1
(a,b)	8	(a,b)	2	(a,b)	3
(b,c)	12	(b,c)	10	(b,c)	7
(a,c)	17	(a,c)	16	(a,c)	9

Table 8. Created Cayley table of the exceptional quasigroup Q_6

	e	(a,b,c)	(a,c,b)	(a,b)	(b,c)	(a,c)
e	e	(a,b,c)	(a,c,b)	(a,b)	(b,c)	(a,c)
(a,b,c)	(a,b,c)	(a,c,b)	e	(b,c)	(a,c)	(a,b)
(a,c,b)	(a,c,b)	e	(a,b,c)	(a,c)	(a,b)	(b,c)
(a,b)	(a,b)	(a,c)	(b,c)	(a,b,c)	e	(a,c,b)
(b,c)	(b,c)	(a,b)	(a,c)	(a,c,b)	(a,b,c)	e
(a,c)	(a,c)	(b,c)	(a,b)	e	(a,c,b)	(a,b,c)

of points in $\mathcal{P}$ are 13 and 2. Now, to define the product of the multiplication $(a,b,c)\cdot(a,b)$ we have to find the unique line in $\mathcal{L}$ that contains the points 13 and 2. Table 6 shows that this is the line No. 6 with points $\{2,7,13\}$. Hence the product of $(a,b,c)\cdot(a,b)$ is point No. 7 which corresponds to the element (b,c) as Table 7 shows. Repeating this procedure for all combinations of rows and columns we obtain the Cayley table of our exceptional quasigroup Q_6 as it is given in Table 8.

Note that in the framework of our entire presentation it is natural to use elements in S_3 as labels (in fact $S_3 \cong D_3$, cf. the notation in the next section). The reader should check that up to the used notation we obtain exactly our starting Cayley table for Q_6. So, in a sense, our Q_6 is interpreted as a twisted dihedral group D_3 of order 6. It remains to describe the form of this twist.

Denote by $x \circ y$ the multiplication in Q_6, while xy is the usual multiplication in D_3. Let us select a "standard" permutation $s = (a,b,c)$, a generator of a cyclic group of order 3. The permutation s will play the role of a distinguished permutation from D_3. Then we obtain

$$x \circ y = \begin{cases} xy, & \text{if } \operatorname{sign}(x) = \operatorname{sign}(y) = 1, \\ xy, & \text{if } \operatorname{sign}(x)\cdot\operatorname{sign}(y) = -1, \\ xys, & \text{if } \operatorname{sign}(x) = \operatorname{sign}(y) = -1. \end{cases}$$

Recall that $\operatorname{sign}(x) = 1$, if x is even and $\operatorname{sign}(x) = -1$ otherwise.

We have now a simple explanation of Q_6.

9 More Examples

In this section we will outline an infinite series of examples of proper quasigroups, each of which provides a positive answer on the question of Barlotti-Strambach. Such a quasigroup Q_{2p} will be described for each prime p, $p \equiv 3 \pmod 4$. Our previous example Q_6 is the first member of the series corresponding to $p = 3$. The methodology of our presentation follows exactly the line which was developed for Q_6. We will not try to reach the complete level of rigor, omitting part of details in a few of the steps if they require some routine computations or more sophisticated reasonings. We will also illustrate each step by a full-length consideration of the second member of the series corresponding to $p = 7$.

9.1 Some Preliminary Observations

Throughout this section p as a rule denotes a prime such that $p \equiv 3 \pmod 4$. Let $\mathbb{Z}_{3p}$ be an additive cyclic group of order $3p$ which is defined by the addition modulo $3p$. By $(\mathbb{Z}_{3p}, \mathbb{Z}_{3p})$ we mean a regular representation of $\mathbb{Z}_{3p}$ which is generated by a permutation $g_1 : x \to x+1 \pmod{3p}$. We denote by C_{3p} a cycle of length $3p$, that is a connected graph of valency 2 with $3p$ vertices. As a rule, we mean a canonical copy of C_{3p}, that is a Cayley graph $Cay(\mathbb{Z}_{3p}, \{1, 3p-1\})$. It is clear that $\mathrm{Aut}(C_{3p}) = D_{3p}$, where D_{3p} is a dihedral group of order $6p$ and degree $3p$. We also have $D_{3p} = \langle g_1, t\rangle$, where $t : x \to -x \pmod{3p}$.

It turns out that $\mathbb{Z}_{3p}$ and D_{3p} have an imprimitivity system consisting of three blocks of size p, namely X_0, X_1, X_2, where $X_i = \{x \in \mathbb{Z}_{3p} | x \equiv i \pmod 3\}$ for $i = 0, 1, 2$. For example, for $p = 7$ we get the blocks $X_0 = \{0, 3, 6, 9, 12, 15, 18\}$, $X_1 = \{1, 4, 7, 10, 13, 16, 19\}$, $X_2 = \{2, 5, 8, 11, 14, 17, 20\}$. Let $\Sigma = Cay(\mathbb{Z}_{3p}, \{3, 3p-3\})$. It is easy to see that Σ is a disconnected regular graph of valency 2 consisting of three connected components, each of which is isomorphic to a cycle C_p. For example, in the case of $p = 7$, Σ is depicted in Fig. 16.

Clearly, $\mathrm{Aut}(\Sigma) \cong S_3 \wr D_p$ is a wreath product of the groups S_3 and D_p, a group of order $6 \cdot (2p)^3 = 48p^3$.

Let $\Delta = Cay(\mathbb{Z}_{3p}, X_1 \cup X_2)$ be a complete regular 3-partite graph of valency $2p$. The graph $\overline{\Delta}$ is isomorphic to $3 \circ K_p$, i.e. the disjoint union of three complete graphs with p vertices. Note that for $p > 3$ the group $\mathrm{Aut}(\Delta) = \mathrm{Aut}(\overline{\Delta}) \cong S_3 \wr S_p$, and thus is strictly larger than $S_3 \wr D_p$.

It is important to note that D_{3p} is a subgroup of $S_3 \wr D_p$. Note also that the permutation t is an involution which has $\frac{3p-1}{2}$ cycles of length 2. Thus, t is an even permutation, as well as g_1. Therefore, D_{3p} consists of even permutations only. On the other hand, the group D_p and therefore $S_3 \wr D_p$ contain odd permutations, for example, each involution in D_p has $\frac{p-1}{2}$ cycles of length 2. Therefore, $G = (S_3 \wr D_p)^{\mathrm{pos}}$, a subgroup of all even permutations in $\mathrm{Aut}(\Sigma)$,

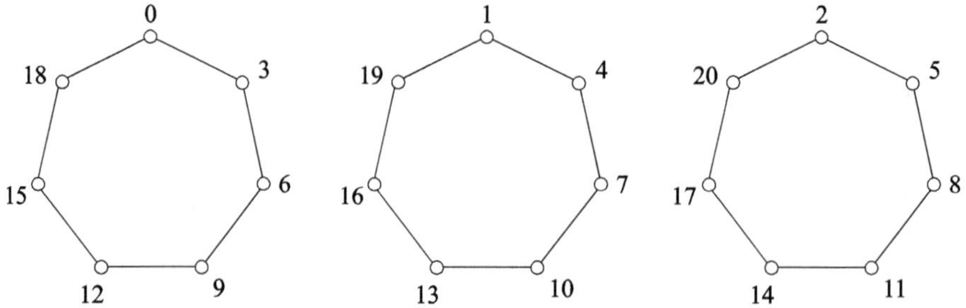

Fig. 16. Graph Σ for $p = 7$

has index 2. This group G of order $24p^3$ is one of the main heroes of our presentation in this section.

9.2 Defining Points and Lines

Now we have to make some combinatorial preparations in order to define our future incidence system $\mathcal{S} = \mathcal{S}_p$ which will turn out to be $TD(3, 2p)$. Let us consider a subgraph F of the canonical graph C_{3p} which consists of all edges in C_{3p} that join a vertex from X_0 to a vertex of X_1. For $p = 7$ such a subgraph is depicted in Fig. 17.

Note that $\mathrm{Aut}(F) \cap D_{3p}$ is a subgroup isomorphic to $D_p \leq D_{3p}$ which has index 3 in D_{3p}. Here $D_p = \langle g_1^3, tg_1 \rangle$. We must now determine $\mathrm{Aut}(F) \cap G$. If, for example, $p = 7$ we get $\mathrm{Aut}(F) \cap G = \langle g_1^3, tg_1, z_2, i_2 \rangle$, where $z_2 = (2, 5, 8, 11, 14, 17, 20)$, $i_2 = (3, 18)(6, 15)(9, 12)(4, 19)(7, 16)(10, 13)$. (Index 2 in z_2 and i_2 refers to X_2, which plays a special role in the definition of these permutations.) In general, defining similar permutations, we get that $\mathrm{Aut}(F) \cap G$ is a group of order $4p^2$.

At this stage we define $\mathcal{P} = F^G$ as the orbit of the action of G on all images of F under $g \in G$, that is $\mathcal{P} = \{F^g | g \in G\}$. Because the action $(G, \mathcal{P})$ is transitive we see that

$$|\mathcal{P}| = \frac{|G|}{|\mathrm{Aut}(F) \cap G|} = \frac{24p^3}{4p^2} = 6p.$$

It is clear that the union of all 1-factors from $\mathcal{P}$ coincides with the edge set of the graph Δ. This set consists of $\frac{1}{2} \cdot 3p \cdot 2p = 3p^2$ edges. Each 1-factor has exactly p edges. Thus, we finally prove that each edge of Δ appears in exactly two 1-factors from $\mathcal{P}$.

Now we define the set $\mathcal{L}$ of lines. Again we consider an image of a certain structure under the action of G. This gives us $\mathcal{L} = C_{3p}^G = \{C_{3p}^g | g \in G\}$. Taking into account that $\mathrm{Aut}(C_{3p}) = D_{3p} \leq G$, we conclude that

$$|\mathcal{L}| = \frac{|G|}{|D_{3p}|} = \frac{24p^3}{6p} = 4p^2.$$

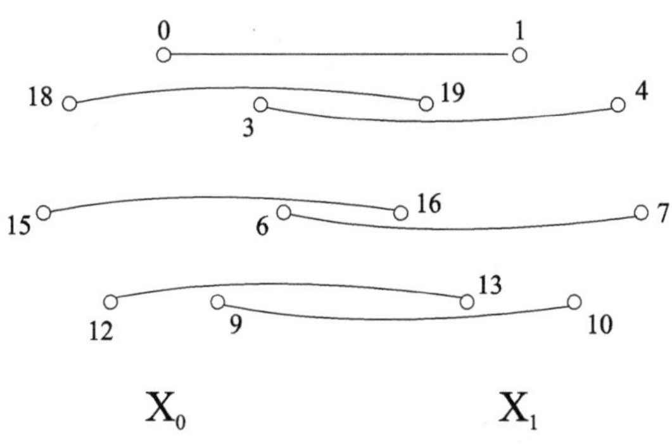

Fig. 17. Subgraph F of C_{21}

9.3 Constructing a Transversal Design

The definition of an incidence system $\mathcal{S} = \mathcal{S}_p$ is now evident. $\mathcal{P}$ is the set of points, $\mathcal{L}$ is the set of lines, the incidence coincides with the set-theoretical inclusion of 1-factors to a cycle. It is clear that $\mathcal{P}$ is naturally partitioned into three sets, namely $\mathcal{P}_{01}, \mathcal{P}_{12}$ and $\mathcal{P}_{02}$ where $\mathcal{P}_{ij}$ is the set of all 1-factors in $\mathcal{P}$ between vertex sets X_i and X_j.

Proposition 6. *The incidence structure $\mathcal{S}_p$ is a transversal design $TD(3, 2p)$.*

Proof. We have $|\mathcal{P}| = 6p, |\mathcal{L}| = 4p^2$ and it is clear that each element of $\mathcal{L}$ can be regarded as a 3-element subset of $\mathcal{P}$. By definition, there is no line which goes through two distinct points in the same group. Due to vertex-transitivity of G, each point is incident to the same number of lines, namely to $\frac{4p^2 \cdot 3}{6p} = 2p$ lines.

It remains then to check that through any two points from distinct groups there is exactly one line which goes through these groups. Note that the group G acts transitively on the ordered pairs of 1-factors from distinct groups. To prove this, consider any concrete 1-factor F, the group $\mathrm{Aut}(F) \cap G$ of order $4p^2$, an arbitrary factor F' in another group and show that the group $\mathrm{Aut}(F) \cap \mathrm{Aut}(F') \cap G$ has order p and, thus, index $4p$ in $\mathrm{Aut}(F) \cap G$. For example, in the case $p = 7$ we obtain $\mathrm{Aut}(F) \cap \mathrm{Aut}(F') \cap G = \langle g_1^3 \rangle$, where F and F' are depicted in Fig. 18 (F contains $\{0, 1\}$, while F' contains $\{1, 2\}$).

Hence, for the proof it is enough to consider a canonical pair of 1-factors. Let F be the 1-factor of C_{3p} which contains the edge $\{0, 1\}$ while F' is a 1-factor of C_{3p} that contains edge $\{1, 2\}$. (These 1-factors are depicted in

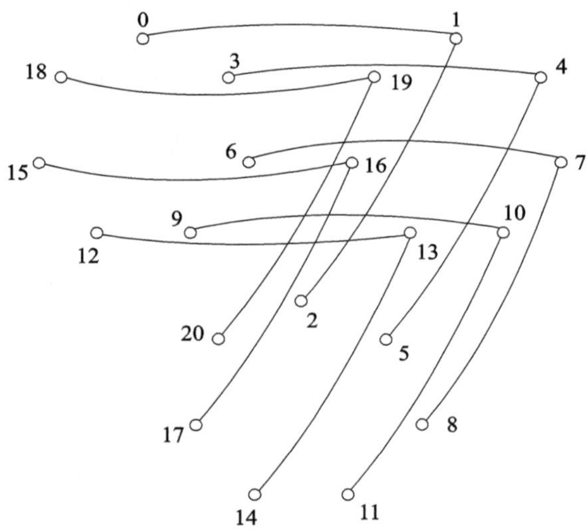

Fig. 18. 1-factors F and F' of Δ

Fig. 18.) Clearly, the canonical cycle C_{3p} goes through F and F'. Our goal is to prove that C_{3p} is the only one in $\mathcal{L}$.

Let $H = \{g \in G | (F^g \in \{F, F'\}) \wedge ((F')^g \in \{F, F'\})\}$ be the stabilizer of the subset $\{F, F'\}$ in G. It follows from the above considerations that $|H| = \frac{|G|}{\frac{1}{2}6p \cdot 4p} = \frac{24p^3}{12p^2} = 2p$.

Now consider the subgroup $\langle g_1^3, g_1^{-1}tg_1 \rangle$. Denote $t_1 = g_1^{-1}tg_1$ and check that $t_1 : x \to 2 - x$ (mod 3p). From this it follows that $\{0,1\}^{t_1} = \{1,2\}$, $\{1,2\}^{t_1} = \{0,1\}$ and more generally, that F and F' are permuted by t_1. Thus $\langle g_1^3, t_1 \rangle$ is a subgroup of H. Taking into account that $|\langle g_1^3, t_1 \rangle| = 2p$, we see that $H = \langle g_1^3, t_1 \rangle$.

Suppose now that $C' \in \mathcal{L}$ is another cycle from $\mathcal{L}$ which goes through F and F'. Then, according to the definition of $\mathcal{L}$, there exists a permutation $g \in G$ such that $(C_{3p})^g = C'$. Note that $C_{3p} = F \cup F' \cup F''$, while $C' = F \cup F' \cup F'''$ for suitable F''' (here F'' is the remaining 1-factor of the canonical C_{3p} which includes the edge $\{2,3\}$). Without loss of generality we may assume, that g preserves the subset $F \cup F'$. Therefore, g belongs to H. Now it easy to check that $H \leq D_{3p} = \mathrm{Aut}(C_{3p})$. Thus, if g preserves C_{3p} and $(F \cup F')^g = F \cup F'$ we have that $(F'')^g = F''$. Therefore, $F''' = F''$ and $C' = C_{3p}$. (We invite the reader to consider the case $p = 7$ with the permutation $t_1 = (0,2)(3,20)(4,19)(5,18)(6,17)(7,16)(8,15)(9,14)(10,13)(11,12)$.) This shows that the incidence structure $\mathcal{S}$ is indeed a transversal design $TD(3, 2p)$. □

9.4 Automorphisms of the Design

We are now describing a simple combinatorial procedure, which is a generalization of the process of construction of the auxiliary structure presented in Sect. 7.5 in Fig. 15. It can be applied for an arbitrary transversal design dual to a 3-net.

Assume $\{A, B, C\}$ is the partition into the groups of the point set of an arbitrary transversal design $TD(3, 2p)$. Let $P_\alpha \in A, P_\beta \in B, P_\gamma \in C$. First we assign to a 3-element subset $\{P_\alpha, P_\beta, P_\gamma\}$ a monochromatic undirected graph $M = M(\{P_\alpha, P_\beta, P_\gamma\})$ with the vertex set $\mathcal{P}$ as follows: $\{P_x, P_y\}$ is an edge in M if and only if $\{P_x, P_y, P_z\}$ is a line in $TD(3, 2p)$ for $z \in \{\alpha, \beta, \gamma\}$. We can also color the vertices and edges of M by the colors from $\{A, B, C\}$. To a vertex of $\mathcal{P}$ we assign the color corresponding to the group containing this vertex. To an edge $\{P_x, P_y\}$ we assign the color of P_z, where $\{P_x, P_y, P_z\}$ is as above. The resulting coloring of the auxiliary graph M will be denoted by $C = C(P_\alpha, P_\beta, P_\gamma)$. The next lemma follows easily from our definitions.

Lemma 4.

(a) The graph $M(\{P_\alpha, P_\beta, P_\gamma\})$ is a regular graph of valency 2, therefore each connectivity component is a cycle.

(b) The graph $M(\{P_\alpha, P_\beta, P_\gamma\})$ is invariant with respect to the setwise stabilizer of the subset $\{P_\alpha, P_\beta, P_\gamma\}$ in the group $\mathrm{Aut}(TD(3, 2p))$.

(c) *The colored graph $C(P_\alpha, P_\beta, P_\gamma)$ is invariant with respect to the pointwise stabilizer of the points $P_\alpha, P_\beta, P_\gamma$ in the group G.*

Remark.

1. Of course, in the formulation of Lemma 4 the number p is an arbitrary integer.
2. The graphs M and C may be defined and exploited in a similar manner for an arbitrary partial linear space with lines of size 3. In particular, in the case of Steiner triple systems, the graphs M form a set of important invariants, see Sect. 4.3.2 of [30].

Our goal is now to describe the full automorphism group $\overline{G} = \mathrm{Aut}(\mathcal{S})$, where $\mathcal{S}$ is the transversal design defined in Sect. 9.2. Recall that the group G, and therefore also $\overline{G}$ acts transitively on the vertices and lines of $\mathcal{S}$.

Moreover, by definition the stabilizer of an arbitrary line from $\mathcal{S}$ in the group G is the dihedral group D_{3p} of order $6p$. In its action on points the group D_{3p} has one orbit of length $3p$ and p orbits of length 3. (To observe this fact, note that each edge of a canonical copy of C_{3p} is contained in exactly one of the above defined factors F, F', F'', as well as in one more factor. These $3p$ factors form the orbit of length $3p$.) One of the orbits of length 3 exactly consists of the three points forming the considered line. The pointwise stabilizer of these three points in G is the unique cyclic group of order p which is contained in D_{3p}. This cyclic group has in its action on $\mathcal{P}$ three orbits of length p and $3p$ fixed points, including the points F, F', F''.

More advanced reasonings show that the auxiliary graph $M(l)$ for an arbitrary line l from $\mathcal{S}$, and in particular for $l = \{F, F', F''\}$ has one cycle of length $3p$ (the one induced by the above orbit of length $3p$) and $\frac{p-1}{2}$ cycles of length 6 (each cycle corresponds to two of the above mentioned orbits of length 3).

To avoid further sophistication we will now restrict ourselves to the consideration of the case $p = 7$. Using COCO, for a certain set of generators of the group G we create the following list of 1-factors from $\mathcal{P}$ (see Table 9). For each of the following pictures we will use the labeling of the elements from $\mathcal{P}$ created by COCO. In particular, the labels $0, 1, 2$ correspond to our 1-factors F, F', F'' respectively.

First we depict the auxiliary graph $C(F, F', F'')$ (see Fig. 19 on page 47, conventions about the colors are the same as in Fig. 15).

It is easy to see that the automorphism group $\mathrm{Aut}(C(F, F', F''))$ is isomorphic to the direct product of the group of order 7 acting semiregularly on the vertex set of the cycle of length 21 with the group of order 48 acting on the vertices of the three hexagons. The latter group is intransitive with three orbits of length 6 corresponding to the subsets (groups) A, B, C of the vertex set $\mathcal{P}$.

We now take vertex 12 from group $C = \mathcal{P}_{02}$ and describe the pointwise stabilizer $\overline{G}_{0,1,12}$. For this purpose we will depict one more auxiliary graph $C(F, F', F''')$, where F''' is the 1-factor with label 12 in our list, see Table 9.

Table 9. Points and lines of $\mathcal{S}$ in the case $p = 7$

	Points of $\mathcal{S}$
0.	$\{\{0,1\},\{3,4\},\{6,7\},\{9,10\},\{12,13\},\{15,16\},\{18,19\}\}$
1.	$\{\{1,2\},\{4,5\},\{7,8\},\{10,11\},\{13,14\},\{16,17\},\{19,20\}\}$
2.	$\{\{0,20\},\{2,3\},\{5,6\},\{8,9\},\{11,12\},\{14,15\},\{17,18\}\}$
3.	$\{\{0,19\},\{1,3\},\{4,6\},\{7,9\},\{10,12\},\{13,15\},\{16,18\}\}$
4.	$\{\{0,1\},\{3,19\},\{4,18\},\{6,16\},\{7,15\},\{9,13\},\{10,12\}\}$
5.	$\{\{1,17\},\{2,16\},\{4,14\},\{5,13\},\{7,11\},\{8,10\},\{19,20\}\}$
6.	$\{\{0,17\},\{2,6\},\{3,20\},\{5,9\},\{8,12\},\{11,15\},\{14,18\}\}$
7.	$\{\{1,20\},\{2,4\},\{5,7\},\{8,10\},\{11,13\},\{14,16\},\{17,19\}\}$
8.	$\{\{0,2\},\{3,5\},\{6,8\},\{9,11\},\{12,14\},\{15,17\},\{18,20\}\}$
9.	$\{\{0,16\},\{1,6\},\{3,19\},\{4,9\},\{7,12\},\{10,15\},\{13,18\}\}$
10.	$\{\{0,19\},\{1,18\},\{3,16\},\{4,15\},\{6,13\},\{7,12\},\{9,10\}\}$
11.	$\{\{1,2\},\{4,20\},\{5,19\},\{7,17\},\{8,16\},\{10,14\},\{11,13\}\}$
12.	$\{\{0,20\},\{2,18\},\{3,17\},\{5,15\},\{6,14\},\{8,12\},\{9,11\}\}$
13.	$\{\{0,4\},\{1,3\},\{6,19\},\{7,18\},\{9,16\},\{10,15\},\{12,13\}\}$
14.	$\{\{0,4\},\{1,18\},\{3,7\},\{6,10\},\{9,13\},\{12,16\},\{15,19\}\}$
15.	$\{\{0,14\},\{2,9\},\{3,17\},\{5,12\},\{6,20\},\{8,15\},\{11,18\}\}$
16.	$\{\{1,20\},\{2,19\},\{4,17\},\{5,16\},\{7,14\},\{8,13\},\{10,11\}\}$
17.	$\{\{1,17\},\{2,7\},\{4,20\},\{5,10\},\{8,13\},\{11,16\},\{14,19\}\}$
18.	$\{\{0,5\},\{2,18\},\{3,8\},\{6,11\},\{9,14\},\{12,17\},\{15,20\}\}$
19.	$\{\{0,13\},\{1,9\},\{3,16\},\{4,12\},\{6,19\},\{7,15\},\{10,18\}\}$
20.	$\{\{0,16\},\{1,15\},\{3,13\},\{4,12\},\{6,10\},\{7,9\},\{18,19\}\}$
21.	$\{\{0,2\},\{3,20\},\{5,18\},\{6,17\},\{8,15\},\{9,14\},\{11,12\}\}$
22.	$\{\{0,5\},\{2,3\},\{6,20\},\{8,18\},\{9,17\},\{11,15\},\{12,14\}\}$
23.	$\{\{1,5\},\{2,4\},\{7,20\},\{8,19\},\{10,17\},\{11,16\},\{13,14\}\}$
24.	$\{\{0,17\},\{2,15\},\{3,14\},\{5,12\},\{6,11\},\{8,9\},\{18,20\}\}$
25.	$\{\{0,7\},\{1,6\},\{3,4\},\{9,19\},\{10,18\},\{12,16\},\{13,15\}\}$
26.	$\{\{1,5\},\{2,19\},\{4,8\},\{7,11\},\{10,14\},\{13,17\},\{16,20\}\}$
27.	$\{\{0,7\},\{1,15\},\{3,10\},\{4,18\},\{6,13\},\{9,16\},\{12,19\}\}$
28.	$\{\{0,11\},\{2,12\},\{3,14\},\{5,15\},\{6,17\},\{8,18\},\{9,20\}\}$
29.	$\{\{1,14\},\{2,10\},\{4,17\},\{5,13\},\{7,20\},\{8,16\},\{11,19\}\}$
30.	$\{\{0,8\},\{2,15\},\{3,11\},\{5,18\},\{6,14\},\{9,17\},\{12,20\}\}$
31.	$\{\{0,10\},\{1,12\},\{3,13\},\{4,15\},\{6,16\},\{7,18\},\{9,19\}\}$
32.	$\{\{0,13\},\{1,12\},\{3,10\},\{4,9\},\{6,7\},\{15,19\},\{16,18\}\}$
33.	$\{\{0,8\},\{2,6\},\{3,5\},\{9,20\},\{11,18\},\{12,17\},\{14,15\}\}$
34.	$\{\{0,14\},\{2,12\},\{3,11\},\{5,9\},\{6,8\},\{15,20\},\{17,18\}\}$
35.	$\{\{1,14\},\{2,13\},\{4,11\},\{5,10\},\{7,8\},\{16,20\},\{17,19\}\}$
36.	$\{\{1,8\},\{2,7\},\{4,5\},\{10,20\},\{11,19\},\{13,17\},\{14,16\}\}$
37.	$\{\{0,10\},\{1,9\},\{3,7\},\{4,6\},\{12,19\},\{13,18\},\{15,16\}\}$
38.	$\{\{1,8\},\{2,16\},\{4,11\},\{5,19\},\{7,14\},\{10,17\},\{13,20\}\}$
39.	$\{\{1,11\},\{2,13\},\{4,14\},\{5,16\},\{7,17\},\{8,19\},\{10,20\}\}$
40.	$\{\{0,11\},\{2,9\},\{3,8\},\{5,6\},\{12,20\},\{14,18\},\{15,17\}\}$
41.	$\{\{1,11\},\{2,10\},\{4,8\},\{5,7\},\{13,20\},\{14,19\},\{16,17\}\}$

Table 9. (Continued)

Lines of $\mathcal{S}$				
0. {0,1,2}	40. {5,9,12}	80. {0,28,29}	120. {13,24,39}	160. {13,30,36}
1. {1,3,6}	41. {9,11,22}	81. {6,27,29}	121. {21,27,41}	161. {12,37,38}
2. {2,4,5}	42. {4,7,21}	82. {2,23,37}	122. {16,27,40}	162. {13,34,38}
3. {2,7,14}	43. {9,16,21}	83. {2,19,39}	123. {16,28,32}	163. {21,37,39}
4. {1,8,14}	44. {7,10,12}	84. {12,19,35}	124. {26,27,30}	164. {10,33,38}
5. {1,9,15}	45. {7,13,22}	85. {11,19,33}	125. {26,28,31}	165. {9,34,41}
6. {5,8,10}	46. {10,17,24}	86. {5,19,21}	126. {18,32,35}	166. {22,32,39}
7. {0,11,12}	47. {1,12,20}	87. {17,22,25}	127. {13,28,41}	167. {24,31,36}
8. {5,6,13}	48. {1,13,33}	88. {17,20,34}	128. {9,17,30}	168. {23,31,34}
9. {0,8,26}	49. {0,5,34}	89. {16,19,22}	129. {3,33,36}	169. {30,35,37}
10. {0,6,7}	50. {4,11,18}	90. {7,20,24}	130. {4,34,39}	170. {8,9,39}
11. {2,13,16}	51. {13,17,21}	91. {8,23,25}	131. {14,22,41}	171. {8,19,29}
12. {2,3,26}	52. {4,26,33}	92. {7,25,33}	132. {11,15,32}	172. {19,23,40}
13. {2,17,27}	53. {12,13,29}	93. {1,24,32}	133. {14,15,29}	173. {20,29,40}
14. {1,18,27}	54. {10,22,26}	94. {1,25,40}	134. {14,30,38}	174. {10,23,28}
15. {1,19,28}	55. {12,27,36}	95. {5,14,40}	135. {15,31,38}	175. {10,15,36}
16. {5,18,20}	56. {11,27,34}	96. {3,5,24}	136. {8,32,41}	176. {24,25,38}
17. {3,12,16}	57. {5,28,37}	97. {13,18,23}	137. {3,18,39}	177. {21,32,38}
18. {3,11,21}	58. {13,15,35}	98. {12,26,32}	138. {4,28,36}	178. {25,30,41}
19. {1,4,22}	59. {7,8,27}	99. {13,26,40}	139. {3,29,30}	179. {17,18,19}
20. {4,12,17}	60. {7,9,28}	100. {0,35,40}	140. {0,33,41}	180. {11,20,28}
21. {12,14,23}	61. {10,16,18}	101. {4,23,30}	141. {6,23,32}	181. {9,36,40}
22. {11,14,24}	62. {3,22,23}	102. {21,25,29}	142. {6,31,39}	182. {10,39,40}
23. {5,15,25}	63. {4,24,29}	103. {20,21,26}	143. {2,32,36}	183. {6,20,36}
24. {4,6,35}	64. {1,10,21}	104. {3,34,35}	144. {21,31,35}	184. {22,27,35}
25. {3,7,15}	65. {14,21,36}	105. {4,38,40}	145. {22,29,37}	185. {15,27,39}
26. {4,8,16}	66. {14,16,34}	106. {12,25,39}	146. {17,32,40}	186. {27,28,38}
27. {0,21,23}	67. {15,16,37}	107. {20,22,38}	147. {5,22,31}	187. {18,37,41}
28. {0,16,24}	68. {14,18,26}	108. {12,31,41}	148. {18,25,36}	188. {9,18,29}
29. {6,16,25}	69. {15,19,26}	109. {11,31,40}	149. {17,33,37}	189. {3,40,41}
30. {6,9,26}	70. {8,20,35}	110. {23,24,27}	150. {16,31,33}	190. {15,20,23}
31. {2,10,35}	71. {3,8,38}	111. {25,28,35}	151. {7,32,34}	191. {14,28,39}
32. {0,18,38}	72. {4,15,41}	112. {8,17,31}	152. {8,36,37}	192. {29,32,33}
33. {0,15,17}	73. {3,17,28}	113. {7,18,31}	153. {7,37,40}	193. {19,24,41}
34. {6,14,17}	74. {8,11,13}	114. {7,19,30}	154. {1,34,37}	194. {20,33,39}
35. {2,11,25}	75. {0,22,36}	115. {16,20,30}	155. {5,27,33}	195. {19,34,36}
36. {2,9,38}	76. {6,11,37}	116. {9,23,33}	156. {9,24,35}	
37. {2,29,31}	77. {6,19,38}	117. {10,29,34}	157. {24,26,37}	
38. {1,30,31}	78. {2,20,41}	118. {10,11,30}	158. {25,26,34}	
39. {5,30,32}	79. {0,30,39}	119. {6,10,41}	159. {14,33,35}	

The graph $C(F,F',F''')$ is depicted in Fig. 20 on page 48. Its automorphism group has order $3! \cdot 4^3 \cdot 2$. We are now interested in the group $\overline{G}_{0,1,12}$. This group fixes the points 0, 1, and therefore also the point 2. By this reason $\overline{G}_{0,1,12} \leq \mathrm{Aut}(C(F,F',F''))$. On the other hand, we clearly have

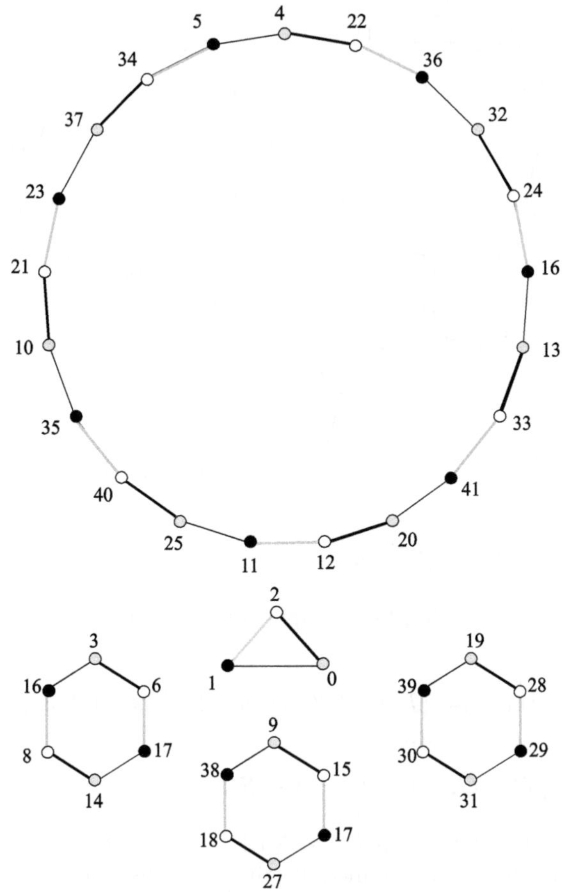

Fig. 19. Graph $C(F, F', F'')$

$\overline{G}_{0,1,12} \leq \mathrm{Aut}(C(F, F', F'''))$. We now need to determine information about the intersection of the two groups $\mathrm{Aut}(C(F, F', F''))$ and $\mathrm{Aut}(C(F, F', F'''))$. This may be done by routine simultaneous inspection of both figures.

First we observe that all vertices of the cycle of length 21 remain in their place. This immediately implies that all vertices in the second graph also remain on their place. In other words, $\mathrm{Aut}(C(F, F', F'')) \cap \mathrm{Aut}(C(F, F', F'''))$ is a group of order 1 and therefore we obtain $|\overline{G}_{0,1,12}| = 1$.

Summing up our reasoning, we formulate the following lemma which is essential for our description of the group $\overline{G}$.

Lemma 5. *Let F, F', F'' be the points of the transversal design $\mathcal{S} = TD(3, 2p)$ as they were defined above. Let F''' be a vertex which belongs to the largest connectivity component of the graph $M(\{F, F', F''\})$. Then for the group $\overline{G} = \mathrm{Aut}(\mathcal{S})$ we have $\overline{G}_{F,F',F'''} = \{e\}$.*

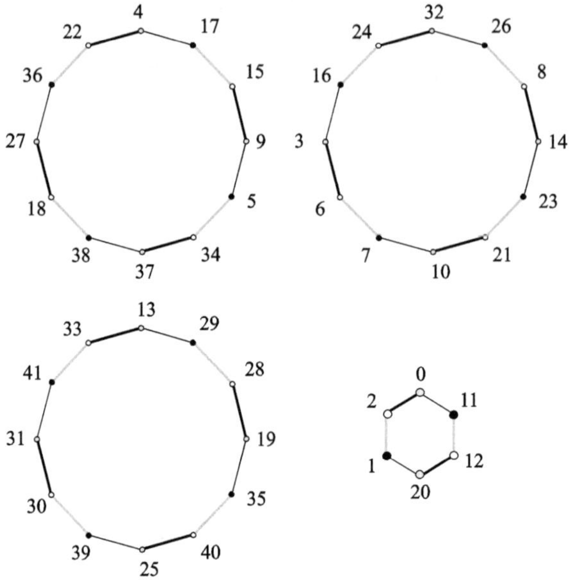

Fig. 20. Graph $C(F, F', F''')$

Note that in this case $\{F, F', F'''\}$ is called a *base* of $\mathrm{Aut}(\mathcal{S})$.
We are now ready to prove

Proposition 7. *$\overline{G} = G$, where $\overline{G} = \mathrm{Aut}(\mathcal{S})$.*

Proof. We consider the group $\overline{G} = \mathrm{Aut}(\mathcal{S})$ in its action on a set $\mathcal{P}$. Let F, F', F''' be as they were defined in the previous lemma. We know that $G \leq \overline{G}$ and $\{F, F', F'''\}$ forms a base of $\overline{G}$. Using, where it is convenient, orbits of G instead of orbits of $\overline{G}$, we obtain:

$$\begin{aligned}
|\overline{G}| &= 6p \cdot |\overline{G}_F| = 6p \cdot |(F')^{\overline{G}_F}| \cdot |\overline{G}_{F,F'}| \\
&= 6p \cdot |(F')^{G_F}| \cdot |\overline{G}_{F,F'}| = 6p \cdot 4p \cdot |\overline{G}_{F,F'}| \\
&= 6p \cdot 4p \cdot |(F''')^{\overline{G}_{F,F'}}| \cdot |\overline{G}_{F,F',F'''}| = 24p^2 \cdot |(F''')^{G_{F,F'}}| \cdot 1 \\
&= 24p^2 \cdot p \cdot 1 = 24p^3.
\end{aligned}$$

Taking into account that $G \leq \overline{G}$ and $|\overline{G}| = 24p^3$, we get that $G = \overline{G} = \mathrm{Aut}(\mathcal{S})$. □

Recall a simple well-known result from group theory (see, e.g. [85], Theorem 4.19):

Proposition 8. *If p is an odd prime, then every group of order $2p$ is either cyclic $\mathbb{Z}_{2p}$ or dihedral D_p.*

Now we can easily see that $\mathrm{Aut}(\mathbb{Z}_{2p}) \cong \mathbb{Z}_{2p}^* \cong \mathbb{Z}_p^*$ is a cyclic group of order $p-1$. A more interesting exercise is to find $\mathrm{Aut}(D_p)$. Here we obtain a group of order $p \cdot (p-1)$, see for details [84].

Thus, we have that

$$|\mathrm{Aut}(SRG(\mathbb{Z}_{2p}))| = 3! \cdot (2p)^2 \cdot (p-1) = 24p^2(p-1) \quad \text{and}$$
$$|\mathrm{Aut}(SRG(D_p))| = 3! \cdot (2p)^2 \cdot p(p-1) = 24p^3(p-1).$$

Finally, with the Proposition 7 we have proven the following proposition:

Proposition 9. *Our transversal design $\mathcal{S}$ is not isomorphic to a transversal design coming from a group.*

Our goal is now to point out a translation group H in $G = \mathrm{Aut}(\mathcal{S})$.

It is clear that H should be a group of order $4p^2$, such that each non-identity element of H does not fix any element of $\mathcal{L}$.

For this purpose it is more convenient to work with the action of G on a small set of $3p$ points. Recall that in terms of this action each element of $\mathcal{L}$ is a copy of a cycle C_{3p} of length $3p$. Such a cycle can be fixed by permutations only from D_{3p}, these permutations (non-identity) have at most one fixed point on the vertices of $\mathbb{Z}_{3p}$.

The group G has a subgroup $(D_p \times D_p \times D_p)^{\mathrm{pos}}$ of order $\frac{1}{2} \cdot (2p)^3$ acting intransitively on $\mathbb{Z}_{3p}$. Let us take any involution i from the third copy of D_p and let us consider a group $H = (D_p \times D_p \times \langle i \rangle)^{\mathrm{pos}}$ which has order $\frac{1}{2} \cdot 2p \cdot 2p \cdot 2 = 4p^2$. Clearly, as an abstract group we have that $H \cong D_p \times D_p$.

It is easy to check that a non-identity element x of H fixes at least two distinct elements of $\mathbb{Z}_{3p}$. Thus, such an element x does not belong to D_{3p}. Therefore, H is a desired translation subgroup of G, that is a subgroup of G which acts regularly on the $4p^2$-element set $\mathcal{L}$.

9.5 Analyzing the Results

We can now claim that the desired result is obtained:

Theorem 2. *Let p be a prime, $p \equiv 3 \pmod 4$ and $\mathcal{S}_p = \mathcal{S}$ be a structure as described above. Then*

(a) $\mathcal{S}$ is a (resolvable) translation design $TD(3, 2p)$.
(b) $G = \mathrm{Aut}(\mathcal{S}) \cong (S_3 \wr D_p)^{\mathrm{pos}}$ is a transitive group of order $24p^3$.
(c) G as a transitive group of degree $4p^2$ in its action on the set $\mathcal{L}$ has a regular subgroup $H \cong (D_p)^2$ of order and degree $4p^2$.
(d) The dual structure to $\mathcal{S}$ is a 3-net which is not coming from a group of order $2p$.

The result of Theorem 2 may be reformulated in terms of partial difference sets, that is there exists a "non-standard" Latin square partial difference set of size $k = 3(2p-1)$ over a group $(D_p)^2$ of order $4p^2$, where $p \equiv 3 \pmod 4$ is a prime.

First of all, we recall that a standard partial difference set X_s corresponds to the $SRG(D_p)$. Here $X_s = \{(g,e),(e,g),(g,g^{-1}) | g \in D_p, g \neq e\}$, where (x,y) is a generic notation for an arbitrary element from the direct square $(D_p)^2$ of two copies of the dihedral group D_p.

We obtain also the existence of a non-standard difference set X_n analyzing the Latin square graph Γ which exists according to Theorem 2.

For example, in case of $p = 7$ we have the following generators for $H = (D_7)^2$:

$$a = (0,3,6,9,12,15,18), \qquad c = (0,3)(6,18)(9,15)(2,5)(8,20)(11,17),$$
$$b = (1,4,7,10,13,16,19), \qquad d = (1,4)(7,19)(10,16)(2,5)(8,20)(11,17).$$

We now have to construct our design $\mathcal{S} = (\mathcal{P}, L)$ and to find a canonical set of $3(2 \cdot 7 - 1) = 39$ permutations from H which move a subgraph F to all its possible copies in $\mathcal{P}$. As such we obtain

$$X_n = \{a, a^2, a^3, a^4, a^5, a^6, d, ad, a^2d, a^3d, a^4d, a^5d, a^6d, b, b^2, b^3, b^4, b^5, b^6, c, bc, b^2c, b^3c, b^4c, b^5c, b^6c, ab, a^2b^2, a^3b^3, a^4b^4, a^5b^5, a^6b^6, acd, a^2bcd, a^3b^2cd, a^4b^3cd, a^5b^4cd, a^6b^5cd, b^6cd\}.$$

In principle, as soon as X_n is found, we can confirm that X_n indeed provides a partial difference set by direct computations in the group ring over H, see also Sect. 10.

We hope that the reader will be able to generalize our construction of X_n from $p = 7$ to the general case considered in this text.

It now remains to describe a Cayley table of the quasigroup Q_{2p}, which is implied by the existence of a transversal design $\mathcal{S} = \mathcal{S}_p$.

For this purpose we identify the element set of Q_{2p} by that of D_p. In this context it is convenient to consider D_p in its canonical action of degree p on the set $\mathbb{Z}_p = \{0, 1, \ldots, p-1\}$. Recall that half of the permutations from D_p are even and half are odd. Let also $a = (0, 1, \ldots, p-1)$ be a canonical generator of $\mathbb{Z}_p$. We define Q_{2p} as follows (here $\circ$ means the new binary operation in Q_{2p}, while $x \cdot y$, or more precisely xy means the usual multiplication in D_p):

$$(*) x \circ y = \begin{cases} xya, & \text{if } x \text{ and } y \text{ are odd,} \\ xy, & \text{otherwise.} \end{cases}$$

The proof of the fact that Q_{2p} is exactly defined by $(*)$ is omitted, though it may be obtained using the methodology described in this paper and with the aid of the partial difference set X_n presented above.

Finally, when our goal was achieved we were surprised by the simple rule $(*)$, which was obtained. A search in the literature, see details in Sect. 10, allowed us to understand that the quasigroup Q_{2p} is known quite well in the quasigroup theory.

We claim that we have succeeded in establishing new properties of Q_{2p}, and more important, to design a technology that "automatically" brings us to the desired result.

10 Conclusion

In this section we present various facts which did not fit naturally into the main part of this text, but which are important enough to be mentioned in the paper.

10.1 Partial Difference Sets and S-rings

Schur rings (briefly S-rings) provide an alternative way for considering Latin square graphs which admit a regular group of automorphisms. We refer, e.g., to [102] and to [73] for basic facts about S-rings and their relations to Cayley graphs. The following proposition which can be easily proved (cf. [96]), provides a simple and efficient criterion. (Note that Sprague attributes its formulation to W. McWorter.)

Proposition 10. *Let K be a group of order n^2. Then the following conditions are equivalent:*

(1) There exists a Latin square graph Γ over a suitable quasigroup Q of order n such that K is a regular subgroup of $\mathcal{T}(Q)$.
(2) The group K contains three subgroups X_1, X_2, X_3 of order n, such that any two of these subgroups have an intersection of size 1.

Remark. In this case we have $X = X_1 \cup X_2 \cup X_3 \setminus \{e\}$, where e is the identity element of K. X is the connection set of a Cayley graph over K which is isomorphic to Γ. Such a connection set is called a partial difference set over K.

Note that the advantage of this criterion is that it provides the most natural way to detect a desired partial difference set in a suitable group K. However, it is not universal, because it may happen that $\mathcal{T}(Q)$ does not contain a regular subgroup, while $\Sigma(Q)$ does.

10.2 Pseudo-geometrical Graphs

Proposition 2 in Sect. 3.2 is, in fact, a particular case of a more general question which was considered by Bose for partial geometries and by Bruck for a particular class of partial geometries, namely, k-nets. Therefore, arguments which provide the bound $n > 23$ are of a general nature. It may happen that for our case of 3-nets a better bound may be found.

Table 10 shows how drastically the number $p(n)$ of pseudo-Latin square graphs on n^2 vertices increases for small n in comparison with the number $m(n)$ of main classes (the value of $p(6)$ is borrowed from [95]).

Table 10. Number of main classes and pseudo Latin square graphs

n	4	5	6
$m(n)$	2	2	12
$p(n)$	2	15	32548

10.3 The Group Case

We believe that Corollary 1 in Sect. 3.4 was first formulated and proved in [90], though in different terms. Theorem 1 was formulated in [45] but without proof (the authors at that time were not aware of [90]). A brief proof of both claims is given in [32]. Certain variants of propositions, which are equivalent to Corollary 1 or treat some of its particular cases, appear in the literature sporadically. It seems that some of the corresponding authors were also not aware of [90]. Here we refer to [86] as one of the earliest alternatives to [90].

10.4 Hamming Graphs and Latin Squares

We regard the model of a Latin square as a coclique in the Hamming graph, which was presented in Sect. 4 on a level of a few examples, as a methodological innovation. Note that implicitly similar links were mentioned in the literature, see e.g. [16, 11, 93].

Unfortunately, this methodological approach at the present time does not seem to be very practical, when we face a problem of the constructive enumeration of Latin squares of order n for relatively large values of n.

10.5 Loops and Permutations

In understanding a permutation of n elements as a bijection of a set $[1, n]$ onto itself, we say that two permutations are *discordant* if no symbol has the same image under both permutations. A complete set of discordant permutations of the set $[1, n]$ consists of n mutually discordant permutations. Such a set may be naturally associated to the rows of a Latin square Q. A corresponding representation of Q is called by Dénes and Keedwell [35] a Cayley-Schönhardt representation (or CS-representation) with credits to [28] and [90]. We were using CS-representations in Sect. 4.6 in order to distinguish isomorphism classes of loops.

Interesting applications of CS-representations for the evaluation of the number of isomorphically distinct 1-factorizations of the complete directed graph may be found in [36].

10.6 Order 5

Euler was the first who, using an exhaustive process, completed the enumeration of all 56 normalized Latin squares of order 5 [37]. Cayley used what is now called the CS-representation of Latin squares in order to reach a more refined classification [28].

Besides obvious interest to Latin squares of order 5 an expert meets many other attractive reasons of investigation. On the one hand, order 5 is the smallest case where we face the occurrence of the existence of non-geometrical strongly regular graphs with the parameters of a graph $SRG(Q)$, see again Table 10.

On the other hand, in order 5 we meet for the first time proper loops, as they were presented in Sect. 4.6. In fact, these loops provided an attractive lead for generations of mathematicians. A partial list of important references includes such names as:

- A. Albert – enumeration of all loops of order 5, [2];
- Bruck – discussion in [23] of a loop of order 5 with a transitive automorphism group;
- D. A. Norton, S. K. Stein – discussion of two examples of quasigroups of order 5 in the context of a developed concept of a cycle in an arbitrary Latin square, [75];
- R. Artzy – demonstration of a loop (L_1 from Sect. 4.6) as an example of a crossed-inverse loop that yields a principal isotope which is not crossed-inverse [3];
- R. Burn – an expository paper, in which the same loop L_1 serves as an example of a loop which does not satisfy the Lagrange theorem, see [27];
- A. D. Keedwell – presents a symmetric idempotent quasigroup of order 5 which provides a decomposition of the graph K_5^* into quadrangles that satisfy certain prescribed properties, [57];
- A. V. Kuznetsov, E. A. Kuznetsov – consideration of a quasigroup of order 5 in the context of constructions of projective planes, [63];
- T. Bier, P. K. Chua – various examples of small loops (starting from order 5) that serve as a motivation for the consideration of a new (at that time) class of numerical invariants of SRG's, called *totient numbers*, which were introduced as a generalization of invariants of nets considered by Bruck in his classical paper [22], see [18];
- R. Peeters – characterization of the Paley graph $P(25)$ in totality of 15 non-isomorphic $\mathrm{srg}(25, 12, 6, 6)$ with the aid of a 5-rank linear combination of matrices A and I, [78].

10.7 Order 6

The investigation of Latin squares and loops of order 6 is a topic that requires the special attention of experts in the history of combinatorics. One of its subdivisions is a proof of the fact that two orthogonal Latin squares of order 6 do not exist.

G. Tarry [98] was the first who established 17 types in order 6 and obtained the value $R_6 = 9408$.

A more detailed classification of Latin squares of order 6 was given by Schönhardt in his classical paper [90]. He provides a reasonably detailed bibliography, giving credit to many predecessors, in particular to Tarry. Unfortunately, for a few decades this classical paper was unknown to many researchers.

It seems that R. A. Fisher and F. Yates were not aware of Tarry's results. The motivation for them to enumerate all Latin squares of order 6 came from a paper of S. M. Jacob (1930) who managed to find just 8192 normalized squares of order 6. Note that Fisher and Yates found in [42] the correct numbers 12, 17 and 9408 corresponding to row 6 in Table 1 of Sect. 4.

Following [38], we distinguish constructive and analytical enumeration of combinatorial objects. In this context almost all of the cited results are referred to the constructive enumeration. P. A. MacMahon presented in [66] a general method for an analytical enumeration of Latin squares of arbitrary order n based on the use of differential operators. MacMahon demonstrated his method on $n = 4$, saying that it became impractical for higher orders. Note that MacMahon, who has the merited reputation as one of the founders of combinatorial analysis, made a mistake (working 150 years later than Euler) in the number R_5 as the result of his constructive search (as it was reported in [42]).

An impressive resurrection of MacMahon's method was undertaken by P. N. Saxena. Using and developing the analytical techniques of the great predecessor Saxena found in [88], the correct value for R_6. Moreover, in [89] the same method was used to provide a correct number $R_7 = 1694200$.

We refer to [4] who speaks about 22 isotopy classes of order 6, and to [26] for the correct number (equal to 109) of loops of order 6. A detailed and helpful catalogue of all quasigroups of order up to 6 was given by Sade in [87].

Independently of [90] the description of the automorphism groups for all 17 types of order 6 is given in [67].

D. Betten in [16] gives a self-contained classification of main classes of order 6. He pays special attention to a Latin square that is mostly related to the problem of the existence of a pair of orthogonal squares. This Latin square belongs to the main class introduced in Sect. 4.6. Betten gives a strict description of the square in terms of the 30 edges of the Dodecahedron and the group A_5.

An interesting analysis of the 12 main classes of order 6 was done with the aid of a computer in [41] and [60], classifying all mutually orthogonal partitions of these squares.

Latin squares of order 6, as well as of other small orders, were used in diverse publications for illustrative purposes. Here we provide just a few of corresponding references.

A. Suschkewitsch in [97] was using a kind of local switching of the Cayley table of the symmetric group of order 6 to obtain examples of quasigroups, satisfying certain postulates introduced in his paper.

E. T. Parker [76] and K. Heinrich [49] considered certain approaches in order to measure how far one may approximate (non-existing) pairs of orthogonal Latin squares of order 6 by their analogues.

10.8 Vertex-transitive Graphs

It was proved in [80] that every abstract group is isomorphic to the automorphism group of a suitable Latin square graph. Another paper of Phelps [81] established that for $n \geq 7$ there exists a Latin square graph with the automorphism group of order 1 and that the number of such graphs goes to infinity with n.

Arguments by Cameron, briefly mentioned in the discussion section of [8], justify the fact that with the growing n almost all Latin square graphs have such a property. The other extremum is related to a question: which Latin square graphs have a transitive automorphism group? Clearly, this is fulfilled for the Cayley table of groups. There exists, however, many examples of proper loops, for which the Latin square graph has a transitive automorphism group, see Sect. 13 of [14].

As we mentioned in the Introduction, the same mathematical objects and propositions concerning Latin squares appear in different fashion and terminology in the statistical design of experiments, graph theory, general algebra, group theory, geometry and algebraic combinatorics. As a result, experts in different areas are sometimes not aware of the achievements of others. We hope that this paper will help build new bridges between all the mentioned areas.

10.9 Q_6

The loop Q_6 in fact was attracting special attention among many experts, especially due to the properties of the group $\mathrm{Aut}(SRG(Q_6))$ and some of its subgroups. This loop was also considered in [8]. Among other recent interesting discussions of the properties of Q_6 and its associated groups we mention [74] and [62].

However, a very significant group-theoretical analysis of the net corresponding to Q_6 was already presented by Sprague in the article [96], a publication that was inspired by [55]. (In fact, the paper [96] was written in 1978–1979 and was initially entitled "Nets with Singer group fixing each parallel class".)

Sprague denotes Q_6 by O, and shows that the net corresponding to it contains a vertex transitive group of order 36 isomorphic to $S_3 \times S_3$. He also presents a corresponding partial difference set in $S_3 \times S_3$. Nevertheless, the complete group $\mathrm{Aut}(SRG(Q_6))$ is not investigated. It is also mentioned that $\mathbb{Z}_6, S_3$, and O are the only loops which imply Latin square graphs corresponding to translation nets of order 6.

Though Sprague's paper was cited e.g., in [12, 56, 48], it seems as though it has not received the credit it deserves.

10.10 Q_{2p}

Loops, which are isomorphic to all their loop isotopes, are usually called G-loops, or loops with the isotopy-isomorphy property. Each group is a G-loop, but there are many proper G-loops. First stages in the development of the theory of G-loops may be attributed, in particular, to [25] and [15]. Due to a pioneering result of E. Wilson [103], which essentially used [26] in order to prove that a prescribed loop L is a G-loop, it is enough to concentrate on two special kinds of isotopes.

An essential breakthrough in the theory of G-loops was done by R. L. Wilson Jr. in his Ph.D thesis (1969), written at the University of Wisconsin under direction of Hans Schneider, and two subsequent papers. In [104], using short combinatorial arguments, and based again on [26], it was proved that a loop of prime order is a G-loop if and only if it is a cyclic group.

In the next paper [105] it is proved that for each even number $2n$ there exist a loop of order $2n$ (we prefer to denote it by Q_{2n}, it is commonly called the Wilson loop) which turns out to be a proper G-loop. In fact, for a prime p the Wilson loop of order $2p$ strictly coincides with the loop Q_{2p} rediscovered by us in this paper as a theoretical generalization of the results of our computer algebra experimentation. This class of Wilson loops has been considered in the literature many times, see e.g., [46, 47].

There is a specific variety of G-loops which is called the conjugacy closed loops, briefly CC-loops. This class was introduced in [47] as loops, in which the right and left multiplications are closed under conjugations. An equivalent definition is suggested in [62], namely a loop should satisfy two identities

$$\text{RCC: } z(yx) = ((zy)/z)(zx) \quad \text{and} \quad \text{LCC: } (xy)z = (xz)(z \setminus (yz)).$$

In this paper Kunen characterizes the loops Q_{2p}: If p is an odd prime, then there are exactly three CC-loops of order $2p$, namely two groups and the loop Q_{2p}. Our new input into the theory of loops Q_{2p} is the precise description of the automorphism groups of $\Gamma = SRG(Q_{2p})$, as well as detecting the regular subgroups in $\mathrm{Aut}(\Gamma)$.

Kunen presents in another paper [61] a wide and deep investigation of various permutation groups related to G-loops. We have no doubt that a careful analysis of the results in [61] may suggest an alternative way to describe the groups $\mathrm{Aut}(\Gamma)$ for Q_{2p}.

10.11 Erich Schönhardt

The article [90] by E. Schönhardt from Tübingen is certainly an outstanding event in the history of combinatorial mathematics in the 20th century. We have already discussed a few significant achievements presented in it. Contrary to many future experts in the theory of Latin squares, Schönhardt was widely knowledgeable of the results of his predecessors; here is (an incomplete) list of names getting credit in [90]: Euler, Cayley, E. Netto, M. Frolov,

MacMahon, Tarry, E. Skolem, E. Coursat, A. Speiser, H. Brandt, W. Burnside, G. Frobenius, G. A. Miller and others (the ordering is according to the first appearance of the name in the text).

The paper consists of three chapters. The first chapter deals with Latin squares and various groups associated to it. The second chapter provides a systematical investigation of various relations between Latin squares, quasigroups and associated groups. The last chapter presents a complete catalogue of Latin squares of order 5 and 6 together with certain groups associated to them.

Unfortunately, a reader (even one who was born and raised in Germany, as one of us) faces essential difficulties in the understanding parts of the text due to the non-standard terminology used by Schönhardt. As it was mentioned, generations of researchers, not familiar with [90], were forced to create their own terminology which has remained very diverse for many decades.

We are sure that a careful translation of [90] will still provide a brilliant source of fresh scientific information.

10.12 Article by Brendan McKay et al.

The article [69] is definitely a landmark in the enumeration theory of Latin squares, and probably the most significant event in its history since the time of Schönhardt.

For many decades this mathematical theory suffered from excessive and idiosyncratic terminology and notation, that come from diverse traditions of many adjacent mathematical domains (from loop theory to combinatorial designs). Based mostly on the practice of Sade, the authors in [69] attempt to unify the notation. This standardization will hopefully influence researchers in the future.

Our Table 1 in Sect. 4 was strongly influenced by the acquaintance with a preliminary version of the paper by McKay et al. We were also trying to make our bibliography complementary to their list of 55 references.

10.13 A Few Books

Below we list a few books which may be helpful to the reader.

The book [65] acquaints with Latin squares, strongly regular graphs, nets and association schemes.

A detailed consideration of transversal designs and nets may be found in [17].

The classical book [34] has served for many decades as a comprehensive source on Latin squares, it has a very detailed and annotated bibliography.

The monograph [15] is one of the first attempts to address the foundations of loops and quasigroups. (One of the authors, who was personally familiar with V. D. Belousov, will forever recall his fascinating features of an algebraist and a person.)

A more modern text [79] provides a very friendly introduction to loops and quasigroups, it is filled with examples and helpful exercises.

The monograph [93] contains a helpful chapter dealing with groups and quasigroups, considering, in particular, loops, nets and isotopy.

The book [10] provides comprehensive information about applications of association schemes in statistics, in particular those schemes which are related to Latin squares.

Finally, we highly recommend one more introductory text, namely [64]. It covers many topics, in particular, some links of Latin squares with groups and graphs.

10.14 More References

As was already mentioned in Sect. 4.5, the case of strongly regular graphs with the parameter set $(v, k, \lambda, \mu) = (25, 12, 5, 6)$ is the smallest case, where we face significant difficulties in developing complete constructive enumeration of all objects. This is why this problem became one of the leads attracting for investigators.

In fact, four teams of researchers attacked this problem more or less at the same time: Shrikhande and Bhat [92]; Arlasarov, Leman and Rosenfeld ([83, 5], see also [101]); Paulus [77]; and finally Corneil and Mathon [31]. The latter paper provides the most complete bibliography and many interesting details about all 15 graphs. In particular, the order and the generators of the groups of the graphs on 25 vertices, which appear in Sect. 4, were already presented in [31]. The automorphism groups of all 3-nets on 25 points are presented on the homepage of Moorhouse [71].

During the last few years a number of very interesting strong results related to Latin squares were obtained by Ian M. Wanless, a former student of McKay. We refer here e.g. to [70] and [100].

There are a few attempts in the literature documenting the efforts to find a closed formula for the number of distinct Latin squares of order n (see Sect. 10.7 as well as the discussion in [69]). In this context we attract the reader's attention to the paper [54], which seems to have gone unnoticed by researchers. Here the author suggests an approach that reduces the problem to the computation of some structure constants in a certain algebra of double cosets of the symmetric group of order n^2 with respect to its intransitive subgroup. No attempts are known to investigate how practical such an approach may be for small values of n.

Section 2.2 of the Thesis [29] discusses helpful isotopy invariants of loops that are formulated in terms of corresponding association schemes.

A number of publications were devoted to the investigation of Latin squares with high symmetry regarded in diverse senses. The articles [19] and [33] serve as nice patterns of such investigations.

For infinite quasigroups and loops that are strictly related to the investigations of 3-webs in classical geometry, see the recent survey [1].

And for description of the automorphism groups of dihedral groups see besides [84], paper [99] and references in it.

10.15 The Presented Project

The project described in this paper has been ongoing since 2001. Its first version was presented at the AMS meeting #991 at Chapel Hill (October 24–25, 2003), see [58]. Since that time we discovered how Q_6 may be extended to an infinite series Q_{2p}. A preliminary version of this result was announced in [51].

Acknowledgments

Conceptually, this project goes back to the Moscow school, in particular to the collaboration of M. K. with I. A. Faradžev, A. V. Ivanov, and especially, with the late Ja. Ju. Gol'fand. More recent collaborations with E. van Dam, M. Muzychuk and A. Woldar have been very helpful.

We also acknowledge R. Bailey, D. Keedwell, N. Kriger, R. Mathon, B. McKay, W. Myrvold, Ch. Praeger, S. Reichard, H. Schneider, E. Spence, A. Sprague, and R. L. Wilson for valuable assistance, information and suggestions.

This project was initiated in August 2001, when M. K. was visiting Alex Rosa at McMaster University, Hamilton, Ontario. Long discussions with Alex on various issues related to Latin squares were very productive and stimulating.

The main part of this contribution was prepared when A. H. was assistant at the Institute of Mathematics, Universität Augsburg, Germany. At that time M. K. was partially supported by Department of Mathematical Sciences, University of Delaware, Newark, DE 19716, USA.

Finally, we would like to thank Ted Eisenberg for his comments on the presentation of this paper.

References

1. M. A. Akivis and V. V. Goldberg, Local algebras of a differential quasigroup, *Bull. Amer. Math. Soc. (N.S.)* (2), **43** (2006), 207–226. Erratum, (3), **43** (2006), 397.
2. A. A. Albert, Quasigroups. II, *Trans. Amer. Math. Soc.*, **55** (1944), 401–419.
3. R. Artzy, Crossed-inverse and related loops, *Trans. Amer. Math. Soc.*, **91** (1959), 480–492.
4. R. Artzy, Isotopy and paratopy of quasigroups, *Proc. Amer. Math. Soc.*, **14** (1963), 429–431.

5. V. L. Arlazarov, A. A. Leman, and M. Z. Rosenfeld, *The Construction and Analysis by a Computer of the Graphs on 25, 26 and 29 Vertices, Preprint*, 58 pp. Institute of Control Theory, Moscow, 1975 (in Russian).
6. V. L. Arlazarov, A. M. Baraev, Ja. Ju. Gol'fand, and I. A. Faradžev, Construction with the use of a computer of all Latin squares of order 8, in *Algorithmic Studies in Combinatorics*, pp. 129–141, Nauka, Moscow, 1978. Collection of papers (in Russian).
7. L. Babai, Automorphism groups, isomorphism, reconstruction, in R. L. Graham, et al. (eds.) *Handbook of Combinatorics*, pp. 1447–1540, Elsevier, Amsterdam, 1995.
8. R. A. Bailey, Strata for randomized experiments, *J. R. Stat. Soc. Ser. B (Methodol.)*, **53** (1991), 27–78.
9. R. A. Bailey, Orthogonal partitions in designed experiments, *Des. Codes Cryptogr.* (3), **8** (1996), 45–77.
10. R. A. Bailey, *Association Schemes. Designed Experiments, Algebra and Combinatorics*, Cambridge Studies in Advanced Mathematics, Vol. 84, Cambridge University Press, Cambridge, 2004.
11. R. A. Bailey and P. Cameron, *Latin squares in experimental design.* http://designtheory.org/library/encyc/latinsq/e/.
12. R. A. Bailey and D. Jungnickel, Translation nets and fixed-point-free group automorphisms, *J. Comb. Theory, Ser. A*, **55** (1990), 1–13.
13. E. Bannai and T. Ito, *Algebraic Combinatorics I. Association Schemes*, Benjamin/Cummings, Menlo Park, 1984.
14. A. Barlotti and K. Strambach, The geometry of binary systems, *Adv. Math.*, **49** (1983), 1–105.
15. V. D. Belousov, *Foundations of the Theory of Quasigroups and Loops*, Nauka, Moscow, 1967 (in Russian).
16. D. Betten, Die 12 lateinischen Quadrate der Ordnung 6, *Coxeter Festschrift I, Mitt. Math. Semin. Gießen*, **163** (1984), 181–188. [The 12 Latin squares of order 6].
17. T. Beth, D. Jungnickel, and H. Lenz, *Design Theory*, Cambridge University Press, Cambridge, 1993.
18. T. Bier and P. K. Chua, Numerical invariants of strongly regular graphs, *J. Comb. Theory, Ser. A*, **49** (1988), 145–171.
19. A. Bonisoli, On 2-transitive 3-nets, *J. Geom.* (1–2), **41** (1991), 42–57. Addendum, (1–2), **42** (1991) 41.
20. R. C. Bose, Strongly regular graphs, partial geometries and partially balanced designs, *Pac. J. Math.*, **13** (1963), 389–419.
21. A. E. Brouwer, A. M. Cohen, and A. Neumaier, *Distance-Regular Graphs*, Springer, Berlin, 1989.
22. R. H. Bruck, Finite nets I. Numerical invariants, *Can. J. Math.*, **3** (1951), 94–107.
23. R. H. Bruck, Loops with transitive automorphism groups, *Pac. J. Math.*, **1** (1951), 481–483.

24. R. H. Bruck, Finite nets II. Uniqueness and embedding, *Pac. J. Math.*, **13** (1963), 421–457.
25. R. H. Bruck, What is a loop? in A. A. Albert (ed.) *Studies in Modern Algebra*, M.A.A. Studies in Mathematics, Vol. 2, pp. 59–99, Prentice Hall, New York, 1963.
26. B. F. Bryant and H. Schneider, Principal loop-isotopes of quasigroups, *Can. J. Math.*, **18** (1966), 120–125.
27. R. P. Burn, Cayley tables and associativity, *Math. Gaz.* (422), **62** (1978), 278–281.
28. A. Cayley, On Latin squares, *Oxford Cambridge Dublin Messenger Math.*, **19** (1890), 135–137.
29. Y. Chang, *Imprimitive Symmetric Association Schemes of Rank 4*, Ph.D. Thesis, University of Michigan, USA, 1994.
30. Ch. J. Colbourn and A. Rosa, *Triple Systems*, Clarendon, Oxford, 1999.
31. D. G. Corneil and R. A. Mathon, Algorithmic techniques for the generation and analysis of strongly regular graphs and other combinatorial configurations, *Ann. Discrete Math.*, **2** (1978), 1–32.
32. E. van Dam, M. Klin, M. Muzychuk, and A. Woldar, Some Implications on Amorphic Association Schemes, in preparation.
33. A. Devillers and J. I. Hall, Rank 3 Latin square designs, *J. Combin. Theory Ser. A* (5), **113** (2006), 894–902.
34. J. Dénes and A. D. Keedwell, *Latin Squares and Their Application*, Academic Press, New York, 1974.
35. J. Dénes and A. D. Keedwell, Latin squares and 1-factorizations of complete graphs. I: Connections between the enumeration of Latin squares and rectangles and r-factorizations of labelled graphs, *Ars Comb.*, **25A** (1988), 109–126.
36. J. Dénes and A. D. Keedwell, Latin squares and one-factorizations of complete graphs. II: Enumerating one-factorizations of the complete directed graph K_n^* using MacMahon's double partition idea, *Util. Math.*, **34** (1988), 73–83.
37. L. Euler, Recherches sur une nouvelle espèce de quarrés magiques, *Verh. Zeeuwsch Gennot. Weten Vliss*, **9** (1782), 85–239.
38. I. A. Faradžev, Constructive enumeration of combinatorical objects, *Colloq. Intern. CNRS*, **260** (1978), 131–135.
39. I. A. Faradžev and M. H. Klin, Computer package for computations with coherent configurations, in *Proc. ISSAC-91*, pp. 219–223, ACM Press, Bonn, 1991.
40. I. A. Faradžev, M. H. Klin, and M. E. Muzichuk, Cellular rings and groups of automorphisms of graphs, in I. A. Faradžev, et al. (eds.) *Investigations in Algebraic Theory of Combinatorial Objects*, pp. 1–152, Kluwer Academic, Dordrecht, 1994.
41. D. J. Finnay, Some enumerations for the 6×6 Latin squares, *Util. Math.*, **21A** (1982), 137–153.

42. R. A. Fisher and F. Yates, The 6×6 Latin squares, *Proc. Cambridge Philos. Soc.*, **30** (1934), 492–507.
43. The GAP Group, GAP – Groups, Algorithms, and Programming, Version 4.2; Aachen, St. Andrews, 1999, http://www-gap.dcs.st-and.ac.uk/~gap.
44. C. D. Godsil, *Algebraic Combinatorics*, Chapman & Hall, New York, 1993.
45. Ja. Ju. Gol'fand, A. V. Ivanov, and M. Klin, Amorphic cellular rings, in I.A. Faradžev, et al. (eds.) *Investigations in Algebraic Theory of Combinatorial Objects*, pp. 167–186, Kluwer Academic, Dordrecht, 1994. (Translation from the Russian original: Investigations in Algebraic Theory of Combinatorial Objects, pp. 32–38 and 39–49, VNIISI, Moscow, 1985.)
46. E. G. Goodaire and D. A. Robinson, Loops which are cyclic extensions of their nuclei, *Compos. Math.*, **45** (1982), 341–356.
47. E. G. Goodaire and D. A. Robinson, A class of loops which are isomorphic to all loop isotopes, *Canad. J. Math.*, **3** (1982), 662–672.
48. D. Hachenberger and D. Jungnickel, Translation nets: A survey, *Discrete Math.*, **106/107** (1992), 231–242.
49. K. Heinrich, Approximation to a self-orthogonal Latin square of order 6, *Ars Combin.*, **4** (1977), 17–24.
50. A. Heinze, *Applications of Schur Rings in Algebraic Combinatorics: Graphs, Partial Difference Sets and Cyclotomic Schemes*, Ph.D. Thesis, Department of Mathematics, Carl von Ossietzky University Oldenburg, Germany, 2001.
51. A. Heinze and M. Klin, Links between Latin squares, nets, graphs and groups: A work inspired by a paper of A. Barlotti and K. Strambach, *Electron. Notes Discrete Math.*, **23** (2005), 13–21.
52. D. R. Hughes and F. C. Piper, *Design Theory*, Cambridge University Press, Cambridge, 1985.
53. L. K. Jørgensen and M. H. Klin, Switching of edges in strongly regular graphs: I. A family of partial difference sets on 100 vertices, *Electron. J. Combin.* (1), **10** (2003), 31 pp.
54. A.-A. A. Jucys, The number of distinct Latin squares as a group-theoretical constant, *J. Combin. Theory A*, **20** (1976), 265–272.
55. D. Jungnickel, Existence results for translation nets, in P. J. Cameron, J. W. P. Hirschfeld, D. R. Hughes (eds.) *Finite Geometries and Designs*, London Math. Soc. Lecture Notes, Vol. 49, pp. 172–196, Cambridge University Press, Cambridge, 1981.
56. D. Jungnickel, Latin squares, their geometries and their groups. A survey, in D. Ray-Chaudhuri (ed.), Coding Theory and Design Theory. Part I, *IMA Vol. Math. Appl.* **20** (1990) 166–225.
57. A. D. Keedwell, Decomposition of complete graphs defined by quasigroups, *Ann. Discrete Math.*, **12** (1982), 185–192.
58. M. H. Klin and A. Heinze, Amorphic association schemes II. A positive answer on a question of Barlotti-Strambach, Preliminary Report, Abstract 991-05-28, *Abstracts of papers presented to Amer. Math. Soc.* **24** (2003) 594.

59. M. Ch. Klin, R. Pöschel, and K. Rosenbaum, *Angewandte Algebra*, VEB Deutscher Verlag der Wissenschaften, Berlin, 1988.
60. H. C. Kirton, Mutually orthogonal partitions of the 6×6 Latin squares, *Util. Math.*, **27** (1985), 265–284.
61. K. Kunen, G-loops and permutation groups, *J. Algebra*, **220** (1999), 694–708.
62. K. Kunen, The structure of conjugacy closed loops, *Trans. Amer. Math. Soc.* (6), **352** (2000), 2889–2911.
63. A. V. Kuznetsov and E. A. Kuznetsov, Two-generator doubly homogeneous quasigroups. Quasigroups and Latin squares, *Mat. Issled.*, **71** (1983), 34–53 (Russian).
64. C. Y. Laywine and G. L. Mullen, *Discrete Mathematics, Using Latin squares*, Wiley, New York, 1998.
65. J. H. van Lint and R. M. Wilson, *A Course in Combinatorics*, 2nd edn. Cambridge University Press, Cambridge, 2001.
66. P. A. MacMahon, *Combinatory Analysis*, Vol. 1, Cambridge University Press, London, 1915.
67. A. E. Malykh and A. N. Pekhletskaya, Automorphism groups of Latin squares of order 6, *Kombinatorika. Uchenye Zapiski Permskogo Gos. Ped. Instituta*, **152** (1976), 63–77 (Russian).
68. B. D. McKay, *Nauty*, http://cs.anu.edu.au/people/bdm/nauty/.
69. B. D. McKay, A. Meynert, and W. Myrvold, Small Latin squares, quasigroups and loops, *J. Comb. Designs*, **15** (2007), 98–119.
70. B. D. McKay and I. M. Wanless, On the number of Latin squares, *Ann. Comb.*, **9** (2005), 335–344.
71. G. E. Moorhouse, *Nets and Latin squares of order 5.* http://www.uwyo.edu/moorhouse/pub/nets5.
72. G. E. Moorhouse, Bruck nets, codes, and characters of loops, *Des. Codes Cryptogr.* (1), **1** (1991), 7–29.
73. M. Muzychuk, M. Klin, and R. Pöschel, The isomorphism problem for circulant graphs via Schur ring theory, in *Codes and Association Schemes*, DIMACS Ser. Discrete Math. Theoret. Comput. Sci., Vol. 56, pp. 241–264, Amer. Math. Soc., Providence, 2001.
74. P. N. Nagy and K. Strambach, Loops as invariant sections in groups, and their geometry, *Canad. J. Math.*, **46** (1994), 1027–1056.
75. D. A. Norton and S. K. Stein, Cycles in algebraic systems, *Proc. Amer. Math. Soc.*, **7** (1956), 999–1004.
76. E. T. Parker, The maximum number of digraph-distinct ordered quadruples on six marks, *J. Combin. Theory A*, **19** (1975), 245–246.
77. A. J. L. Paulus, *Conference Matrices and Graphs of Order 26*, T.H.-Report 73-WSK-06, Technological University, Dept. of Mathematics, Eindhoven, Netherlands, 1973, 89 pp.
78. R. Peeters, Uniqueness of strongly regular graphs having minimal p-rank, *Linear Algebra Appl.*, **226/228** (1995), 9–31.
79. H. O. Pflugfelder, *Quasigroups and Loops: Introduction*, Sigma Series in Pure Mathematics, Vol. 7, Heldermann, Berlin, 1990.

80. K. T. Phelps, Latin square graphs and their automorphism groups, *Ars Combin.*, **7** (1979), 273–299.
81. K. T. Phelps, Automorphism free Latin square graphs, *Discrete Math.*, **31** (1980), 193–200.
82. D. A. Robinson, The Bryant-Schneider group of a loop, *Ann. Soc. Sci. Bruxelles, Sér. I*, **94** (1980), 69–81.
83. M. Z. Rosenfeld, On the construction and properties of some families of strongly regular graphs, *Uspekhi Mat. Nauk*, **28** (1973), 197–198 (Russian).
84. F. Rotmaler, Automorphism groups of dihedral groups, *Ukr. Math. J.*, **29** (1977), 162–167 (Translation from Russian).
85. J. J. Rotman, *An Introduction to the Theory of Groups*, 4th edn. Graduate Texts in Mathematics, Vol. 148, Springer, New York, 1995.
86. A. Sade, Quasigroupes isotopes. Autotopies d'un groupe, *Ann. Soc. Sci. Bruxelles Sir. III*, **81** (1967), 231–239.
87. A. Sade, Morphisms de quasigroups table, *Revista Fac. Sciencias de Lisboa*, **13** (1971), 149–172.
88. P. N. Saxena, A simplified method of enumerating Latin squares by MacMahon's differential operators. Part I. The 6×6 Latin squares, *J. Indian Soc. Argicultural Stat.*, **2** (1950), 161–188.
89. P. N. Saxena, A simplified method of enumerating Latin squares by MacMahon's differential operators. Part II. The 7×7 Latin squares, *J. Indian Soc. Agricultural Statist.*, **3** (1951), 24–79.
90. E. Schönhardt, Über lateinische Quadrate und Unionen, *J. Reine Angewandte Math.*, **163** (1930), 183–230.
91. S. S. Shrikhande, The uniqueness of the L_2-association scheme, *Ann. Math. Statist.*, **30** (1959), 781–798.
92. S. S. Shrikhande and V. N. Bhat, Graphs derivable from $L_3(5)$ graphs, *Sankhyā Ser. A*, **33** (1971), 315–350.
93. J. D. H. Smith and A. B. Romanovska, *Post-Modern Algebra*, Wiley, New York, 1999.
94. L. H. Soicher, *The GRAPE Package for GAP.* http://www.maths.qmul.ac.uk/~leonard/grape/.
95. E. Spence, *Strongly Regular Graphs on at Most 64 Vertices.* http://www.maths.gla.ac.uk/~es/srgraphs.html.
96. A. P. Sprague, Translation nets, *Mitt. Math. Semin. Gießen*, **157** (1982), 46–68.
97. A. Suschkewitsch, On a generalization of the associative law, *Trans. Amer. Math. Soc.*, **31** (1929), 204–214.
98. G. Tarry, Le problème des 36 officiers, *Assoc. Franc. Paris*, **29** (1900), 170–203.
99. G. L. Walls, Automorphism groups, *Amer. Math. Monthly* (6), **93** (1986), 459–462.
100. I. M. Wanless, About Latin squares based on cyclotomic orthomorphisms, *Electron. J. Combin.*, **12** (2005), R22.
101. B. Weisfeiler (ed.) *On Construction and Identification of Graphs.* With contributions by A. Lehman, G. M. Adelson-Velsky, V. Arlazarov, I.

Faragev, A. Uskov, I. Zuev, M. Rosenfeld, and B. Weisfeiler, Lecture Notes in Mathematics, Vol. 558, Springer-Verlag, Berlin, 1976.

102. H. Wielandt, *Finite Permutation Groups*, Academic Press, New York, 1964. (Translated from the German by R. Bercov.)

103. E. L. Wilson, A class of loops with the isotopy-isomorphy property, *Canad. J. Math.*, **18** (1966), 589–592.

104. R. L. Wilson Jr., Isotopy-isomorphy loops of prime order, *J. Algebra*, **31** (1974), 117–119.

105. R. L. Wilson Jr., Quasidirect products of quasigroups, *Comm. Algebra*, **3** (1975), 835–850.

Siamese Combinatorial Objects via Computer Algebra Experimentation

Mikhail Klin[1], Sven Reichard[2], and Andrew Woldar[3]

[1] Ben-Gurion University of the Negev, Beer Sheva 84105, Israel. klin@cs.bgu.ac.il
[2] University of Western Australia, Crawley 6009, Western Australia. reichard@maths.uwa.edu.au
[3] Villanova University, Villanova, PA 19085, USA. andrew.woldar@villanova.edu

Summary. Following Kharaghani and Torabi [On a decomposition of complete graphs, *Graphs Comb.*, **19** (2003), 519–526], we introduce new concepts of Siamese color graph, Siamese association scheme and Siamese Steiner design. With the aid of a computer, we determine all Siamese objects on 15 points, as well as hundreds on 40 points. As a generalization of accumulated observations, an infinite series of Siamese association schemes related to certain imprimitive actions of the groups $PSL(2, q^2)$ is outlined. Special attention is paid to the spirit of computer-aided activity, namely to algorithms, technical data, successful *ad hoc* tricks, and computer-free interpretations of obtained results.

Key words: Color graph, Coherent configuration, Association scheme, Distance regular graph, Strongly regular graph, Generalized quadrangle, Spread, Steiner system, Siamese color graph, Siamese association scheme, Siamese Steiner design, Computer algebra package

1 Introduction

The starting point for this project was our acquaintance with the paper [35], in which an infinite series of special color graphs was constructed. Quite early on, we realized how one could naturally axiomatize the properties of these constructed objects in order to obtain, in our terminology, a Siamese association scheme – more generally, a Siamese color graph. Certain mono- and bi-chromatic graphs related to Siamese color graphs form distance regular and strongly regular graphs; in particular, the strongly regular graphs have the same parameters as point graphs of generalized quadrangles. If moreover, these strongly regular graphs are geometric, then one additional structure is implied – a Siamese Steiner design.

In such manner, we came about a quite remarkable and hopefully fruitful link between such well studied objects as color graphs, association schemes

M. Klin et al. (eds.), Algorithmic Algebraic Combinatorics and Gröbner Bases, DOI 10.1007/978-3-642-01960-9_2, © Springer-Verlag Berlin Heidelberg 2009

and Steiner designs – all termed by us *Siamese objects*. As a next step, we constructed and investigated many such objects at both the computational and theoretical level.

From the early inception of our project, the use of computers was an inalienable part of all our activities. Indeed, it would be fair to say that virtually all theoretical results presented in this paper are generalizations of observations gleaned through one or another computer experiment.

A brief account of our discoveries in a form very close to the style of an extended abstract is presented in [39]. Elements of the general theory of Siamese objects will be developed in forthcoming papers, particularly in [40, 41]. In the course of our work over these papers, we realized that there is an excellent opportunity to prepare a text of absolutely different genre – one that lies somewhere between the two extreme cases of extended abstract and comprehensive treatise. This is the genre of a tutorial paper, which is presented in the current text.

One of our multiple objectives is to provide a brief and reasonably friendly introduction to various concepts from algebraic combinatorics, including coherent configurations, association schemes, distance regular graphs, strongly regular graphs, etc. It is hoped that this portion of the text will serve not only as an aid to the readers of *this* paper, but to those who have scientific interest in any number of accompanying contributions comprising this collection.

We are, however, attempting to achieve this objective in conjunction with a second important goal – to introduce the reader to the recently formulated notion of Siamese combinatorial objects, with sufficient attention paid to theoretical aspects, striking examples, and relevant links to objects from other areas of combinatorics such as geometry and design theory.

The word "algorithmic," as it appears in the title of the entire collection of papers, clearly dictates what are our remaining goals:

- To describe the main methodological features of our vision of computer algebra experimentation, as it applies to combinatorics.
- To introduce the reader to extant computer packages such as COCO and GAP.
- To share our experience, obtained at the forefront of computational and theoretical exploration.
- To challenge the reader to extend, or otherwise generalize, our results.

We conclude with a brief outline of the balance of the paper. In Sect. 2, all preliminary information is gathered. Computer packages are discussed in Sect. 3, together with general features of our methodology. Section 4 is devoted to main definitions and simple theoretical facts related to Siamese objects, while Sect. 5 provides an overview of our activities. (Section 5 may be regarded as a synopsis of material to appear in [40] and [41].) The main scientific load of the paper occurs in Sects. 6, 7, 8 and 9, where we present all experimental and theoretical results concerning Siamese objects on 15 and 40 points. We conclude with additional remarks in Sect. 10, including some of purely speculative nature.

2 Preliminaries

2.1 Color Graphs

Definition 1. *A color graph Γ is a pair $(V, \mathcal{R})$, where V is a set of vertices and $\mathcal{R}$ a set of (non-empty) disjoint binary relations on V such that $\bigcup_{R\in\mathcal{R}} R = V^2$. We refer to the elements of $\mathcal{R}$ as the colors of Γ, and to the number $|\mathcal{R}|$ of its colors as the rank of Γ.*

In other words, a color graph is an edge-coloring of a complete graph. Note that any function ϕ defined on V^2 defines a color graph.

Given a color graph Γ, we define its *adjacency matrix* to be the $v \times v$ matrix $A = (a_{ij})$ for which $a_{ij} = t$ if $(x_i, x_j) \in R_t$, $R_t \in \mathcal{R}$.

Definition 2. *Let $\Gamma = (V, \mathcal{R})$ and $\Gamma' = (V', \mathcal{R}')$ be color graphs. An isomorphism $\phi : \Gamma \to \Gamma'$ is a bijection of V onto V' which induces a bijection $\psi : \mathcal{R} \leftrightarrow \mathcal{R}'$ of colors. A weak (or color) automorphism is an isomorphism $\phi : \Gamma \to \Gamma$. If, in addition, the induced map ψ is the identity on $\mathcal{R}$ we call ϕ a (strong) automorphism.*

We denote by $CAut(\Gamma)$ and $Aut(\Gamma)$ the groups of weak and strong automorphisms of Γ, respectively. We shall often refer to $Aut(\Gamma)$ as the *group of Γ*, and to $CAut(\Gamma)$ as the *color group of Γ*.

2.2 Coherent Configurations and Association Schemes

A coherent configuration is one of the initial notions which, in principle, we presume to be known (e.g., see [29, 18, 12]). However, to keep our text self-contained we give its definition.

Definition 3. *A color graph $\mathcal{M} = (X, \mathcal{R})$, $\mathcal{R} = \{R_i \mid i \in I\}$, is a coherent configuration if the following conditions are satisfied:*

(1) The identity relation $Id_X = \{(x, x) \mid x \in X\}$ is a union of suitable relations R_i, $i \in I'$, $I' \subset I$.
(2) For each $i \in I$ there exists $i' \in I$ such that $R_i^t = R_{i'}$, where $R_i^t := \{(x, y) \mid (y, x) \in R_i\}$.
(3) For any $i, j, k \in I$, the number p_{ij}^k of elements $z \in X$ for which $(x, z) \in R_i$ and $(z, y) \in R_j$ is constant, provided $(x, y) \in R_k$.

The constants p_{ij}^k appearing in Definition 3 are called *intersection numbers* (or *structure constants*) of $\mathcal{M}$.

Given a coherent configuration $\mathcal{M} = (X, \{R_i\})$ we refer to the relations R_i as *basis relations*. The graphs $\Gamma_i = (X, R_i)$ will be called *basis graphs*, whereas their adjacency matrices $A_i = A(\Gamma_i)$ will be called *basis matrices*. This allows us to switch freely between the languages of matrices, relations and graphs.

Because a coherent configuration is defined as a particular case of a color graph, all notions defined for color graphs apply to coherent configurations as a special case.

Similarly, we may regard association schemes as a particular class of coherent configurations (see [18]). For completeness, we give its definition as well.

Definition 4. *A coherent configuration $\mathcal{M} = (X, \{R_i\})$ is an association scheme if the identity relation Id_X is one of the basis relations of $\mathcal{M}$. Typically, we denote this basis relation by R_0.*

We stress that in our terminology an association scheme is not presumed to be symmetric or commutative, see [1].

In what follows we call the identity relation $Id_X = R_0$ of an association scheme defined over X the *trivial relation*, while all nontrivial relations are called *classes*. We call a basis graph *nontrivial* if its arc set is a class. In a *primitive* association scheme all nontrivial basis graphs are connected; otherwise the scheme is *imprimitive*.

A class of examples of association schemes arises from distance regular graphs:

Definition 5. *Let Γ be a connected graph of diameter d. Suppose that for any pair of vertices (x, y), the number of vertices z at distance i from x and at distance j from y depends only on the distance k of x and y. Then Γ is called a distance regular graph (briefly, drg).*

Given such a graph Γ, we can define relations R_i, $i = 0, \ldots, d$, with $(x, y) \in R_i$ if their distance in Γ is i.

Proposition 1. *The relations R_i defined above form an association scheme with d classes.*

Association schemes coming from distance regular graphs are called *metric*, or sometimes *P-polynomial.*

We will use a number of additional notions, including imprimitive drg, antipodal cover and quotient graph. We refer the reader to [9] for their discussion.

Let $\Gamma = (V, E)$ be a finite simple graph, that is, undirected, without loops or multiple edges. The *order* of Γ is $v = |V|$. The number of neighbors of a vertex $x \in V$ is called the *valency* of x. Recall that Γ is called *regular* if all its vertices have the same valency k. In such case, we further say that Γ has valency k.

Although a strongly regular graph is a particular case of a drg, we prefer to define this notion in a separate self-contained manner.

Definition 6. *Let Γ be a regular graph of order v and valency k. Suppose there exist nonnegative integers λ and μ such that all pairs of adjacent vertices in Γ have λ common neighbors, and all pairs of distinct non-adjacent vertices have μ common neighbors. Then Γ is a strongly regular graph (briefly, srg) with the parameters (v, k, λ, μ). (We sometimes refer to Γ as an srg(v, k, λ, μ).)*

Graphs of the form $m \circ K_n$ (m disjoint copies of the complete graph K_n) comprise the class of srg's with $\mu = 0$. As the complement of an srg is again an srg, we automatically obtain a second class of imprimitive srg's, namely that consisting of the so-called complete multipartite graphs $\overline{m \circ K_n}$.

Example 1. The Petersen graph P is one of the most famous strongly regular graphs. Formally, it can be defined as the complement $\overline{T(5)}$ of the triangular graph $T(5)$. In other words, vertices are 2-element subsets of a fixed 5-element set, with two vertices adjacent if and only if their corresponding subsets are disjoint.

2.3 Incidence Structures

2.3.1 General Definitions

Definition 7. *Let P and B be sets, and let $I \subseteq P \times B$ be a relation, referred to as incidence. Then (P, B, I) is called an incidence structure.*

Usually, the elements of P are called *points*, and those of B, *blocks*. However, sometimes in place of block one uses the more geometric term *line*.

Definition 8. *Let (P, B, I) be an incidence structure. The incidence structure (B, P, I^T) obtained by interchanging blocks and points, and reversing the incidence relation, is called its dual structure.*

Given an incidence structure, we can naturally define three graphs related to it. The incidence graph (or Levi graph) is the bipartite graph defined on $P \cup B$ where a point and a block are adjacent if they are incident. The point graph is defined on P with two points adjacent if they are "collinear" (i.e., incident with a common block). Finally, the block graph is the point graph of the dual structure (i.e., vertices are blocks, and two blocks are adjacent if they are incident to a common point).

2.3.2 Steiner Systems

Definition 9. *Let (P, B) be an incidence structure in which each block contains k points and each set of t points is contained in a unique block. Then (P, B) is called a Steiner system $S(t, k, v)$, where $v = |P|$.*

If $t = 2$, $k = 3$, we speak of a *Steiner triple system*, or $STS(v)$. We refer the reader to [31] for a concise treatment of incidence structures, and to [2] for detailed information on Steiner systems.

Example 2. Consider the 4-dimensional vector space F^4 defined over the finite field F of q elements. Let V and B denote, respectively, the set of all 1-dimensional and 2-dimensional subspaces of F^4. Define incidence to be ordinary inclusion. Then the resulting incidence structure (V, B) provides a classical example of a Steiner system $S(2, q+1, q^3+q^2+q+1)$. We denote it by $PG(3, q)$ and call it *projective 3-space over* F. Elements of V are called *projective points* and those of B, *projective lines.* Note that traditionally $PG(3, q)$ also includes "planes," which correspond to 3-dimensional subspaces of F^4.

2.3.3 Generalized Quadrangles

Definition 10. *A partial geometry with parameters* (K, R, T) *is an incidence structure such that each block (or line) contains* K *points, each point lies on* R *lines, each pair of distinct points lies on at most one line, and for each line* l *and point* P *not on* l*, there exist exactly* T *lines through* P *that intersect* l*.*

Definition 11. *A* generalized quadrangle *(briefly,* GQ*) is a partial geometry with* $T = 1$*. The pair* $(s = K - 1, t = R - 1)$ *is called the* order *of the generalized quadrangle. We will denote a generalized quadrangle of order* (s, t) *by* $GQ(s, t)$*. In the case of* $s = t$*, we will simply speak of a GQ of order* s *and write* $GQ(s)$*.*

Theorem 1. *The point graph of a* $GQ(s, t)$ *is strongly regular, with parameters*

$$v = (s+1)(st+1), \qquad k = s(t+1), \qquad \lambda = s-1, \qquad \mu = t+1.$$

We often refer to the point graph of a $GQ(s, t)$ as a $GQ(s, t)$-graph.

Example 3. A classical example of a $GQ(2)$ is constructed with the aid of a 6-element set X, wherein points correspond to 2-element subsets of X and lines to partitions of X into three parts of equal size. Incidence is containment. Fulfillment of the axioms of a GQ is an easy exercise for the reader. The point graph of this $GQ(2)$ is the complement graph $\overline{T(6)}$ of $T(6)$. It is an srg$(15, 6, 1, 3)$. This structure is a classical mathematical object which goes back to J.J. Sylvester [64], who called it the "duad-syntheme geometry."

We refer the reader to [54] for more information on generalized quadrangles.

2.4 Kramer–Mesner Method and Related Issues

We now briefly discuss a method which, in principle, allows one to construct all incidence structures with prescribed parameters, which are invariant with respect to a given permutation group (H, Ω). It is usually called the *Kramer–Mesner method,* due to its formal presentation in [43].

Table 1. Matrix $M = m(i, j)$ of Example 4

	K_1	K_2	K_3	K_4	K_5	Σ
T_1	1	3	6	3	0	13
T_2	0	0	4	8	1	13

Below we provide an outline of the algorithm as it applies to the special case of Steiner systems. Thus, we presume we are looking for systems $S(t, k, v)$ which are invariant with respect to (H, Ω), where $|\Omega| = v$. Our treatment follows closely the spirit of [5].

Denote by $\mathcal{T}$ the set of all orbits of (H, Ω) in its induced action on the t-element subsets of Ω, and let $a = |\mathcal{T}|$. Similarly, denote by $\mathcal{K}$ the set of all orbits of (H, Ω) on the k-element subsets of Ω, and let $b = |\mathcal{K}|$.

We now form an $a \times b$ matrix M, with rows and columns indexed by the elements of $\mathcal{T}$ and $\mathcal{K}$, respectively, in which the entry $m(i, j)$ of M is defined as follows: Let $X \in T_i$ be an arbitrary t-element set in the i-th orbit T_i of $\mathcal{T}$. Then $m(i, j) = |\{Y \in K_j \mid X \subseteq Y\}|$, where K_j is the j-th orbit of $\mathcal{K}$. It is evident that $m(i, j)$ does not depend on our choice of $X \in T_i$.

Suppose we are able to find a collection $\{j_1, j_2, \ldots, j_s\}$ of columns such that their sum is equal to the all-ones vector of length a. Then, setting $B = \bigcup_{e=1}^{s} K_{j_e}$, we see that (Ω, B) is evidently an $S(t, k, v)$. Conversely, every such Steiner design which is invariant with respect to (H, Ω) arises in this manner.

Example 4. Let us construct an example of an $STS(15)$ using the Kramer–Mesner method. Consider the symmetric group S_6 in its natural action on $X = \{1, 2, 3, 4, 5, 6\}$, and let Ω denote the set of all 2-element subsets of X. Because we are looking for an $S(2, 3, 15)$, we have to describe all orbits of (S_6, Ω) on 2- and 3-element subsets of Ω.

It is well known that such orbits above are in bijective correspondence with the isomorphism classes of graphs on six vertices having two and three edges, respectively. Thus, we switch to graphical language. Let $\mathcal{K}_6$ denote the complete graph with vertex set X. Then the set Ω corresponds to the edge set of $\mathcal{K}_6$, and the entry $m(i, j)$ corresponds to the number of ways one can extend a fixed copy of T_i (as a two-edge subgraph of $\mathcal{K}_6$) to a graph isomorphic to K_j by adding one new edge. Here, T_1 denotes the path of length 2, T_2 the graph with two disjoint edges, K_1 the triangle, K_2 the star, K_3 the path of length 3, K_4 the graph with two components (a two-path and an edge), and K_5 the graph with three disjoint edges.

As the reader may easily verify, we obtain the matrix M in Table 1, represented in tabular form. We also include a control sum Σ, which is nothing more than the ordinary column sum of M. (Note that Σ always coincides with $|\Omega| - t$, in our case $15 - 2 = 13$.)

Now an easy inspection reveals that the first and last columns of M induce an $STS(15)$. In other words, the design will have as its points the 15 edges of the complete graph $\mathcal{K}_6$, and as blocks the 20 triangles of $\mathcal{K}_6$ (contributed from column 1) plus the 15 1-factors of $\mathcal{K}_6$ (contributed from column 5). Thus, we get an $STS(15)$. This design will play an essential role in our presentation.

Remark 1. The $STS(15)$ just constructed is a particular case of a so-called graphical design, see [5] for a systematic discussion.

2.5 Double Cosets

An important class of association schemes, namely schemes of 2-orbits of transitive permutation groups (equivalently, centralizer algebras of such groups), may be formulated in purely group theoretic terms.

Let (G, Ω) be a transitive permutation group, and let $H = G_\alpha$ be the stabilizer in G of a point $\alpha \in \Omega$. Let $2\text{-}orb(G, \Omega) = \{R_0, R_1, \ldots, R_d\}$ be the set of orbits of the induced action of (G, Ω) on Ω^2, where R_0, as usual, denotes the 2-orbit with representative (α, α). Then, for each $0 \leq i \leq d$, the set $\{g \in G \mid (\alpha, \alpha^g) \in R_i\}$ is a so-called *double coset* of H in G, that is a subset of G of the form Hg_iH for a suitable $g_i \in G$.

It turns out that double cosets are in bijective correspondence with the elements of $2\text{-}orb(G, \Omega)$, an observation which was used implicitly throughout the early part of the 20th century. During the last few decades the theory has found numerous applications in computer algebra, e.g., see [18] and [42] for details. We mention that the transitivity assumption is actually not required in the establishment of the correspondence between 2-orbits and double cosets, although it does lead to a marked simplification.

Example 5. Let us illustrate by way of example the just defined notion of double coset. We take $G = \langle (1,2,3,4), (1,3) \rangle = D_4$, the dihedral group of order 8 and degree 4. For convenience, we give an explicit listing of its elements:

$$e, (1,2,3,4), (1,3)(2,4), (1,4,3,2), (2,4), (1,2)(3,4), (1,3), (14)(23).$$

Clearly G is a transitive group, with stabilizer $H = G_1 = \{e, (2,4)\}$. It is now easy to compute: $H = HeH$, $H(1,3)H = \{(1,3)(2,4), (1,3)\}$, and $H(1,2,3,4)H = \{(1,2,3,4), (1,2)(3,4), (1,4,3,2), (1,4)(3,2)\}$. Using these three double cosets, the reader can easily see that the corresponding three 2-orbits of D_4 are those with representatives $(1,1)$, $(1,3)$ and $(1,2)$, respectively.

3 Computer Algebra Tools

3.1 Computations in Combinatorics

The advent of computers in algebraic combinatorics is a relatively recent event. In yesteryear nearly all results were accomplished purely at the theoretical level, but today the role of machine is pervasive, being used to search for interesting objects, to determine their combinatorial and algebraic properties, and to generate conjectures. In particular, the creation of symbolic algebra systems has had profound impact on this theoretical-experimental synergy, which nowadays seems indispensable.

In addition to algorithms designed to perform both general and specific tasks, computer algebra systems come with, in many cases, substantial libraries of combinatorial and algebraic data that have been accumulated by generations of mathematicians.

We distinguish two kinds of such packages. General purpose systems are designed to deal with a broad range of mathematical problems. Examples are GAP [59] and Maple, which include their own programming languages in order to be extensible. In contrast, specialized packages are designed for one specific task. Examples here are COCO [17] for the investigation of coherent configurations, and Discreta for the construction of t-designs.

3.2 COCO

COCO (COherent COnfigurations) is a collection of programs designed specifically for the investigation of coherent configurations. It was developed around 1990 by members of the Moscow Seminar on the Algebraic Theory of Combinatorial Objects, in particular, I. A. Faradžev and M. Klin [17]. Originally written in Fortran-4, it has since been ported to C and adjusted for use on personal computers. The next version, which has been slightly modified and ported to UNIX by A. E. Brouwer, is available from his homepage [8].

In addition to having a modest library of group theoretic data, COCO provides the following facilities:

- **Inducing** Given a permutation group (G, Ω) and a combinatorial structure $\mathfrak{X}$, compute the orbit $\mathfrak{X}^G$ and the action of G on $\mathfrak{X}^G$.
- **Color Graph** Given a permutation group (G, Ω), compute its centralizer algebra $V(G, \Omega)$.
- **Intersection Numbers** Given a coherent configuration W, compute its intersection numbers (i.e., structure constants of its corresponding coherent algebra).
- **Subschemes** Given the intersection numbers of a coherent configuration, determine all mergings of its classes which lead to homogeneous subconfigurations (fusion association schemes).
- **Automorphism Groups** Given a coherent configuration and a set of mergings, compute the automorphism groups of the resulting fusion schemes.

Typically, these five commands are used in the order specified. This allows one to determine all association schemes invariant with respect to a given permutation group (G, Ω), as well as their automorphism groups (i.e., the 2-closed overgroups of (G, Ω) in S_Ω).

Regarding the process of inducing, we mention that orbits are generally easier to compute than coset representatives. Thus, given a (transitive) permutation group $(G, G/H)$ it is always convenient to have a combinatorial structure $\mathfrak{X}$ for which the action of G on the orbit $\mathfrak{X}^G$ is similar to the action of G on G/H. (In such case, H is the stabilizer in G of $\mathfrak{X}$.) This provides a time- and space-efficient way to delineate permutation representations of finite groups.

In general, it may be difficult to verify that H is the full stabilizer of $\mathfrak{X}$ in G. In contrast, in the case where H is maximal in G one may establish this fact by simply showing that $\mathfrak{X}$ is invariant under H but not under G. Thus, it is especially easy to find representations for primitive permutation actions of G. In the case of imprimitive representations, *ad hoc* tricks may be helpful.

3.3 GAP

Whereas COCO is designed for highly specialized tasks, GAP provides solutions for a wide range of problems in discrete mathematics, with a strong focus on group theory. GAP (Groups, Algorithms, Programs) was originally developed during the 1990's at the Rheinisch-Westfälische Hochschule in Aachen, Germany [59]. More recently, its main development has shifted to St. Andrews, Scotland.

The GAP package contains a relatively small kernel written in C, which makes it portable to all current operating systems. This kernel contains basic IO-capabilities, representations of primitive data types such as arbitrary precision integers, finite permutations and finite field elements, and an interpreter for the GAP language. In the most current versions, this language has been enhanced with object-oriented features such as data encapsulation and runtime polymorphism.

What makes GAP so powerful is its vast library of routines and data written in the GAP language. It contains state-of-the-art algorithms for the investigation of permutation groups, matrix groups, rings, finite fields, etc.

One additional feature is the support of so-called *share packages,* which are independent routines, typically written by different authors, to deal with specific algebraic or combinatorial objects.

3.4 GRAPE

One of the share packages mentioned above is GRAPE (GRaph Algorithms using PErmutation groups) developed by L. Soicher [62]. Its basic premise is that each graph is invariant under some group (possibly the trivial group) acting on its vertices. Such a group is stored together with the ambient graph in order to make the representation more compact, and thus speed up calculations. In fact, many of the algorithms rely on backtracking, and the known group is used to perform isomorph rejection.

Among the operations available for graphs are global invariants (diameter, girth, independence number, distance-regularity), structural determinants related to subsets of vertices (induced subgraphs, cliques, connected components), and so on. More elaborate algorithms are provided, for example, to enumerate all partial geometries having a prescribed point graph.

3.5 nauty

Over the last decade, the standard program for graph isomorphism problems is B. McKay's nauty (no automorphisms, yes?) [49]. Its name reflects the fact that, statistically speaking, a random graph has trivial automorphism group. This fact is used heuristically to find invariants which distinguish vertices of the graph; thus the algorithms are especially successful when the automorphism group is small.

For a given graph, nauty computes the automorphism group together with a canonical labeling of the vertices. The premise here is that two canonically labeled graphs are isomorphic if and only if they are equal; hence nauty enables a quick isomorphism check (quadratic in the number of vertices). Though the determination of a labeling is not computationally economic, it need only be performed once for each graph under consideration.

nauty is supplied with its own language and file format; however, there also exists an interface to it, provided courtesy of GRAPE. Thus, one need not learn another language in order to benefit from nauty's most important capabilities, which are consequently available through GAP.

3.6 Computer Algebra Experimentation

As has been mentioned several times, in the course of our investigation of Siamese objects we relied quite extensively on the use of computers. Below we distinguish the methodological stages of their use, which, from our own experience, are consistent with the steps taken in the investigation of general combinatorial structures.

- Implement existing programs to obtain quick results for relatively small objects.
- Determine limitations of existing programs.
- Create special task-oriented algorithms to enhance or replace available programs.
- Implement new algorithms on known objects as a means of measuring their efficacy.
- Search for objects of larger size, as well as establish new properties of known objects.
- Interpret and generalize experimental results on a theoretical level.

3.7 Explanation Versus Interpretation

As we have maintained throughout, the discovery of new combinatorial objects, or of unknown properties of known objects, is an immediate goal of computer algebra experimentation. The construction of a new incidence structure, color graph, strongly regular graph, distance regular graph, spread in a known graph, embedding of one object in another – all these tasks were performed many times during thc course of this project. Additional such tasks include computation of an automorphism group, its rank, its orbit structure in a specified action, isomorphism-testing, and so on.

In all cases the obtained results were computer-dependent, that is, the object or its property was elaborated in terms of some routine data generated by machine. Thus, ultimately, we were faced with the important task of performing *a posteriori* reasoning, in order to get a description of the resulting object or property which was both clear and friendly. We distinguish below two levels of such description.

Suppose, for example, we obtain a computer-generated description of an incidence structure $\mathfrak{S} = (P, B)$. By an *explanation* of $\mathfrak{S}$, we mean a lucid computer-free description of P, B, and the incidence between them. Essential use of a computer, or of additional hand calculations, is not required in this case.

By an *interpretation* of $\mathfrak{S}$, we mean that in addition to an explanation we have also a clear self-contained proof that $\mathfrak{S}$ indeed has the properties it is purported to have (for example, $\mathfrak{S}$ is a Steiner system with parameters t, k, v).

Ideally, an interpretation should be reasonably short, aesthetically pleasing, and methodologically clear. We can also speak about a "conditional interpretation," in which the proof depends on some well established and reliable source of information (for example, a catalog).

In this paper we provide a number of explanations and interpretations. As a rule, we prefer to allow the reader to decide which level of description has been achieved, due to the delicate and subjective nature of the distinction we have set forth.

4 Siamese Objects: Main Definitions

4.1 Siamese Color Graphs

Definition 12. *Let* $W = (V, \{Id_V, S, R_1, R_2, \ldots, R_n\})$ *be a color graph for which*

(1) (V, S) *is an imprimitive disconnected srg, i.e., a partition of* V *into cliques of equal size. In what follows, it will be called a spread.*
(2) For each i*, graph* (V, R_i) *is an imprimitive drg of diameter 3 with antipodal system* S*.*
(3) For each i*,* $(V, R_i \cup S)$ *is an srg.*

Then W is a Siamese color graph. We call S the spread of Γ, and the number n of drg's the Siamese rank of W.

Given a Siamese color graph W we indicate by $(v, k, \lambda, \mu, \sigma)$ its parameter set, where (v, k, λ, μ) is the (common) parameter set of each srg $(V, R_i \cup S)$ and σ is the valency of the spread S. There are obvious necessary conditions which must be satisfied by such parameters, and we refer to any set $(v, k, \lambda, \mu, \sigma)$ which satisfies these conditions as *feasible.*

Siamese color graphs were first studied by Kharaghani and Torabi [35]. The word "Siamese" comes from the observation that any two of the strongly regular graphs share the spread S, so are like conjoined twins. However, after surgical removal of the spread, both "twins" can live an independent life as distance regular graphs.

4.2 Siamese Association Schemes

Definition 13. *An association scheme*

$$W = (V, \{Id_V, S_1, \ldots, S_n, R_1, \ldots, R_k\})$$

is a Siamese association scheme if $(V, \{Id_V, \bigcup S_i, R_1, \ldots, R_k\})$ is a Siamese color graph.

In other words, we allow the spread to be a union of basis relations of the scheme. Alternatively, to any association scheme there corresponds a color graph. In this context, the spread in a Siamese association scheme is allowed to be comprised of many colors, provided these colors do not occur anywhere outside the spread.

Consequently, given a Siamese color graph one may ask whether it is coming from a Siamese association scheme. We say in this case that the Siamese color graph *admits* a Siamese association scheme.

Definition 14. *A Siamese color graph is said to be geometric if each srg $(V, R_i \cup S)$ is the point graph of a suitable generalized quadrangle. It is called pseudo-geometric if its parameter set coincides with that of a geometric Siamese color graph (cf. Theorem 2 below).*

4.3 Siamese Steiner Designs

Proposition 2. *Let W be a Siamese color graph with the parameters*

$$\left(\frac{q^4-1}{q-1}, q(q+1), q-1, q+1, q+1\right).$$

Further assume W is geometric. For each point graph $(V, R_i \cup S)$ construct a corresponding generalized quadrangle. Let B denote the union of all lines in all resulting GQ's. Then the incidence structure

$$\mathcal{S} = (V, B)$$

is a Steiner design

$$\mathcal{S} = S\left(2, q+1, \frac{q^4-1}{q-1}\right).$$

Thus a geometric Siamese color graph provides a Steiner system with a spread, and a partition of the remaining blocks into sets which together with the spread form generalized quadrangles. We will call this a *Siamese partition* of the Steiner system. We further call such a partition *coherent* if the color graph admits a Siamese association scheme.

It is easy to see that a Siamese partition of a Siamese Steiner system provides a geometric Siamese color graph W.

4.4 Pattern of Investigation

Given a Siamese color graph W, we can define many derived combinatorial objects. In order to investigate them, we try to describe their automorphism groups. There are quite a few groups of interest to consider:

- the automorphism group $Aut(W)$,
- the color group $CAut(W)$,
- the normalizer group $N(W)$ (defined to be the normalizer in S_V of $Aut(W)$, where V is the vertex set of W),
- the automorphism groups of the distance regular graphs,
- the automorphism groups of the strongly regular graphs.

If Γ is geometric, we have in addition

- the automorphism groups of the GQ,
- the automorphism group of the Steiner system,
- the automorphism group of its Siamese partition.

Finally, if $s = t$ or $s = t + 2$ then each srg defines a symmetric design, which gives us one more group to consider.

In fact, not all of these groups are distinct. The automorphism group of a GQ coincides with that of its point graph; if W admits a Siamese partition, then the automorphism group of the partition coincides with $CAut(W)$; if W admits a Schurian Siamese association scheme, then $CAut(W)$ coincides with $N(W)$.

In what follows, we restrict ourselves to the case $s = t$.

4.5 Siamese Graphs as Simultaneous Antipodal Covers

Let Γ be a graph. Suppose Γ has an equitable partition Π (e.g., see [9] for a definition) such that

- each class (i.e., partition cell) is a coclique,
- each point is adjacent to at most one point from each class.

In other words, the subgraph induced by any two classes is either empty or a matching (i.e., bipartite and regular of valency 1).

We can now define a graph Δ on Π by joining two classes if there are edges between them. Thus, Δ is the quotient graph Γ/Π. In this case we say that Γ is a *cover* of Δ. It is called an n-fold cover if each class in Π has n elements.

In the particular case in which Γ is an antipodal drg and Π is its antipodal system, we can speak of an *antipodal cover* of Δ. The distance regular graphs that occur in Siamese color graphs are such antipodal covers, see [9] for their detailed study.

A seminal result here is the following theorem due to A. E. Brouwer:

Theorem 2. *Let Γ be a pseudo-geometric $GQ(s,t)$-graph with a spread. Then removing the spread from Γ gives a distance regular graph which is an $(s+1)$-fold cover of the complete graph K_{st+1}. Conversely, any drg which is an $(s+1)$-fold cover of the complete graph K_{st+1} may be obtained by removing a spread from a suitable pseudo-geometric $GQ(s,t)$-graph.*

This gives the following interpretation of Siamese color graphs: The distance regular graphs in a Siamese color graph form simultaneous antipodal covers of K_{st+1} which partition the edges of the complete multipartite graph $\overline{S}$, where S is the spread.

5 Review of Main Results

Part of the results appearing in this paper were for the first time briefly reported in [38]. As previously mentioned, an extended abstract of main accomplishments may be found in [39], while a more comprehensive treatment is given in [55]. Currently, we are in the process of preparing papers [40] and [41], which, we hope, will provide a concise, complete, and rigorous introduction to Siamese combinatorial objects, an area which is developmentally in its infancy. Thus, we make no pretense that the current text serve as a complete account of the subject.

Recall two of our earlier stated goals for this paper: to involve the reader into the history and logistics of our computer-aided search for Siamese objects, and to share the excitement we experienced from the many observations and discoveries which resulted from this search. Naturally, these goals remain intact. They will serve to motivate the remainder of our text, beginning with the following short summary which incorporates as a whole, the results of our main activity.

We started by considering as our initial example the Siamese color graph on 15 points given in [35], which we interpreted group theoretically as a transitive action of the group A_5 of degree 15. As a second example, we considered a suitable action of A_6 on 40 points, and found one more Siamese color graph, which (as was the case for the one on 15 points) admitted a Siamese association scheme.

At this stage, we were already able to extrapolate on an evident observation, namely $A_5 \cong PSL(2,4)$ and $A_6 \cong PSL(2,9)$. Thus, we began an investigation of analogous imprimitive actions of the groups $PSL(2,q^2)$ on $\frac{q^4-1}{q-1}$ points. Subsequent additional experimentation performed on the cases $q = 4,5,7$ soon convinced us that we were on the threshold of an infinite series of Siamese objects related to the groups $PSL(2,q^2)$, the generalized quadrangles $W(q)$, and the classical Steiner designs $PG(3,q)$.

Analyzing various sources of information from algebraic combinatorics [7, 47], finite geometry [30], and group theory [16, 20, 66], we next prepared an outline of a proof of the following:

Theorem 3. *For each prime power q, there exists an imprimitive action of $PSL(2,q^2)$ of degree q^3+q^2+q+1 for which the corresponding association scheme of 2-orbits is a geometric Siamese association scheme on $v = \frac{q^4-1}{q-1}$ points.*

It is easy to show that the resulting Siamese Steiner design is isomorphic to $PG(3,q)$. Also, it is well known that $Aut(PG(3,q)) = P\Gamma L(4,q)$. Due to the inclusion $PSL(2,q^2) \leq P\Gamma L(4,q)$, we next observed that the formerly described action of $PSL(2,q^2)$ on v points (the projective points of $PG(3,q^2)$) induces another action of $PSL(2,q^2)$ on the projective lines of $PG(3,q^2)$. Thus we obtained:

Corollary 1. *Consider the transitive action of $PSL(2,q^2)$ on the points of $PG(3,q)$ which results from the inclusion $PSL(2,q^2) \leq P\Gamma L(4,q)$. Next consider the induced action of $PSL(2,q^2)$ on the lines of $PG(3,q)$. Then the orbits of this latter action admit a Siamese partition of $PG(3,q)$.*

This is briefly how we came about an infinite series of Siamese objects (i.e., color graphs, association schemes, Steiner designs, and coherent Siamese partitions). Note that the list of references provided above is not complete; its function is merely to indicate to the reader the origins of some crucial ideas.

We refer to this series as "classical," despite the fact that its presentation has never appeared before [38] and [55]. Our reasoning is that everything needed for the construction of the series, and all subsequent verification of its correctness, is readily available as classical results, especially those borrowed from group theory.

We also think that the material presented here sheds some new light on classical isomorphisms. Implicitly, a few such isomorphisms will be touched upon in subsequent sections of this paper; however, we make no attempt to treat these isomorphisms in a totally rigorous and comprehensive manner.

6 Initial Example on 15 Points

Recall once more the early part of our story: We worked with the text [35], which presents in evident form the adjacency matrix of a color graph on 15 points. Using GAP, we found the automorphism group of this color graph, which corresponds to the unique (up to similarity) imprimitive representation of A_5 of degree 15. From there we constructed the centralizer algebra of this representation, and proved that a certain merging of its classes produced the color graph from [35]. Below we repeat this portion of the job; in particular, we give an explicit description of the centralizer algebra.

6.1 Data from COCO

Let us start with the following generators of A_5:

$$A_5 = \langle (0, 1, 2, 3, 4), (2, 3, 4) \rangle.$$

We consider the 1-factor of the complete graph on the vertex set $\{0, 1, 2, 3, 4\}$ given by $x = \{\{0, 1\}, \{2, 3\}\}$, with isolated vertex 4. Clearly, $Aut(x) = D_4$, where $D_4 = \langle (0, 2, 1, 3), (0, 1) \rangle$ is the dihedral group of order 8. Note that the intersection $D_4 \cap A_5$ coincides with an elementary abelian group $E_4 = \langle (0, 1)(2, 3), (0, 2)(1, 3) \rangle$ of order 4. Consequently, if we set $\Omega = x^{A_5}$ (where x^{A_5} is the A_5-orbit containing x), then the action of A_5 on Ω is similar to its action on the coset space A_5/E_4. That is, (A_5, Ω) is the desired imprimitive action of degree 15.

The numeration of elements of Ω given in Table 2 was internally generated by COCO, and we shall adopt it throughout. (For example, "0" is COCO's designation for the element $\{\{0, 1\}, \{2, 3\}\} \in \Omega$.)

Starting with the induced generators of the permutation group (A_5, Ω), we get a description of the scheme $\mathcal{M} = (\Omega, 2\text{-}orb(A_5, \Omega))$. Namely $\mathcal{M}$ has five classes R_1, R_2, R_3, R_4, R_5 of respective valencies $4, 4, 4, 1, 1$. Of these, R_4, R_5 form a pair of antisymmetric classes.

Another function of COCO provides a list of the intersection numbers of $\mathcal{M}$, i.e., structure constants of the centralizer algebra $V(A_5, \Omega)$. Note that analysis of these structure constants shows that $V(A_5, \Omega)$ is a non-commutative adjacency algebra, in other words, $\mathcal{M}$ is a non-commutative association scheme.

Table 2. Elements of Ω as internally generated by COCO

0	1	2	3	4
{{0,1},{2,3}}	{{1,2},{3,4}}	{{0,1},{3,4}}	{{0,4},{2,3}}	{{1,3},{2,4}}
5	6	7	8	9
{{0,4},{1,2}}	{{0,1},{2,4}}	{{0,2},{3,4}}	{{0,3},{2,4}}	{{1,4},{2,3}}
10	11	12	13	14
{{0,2},{1,3}}	{{0,3},{1,2}}	{{0,4},{1,3}}	{{0,3},{1,4}}	{{0,2},{1,4}}

Finally, we ask COCO to describe all fusion schemes of $\mathcal{M}$ and their respective automorphism groups. In particular, we get that $Aut(\mathcal{M}) = A_5$ (that is, (A_5, Ω) is a 2-closed permutation group). Regarding proper fusion schemes, we obtain that $(\Omega, R_4 \cup R_5)$ defines an imprimitive strongly regular graph of the form $5 \circ K_3$, while each of the graphs $(\Omega, R_i \cup R_4 \cup R_5)$, $1 \leq i \leq 3$, provides an srg$(15, 6, 1, 3)$ having automorphism group of order 720.

Thus, we are able to prove with the aid of COCO that $\mathcal{M}$ is a (Schurian) Siamese association scheme.

For the reader's convenience, we provide an explicit list of the 2-orbits of (A_5, Ω) as follows: $R_1 = (0,1)^{A_5}$, $R_2 = (0,2)^{A_5}$, $R_3 = (0,4)^{A_5}$, $R_4 = (0,10)^{A_5}$, $R_5 = (0,11)^{A_5}$.

Though all of these results may be obtained *a posteriori* without the aid of a computer, we do not feel it is prudent to challenge the reader to fulfill these computations independently. One of our reasons is that this computer-free task may be accomplished instead by some quite beautiful theoretical considerations. Below we restrict ourselves to only a brief outline of such activities; for details, the reader is referred to [40, 41] and [55].

6.2 Theoretical Interpretation

The existence of the Siamese association scheme $\mathcal{M}$ above implies the existence of a Siamese $STS(15)$ which we denote by $\mathfrak{S}$.

It is well known that $Aut(\mathfrak{S}) \cong A_8 \cong PSL(4,2)$. In fact, one is able to establish the exceptional isomorphism $A_8 \cong PSL(4,2)$ directly, by simultaneous consideration of the classical geometric model, which gives $Aut(PG(3,2)) = PSL(4,2)$, and the sporadic combinatorial model (consisting of an A_8-orbit of affine designs $AG(3,2)$ and all partitions of an 8-element set into two parts of equal size), which gives $Aut(PG(3,2)) = A_8$.

We list here some additional striking facts which are required by us (see also 7.1):

- Up to isomorphism, there exists a unique generalized quadrangle of order 2, hence it is self-dual and isomorphic to $W(2)$.
- The classical model of $W(2)$ is provided by all 2-element subsets of $\{1, 2, 3, 4, 5, 6\}$ (as points) and all partitions of $\{1, 2, 3, 4, 5, 6\}$ into three 2-element subsets (as lines), cf. Example 3.
- The complement $\overline{T(6)}$ of the triangular graph $T(6)$ is the unique srg$(15, 6, 1, 3)$.
- $\overline{T(6)}$ is geometric (namely, it is the point graph of $W(2)$), and $Aut(\overline{T(6)}) = S_6$.
- Up to isomorphism, there exists a unique antipodal drg of valency 4 and diameter 3 on 15 points, namely the line graph of the Petersen graph.

Altogether, this classical information allows us to prove that $\mathfrak{S}$ is isomorphic to $PG(3,2)$. Moreover, we can prove that $\mathfrak{S}$ has $\binom{8}{3} = 56$ different Siamese partitions. Let τ be one such partition. Then $Aut(\tau)$ is isomorphic to the stabilizer of a 3-element subset of $\{1, \ldots, 8\}$ in the natural action $(A_8, \{1, \ldots, 8\})$; in other words, $Aut(\tau)$ is isomorphic to $(S_5 \times S_3)^+$, where we indicate by G^+ the subgroup of all even permutations in group G.

Although a more detailed account of this result is beyond the scope of this paper, we nevertheless are able to present in Sect. 7 an alternative interpretation of the group $(S_5 \times S_3)^+$ which is essentially computational in nature, and which fits very naturally within the frames of our current exposition.

6.3 A Few Words About $STS(15)$

A complete listing of all $STS(15)$ was elaborated quite a long time ago in the paper [13], where a catalog of 80 isomorphism classes of such combinatorial designs was given, together with sufficiently detailed arguments to establish its completeness.

Later on, the same problem was attacked by R. A. Fisher [19] who determined only 79 isomorphism classes. At the dawn of the computer era, however, the result of [13] was confirmed in [27].

Nowadays, information about the 80 isomorphism classes of $STS(15)$ is regarded as one of the classical sources in design theory. The paper [48] provides a detailed catalog of these designs together with their numerous properties. We will refer many times to this classical source, in particular we will henceforth adopt their notation $STS(15)\#1$ for the design $\mathfrak{S} = PG(3, 2)$.

An interesting observation was exploited in [36], where the authors considered a certain kind of switching (sometimes referred to a "Pasch switching," e.g., see [24]) which transforms an initial $STS(15)$ into a second, possibly non-isomorphic, $STS(15)$. If we construct a graph whose vertex set consists of the 80 isomorphism classes of $STS(15)$, with two vertices adjacent if and only if a representative of the first class may be switched to a representative of the second class, then this graph will have two connected components: one of size 79 (which includes $STS(15)\#1$ as a vertex of valency one), and one isolated vertex. It is the class associated with this isolated vertex that totally escaped the attention of Fisher in [19]. Following [48], we denote this "isolated" class by $STS(15)\#80$.

At this point we would like to make a quite important observation. Among all 80 $STS(15)$, there are exactly two which have point-transitive automorphism group: $STS(15)\#1$ (with automorphism group A_8), and $STS(15)\#80$ (with automorphism group of order 60). We will discuss other properties of $STS(15)$ as the need arises in the course of our presentation.

7 Automorphism Group of a Siamese Partition for $STS(15)\#1$

Although the reader may be justified in regarding this section as somewhat tangential to our task, we have included it in order to fulfill a broader mission of our presentation: to acquaint the reader with the methodology of experimentation, and to explain what is meant by computer algebra experimentation "in action." We start by delineating the rules of the game.

We work with the Siamese association scheme $\mathcal{M}$ on 15 points and the corresponding Siamese design $\mathfrak{S} = PG(3,2)$. Denote by $\tau = \tau(\mathfrak{S}) = (P, \{B_1, B_2, B_3, S\})$ a Siamese partition of $\mathfrak{S}$. Here P is the point set of $\mathfrak{S}$ (replacing the earlier notation Ω for the point set of $\mathcal{M}$), S is a spread, and $B_i \cup S$ is the set of lines of $GQ(2)\#i$, $i \in \{1,2,3\}$. Let $N = Aut(\tau)$ be the automorphism group of the partition τ.

We know already from two independent sources (via computer, and via theoretical reasonings) that $Aut(\tau) \cong (S_5 \times S_3)^+$. We agree to postpone a complete theoretical description to the sequel [41], preferring here to give a self-contained "naive" explanation of N.

7.1 Summary of Known Results

We alert the reader that throughout this section the permutation group (A_5, Ω) of Sect. 6.1 will henceforth be denoted by (A_5, P), commensurate with our choice of notation above.

7.1.1 Association scheme $\mathcal{M}$ is clearly obtained as the scheme of 2-orbits of (A_5, P). As a by-product of our activity, we will find an alternative way to establish that $Aut(\mathcal{M}) = A_5$.

By construction, A_5 acts transitively on P. In what follows, let $\left\{ {P \atop 3} \right\}$ denote the set of all 3-element subsets of P.

We wish to evaluate the number t_3 of orbits of $(A_5, \left\{ {P \atop 3} \right\})$. For this goal we use a well known orbit-counting lemma (see [37, 18] for its formulation, and some insightful examples of its combinatorial applications). We denote by $Z(H,X)$ the cycle index polynomial of the permutation group (H,X).

Clearly, $Z(A_5, \{0,1,2,3,4\}) = \frac{1}{60}(x_1^5 + 15x_1x_2^2 + 20x_1^2x_3 + 24x_5)$. From this, it is easy to establish that

$$Z(A_5, P) = \frac{1}{60}\left(x_1^{15} + 15x_1^3x_2^6 + 20x_3^5 + 24x_5^3\right).$$

Thus, we obtain $t_3 = \frac{1}{60}(\binom{15}{3} + 15(1+18) + 20 \cdot 5) = 14$. We conclude that $(A_5, \left\{ {P \atop 3} \right\})$ has 14 orbits.

7.1.2 Clearly, we may identify the elements of P with the edges of the Petersen graph. Moreover, from the indicated representative of R_2 (cf. Sect. 6.1), we immediately conclude that the graph $\Gamma_2 = (P, R_2)$ is none other than the line graph of the Petersen graph. (In fact, one can find this same interpretation of Γ_2 occurring already on page 1 of [9].) Finally, simply recall that $Aut(\Gamma_2) \cong S_5$ (e.g., use the famous Whitney-Jung Theorem, see [28]).

We now wish to repeat our former computation of t_3, only this time replacing A_5 by S_5. This initially gives that $Z(S_5, \{0,1,2,3,4\})$ equals

$$\frac{1}{120}\left(x_1^5 + 10x_1^3x_2 + 20x_2x_3 + 30x_1x_4 + 15x_1x_2^2 + 20x_1^2x_3 + 24x_5\right)$$

from which it follows that

$$Z(S_5, P) = \frac{1}{120}\left(x_1^{15} + 10x_1^3x_2^6 + 20x_3x_6^2 + 30x_1x_2x_4^3 + \cdots\right).$$

From here we proceed to compute

$$t_3\left(S_5, \left\{\begin{matrix}P\\3\end{matrix}\right\}\right) = \frac{1}{120}\left(\binom{15}{3} + 10(1+18) + 20\cdot 1 + 30\cdot 1 \right.$$
$$\left. + 15(1+18) + 20\cdot 5\right) = 9,$$

in other words, group $(S_5, \{\begin{smallmatrix}P\\3\end{smallmatrix}\})$ has 9 orbits.

7.1.3 We know that each srg in our Siamese association scheme $\mathcal{M}$ is isomorphic to $\overline{T(6)}$ and has automorphism group S_6. In this case, P is the vertex set of $\overline{T(6)}$. Listing of the orbits of $(S_6, \{\begin{smallmatrix}P\\3\end{smallmatrix}\})$ has already been accomplished, see Example 4.

7.1.4 If we now permit ourselves a glance at the catalog [48], we easily identify the Steiner design of Example 4 as $STS(15)\#1$; indeed, it is the only $STS(15)$ with automorphism group of order properly divisible by 60. However, it turns out that such identification is not crucial for the fulfillment of our remaining steps.

7.1.5 Because $\mathcal{M}$ is a Schurian association scheme, it is easy to show that N coincides with the normalizer in S_{15} of the 2-closure of (A_5, P). Again, appealing to [48] we see that $Aut(\mathfrak{S})$ contains a cycle of length 15, and we are able to show that such a cycle belongs to N. (Important note: In future considerations of our classical infinite series, existence of a similar cycle in each case may be extracted from the famous Singer Theorem, which states that $Aut(PG(3,q))$ contains a cyclic subgroup which acts transitively on points.)

Nevertheless, we can avoid entirely such considerations here by using GAP, which identifies N as a group of order 360 which is isomorphic (as an abstract group) to $(S_5 \times S_3)^+$. We shall use this information as our real starting point.

Our goal is to better understand the group (N, P). In what follows, we shall regard S_5 and S_3 in their natural actions on $\{0,1,2,3,4\}$ and $\{5,6,7\}$, respectively.

7.2 Other Roads to Group N

Before we continue with our exploration of group N, we would like to mention that there are at least two additional ways to approach its identification as an abstract group, which do not rely on computer:

- Identification of $Aut(\mathfrak{S})$ as A_8, see [40, 41].
- Combined use of two portions of group theoretic information, namely $\mathbb{Z}_{15} \leq N$ and $S_5 \leq N$ (the latter coming from our knowledge of $Aut(\Gamma_2)$). Together, these provide the ingredients for a computer-free derivation of the structure of N.

7.3 Subdirect Products

We need to briefly recall the important group theoretic notion of subdirect product. Let H_1, H_2, and M be groups, and let $f_1 : H_1 \to M$, $f_2 : H_2 \to M$ be homomorphisms onto M. Consider the set $H \subset H_1 \times H_2$ defined by $(a, b) \in H$ if $f_1(a) = f_2(b)$ for $a \in H_1$, $b \in H_2$. Then H is a group, called the *subdirect product* of H_1 and H_2 with respect to the homomorphisms f_1, f_2. The order of H is given by

$$|H| = \frac{|H_1| \cdot |H_2|}{|M|}.$$

More information on subdirect products may be found in [60] and [56].

Example 6. We illustrate by simple example the notion of subdirect product. Consider $H_1 = \langle(1,2)\rangle \cong S_2$ and $H_2 = \langle(3,4)\rangle \cong S_2$ in their natural actions on $\{1,2\}$ and $\{3,4\}$, respectively. Thus, the group $G = H_1 \times H_2$ acts intransitively on $\{1,2,3,4\}$. Let us describe all subgroups of order 2 in G. Evidently, we have the initial subgroups H_1 and H_2 (up to identification with their embeddings), however we have one more group $H = \langle(1,2)(3,4)\rangle$. In fact, group H is a subdirect product, with respect to the trivial homomorphisms from H_1 and H_2 onto S_2.

A more interesting example will be discussed below.

7.4 Faithful Actions of N on 15 Points

We know our group N acts faithfully on 15 points. Here we want to further investigate and interpret this action.

Clearly, such an action is similar to $(N, N/K)$, where N/K is the coset space of an appropriate subgroup $K \leq N$. Moreover, we know $|K| = 24$, and further that K is an anti-invariant subgroup of N (meaning that K does not contain any non-trivial normal subgroup of N). An obvious candidate for K is a group $(S_4 \times S_2)^+$ where, for example, the evident factors act on $\{0,1,2,3\}$ and $\{5,6\}$, respectively.

At first glance it would appear that this candidate fulfills all requirements. Indeed, it is not hard to show that the resulting permutation group $(N, N/K)$ is similar to a subgroup of index two in the direct product of permutation groups S_5 and S_3 (see [37] for a discussion of this notion), and more so, that this subgroup is comprised only of even permutations. However, one can now establish that the degree 15 action of a copy of A_5 in the resulting transitive action of N must be intransitive with three orbits of length 5. As this contradicts the initial action (A_5, Ω), we must seek another candidate for the role of K.

Recall that S_4 has a quite exceptional property among symmetric groups S_n, namely it admits S_3 as a homomorphic image. (Perhaps the easiest way to see this is to consider the set X of all 1-factors of the complete graph K_4. Clearly $S_4 = Aut(K_4)$ acts transitively on X, which consists of three members.) Thus, we can construct a subdirect product K of $S_4 \times S_3$, with respect to homomorphisms $f_1 : S_4 \to S_3$ and $f_2 : S_3 \to S_3$, which has order $|K| = \frac{|S_4|\cdot|S_3|}{|S_3|} = 24$. Moreover, it is easy to see that all permutations in $(K, \{0, \ldots, 7\})$ are even, whence $K \leq (S_5 \times S_3)^+$. Upon further examination, we conclude that K is indeed an ideal candidate.

7.5 Explicit Desired Action of N on 15 Points

There is a notable distinction between natural and induced actions, particularly when describing group generators via computer. Because N is a subgroup of the symmetric group on $\{0, \ldots, 7\}$, we may regard the action $(N, \{0, \ldots, 7\})$ as natural and well understood. This is not true, however, of the induced action $(N, N/K)$. Indeed, to understand this action we need to recruit the aid of a suitable combinatorial structure $\mathfrak{X}$, defined on the base set $\{0, \ldots, 7\}$, for which $Aut(\mathfrak{X}) \cap N = K$. This mirrors a general paradigm for describing induced actions and their generators via machine as implemented in COCO.

Thus, we let $N \cong (S_5 \times S_3)^+$ be given explicitly by

$$N = \langle (0, 1, 2), (0, 1, 2, 3, 4), (5, 6, 7), (0, 1)(5, 6) \rangle,$$

and we define the structure

$$\mathfrak{X} = \{\{0, 1, 5\}, \{2, 3, 5\}, \{0, 2, 6\}, \{1, 3, 6\}, \{0, 3, 7\}, \{1, 2, 7\}\}.$$

The reader is now asked to verify that

$$Aut(\mathfrak{X}) \cap N = \langle (0, 1, 2, 3)(5, 7), (0, 1)(2, 3), (0, 1, 2)(5, 7, 6) \rangle = K.$$

This means that if we now construct the orbit $\Omega' = \mathfrak{X}^N$, then $|\Omega'| = 15$, and the permutation group (N, Ω') contains a subgroup A_5 whose action on Ω' is similar to the desired action (A_5, Ω). Thus, we may identify the elements of Ω with those of Ω'.

More precisely, if we take $A_5 = \langle(0,1,2),(0,1,2,3,4)\rangle$ then $A_5 \cap Aut(\mathfrak{X}) = \langle(0,1)(2,3),(0,2)(1,3)\rangle$ is a subgroup of order 4, in fact it is the same group E_4 which appears in Sect. 6.1. Thus our action (A_5, Ω') is transitive and similar to $(A_5, A_5/E_4)$, in other words, it is similar to (A_5, Ω). This implies that after a suitable identification of elements, (N, Ω) is an overgroup of (A_5, Ω).

Let us briefly explain the genesis of structure $\mathfrak{X}$ above. Recall from Sect. 6.1 the set X of 1-factors of the complete graph on $\{0,1,2,3\}$, namely $X = \{\{\{0,1\},\{2,3\}\},\{\{0,2\},\{1,3\}\},\{\{0,3\},\{1,2\}\}\}$. We label the members of X by 5, 6, 7 in some specified order. Subject to this labeling, we now construct $\mathfrak{X}$ as the set of all $\{a,b,c\}$ for which $\{a,b\}$ is an edge in the 1-factor labeled c. In a similar manner, we perform this construction starting with sets of 1-factors arising from other choices of 4-vertex subsets of $\{0,1,2,3,4\}$ (just as X arose from the subset $\{0,1,2,3\}$).

In a sense, group N "coordinates" this job. The objective difficulty in the consideration of set Ω' is the following: There are $15 \cdot 3! = 90$ ways to label elements of Ω (1-factors of K_5 with one isolated point) by the elements of $\{5,6,7\}$. The orbit $\Omega' = \mathfrak{X}^N$ consists of 1/6 of these labellings.

For convenience, we include below an explicit list of the elements of Ω' as generated by COCO. We believe that the reader will agree, pending its close examination, that the task of selecting these 15 elements from the entire list of 90 options is not so easy.

0. $\{\{0,1,5\},\{0,2,6\},\{0,3,7\},\{1,2,7\},\{1,3,6\},\{2,3,5\}\}$
1. $\{\{0,1,6\},\{0,2,7\},\{0,3,5\},\{1,2,5\},\{1,3,7\},\{2,3,6\}\}$
2. $\{\{1,2,5\},\{1,3,6\},\{1,4,7\},\{2,3,7\},\{2,4,6\},\{3,4,5\}\}$
3. $\{\{0,1,7\},\{0,2,5\},\{0,3,6\},\{1,2,6\},\{1,3,5\},\{2,3,7\}\}$
4. $\{\{1,2,6\},\{1,3,7\},\{1,4,5\},\{2,3,5\},\{2,4,7\},\{3,4,6\}\}$
5. $\{\{0,2,5\},\{0,3,7\},\{0,4,6\},\{2,3,6\},\{2,4,7\},\{3,4,5\}\}$
6. $\{\{0,2,7\},\{0,3,6\},\{0,4,5\},\{2,3,5\},\{2,4,6\},\{3,4,7\}\}$
7. $\{\{0,2,6\},\{0,3,5\},\{0,4,7\},\{2,3,7\},\{2,4,5\},\{3,4,6\}\}$
8. $\{\{1,2,7\},\{1,3,5\},\{1,4,6\},\{2,3,6\},\{2,4,5\},\{3,4,7\}\}$
9. $\{\{0,1,5\},\{0,3,6\},\{0,4,7\},\{1,3,7\},\{1,4,6\},\{3,4,5\}\}$
10. $\{\{0,1,6\},\{0,3,7\},\{0,4,5\},\{1,3,5\},\{1,4,7\},\{3,4,6\}\}$
11. $\{\{0,1,7\},\{0,3,5\},\{0,4,6\},\{1,3,6\},\{1,4,5\},\{3,4,7\}\}$
12. $\{\{0,1,7\},\{0,2,6\},\{0,4,5\},\{1,2,5\},\{1,4,6\},\{2,4,7\}\}$
13. $\{\{0,1,5\},\{0,2,7\},\{0,4,6\},\{1,2,6\},\{1,4,7\},\{2,4,5\}\}$
14. $\{\{0,1,6\},\{0,2,5\},\{0,4,7\},\{1,2,7\},\{1,4,5\},\{2,4,6\}\}$

7.6 Analytic Enumeration of Orbits of $(N, \{{\Omega \atop 3}\})$

We now count the orbits of the action of N on 3-element subsets of Ω. Because this task requires more comprehensive efforts, we organize our computations

Table 3. Cycle index data for group (N, Ω)

#	$g_1 \in S_5$	$g_2 \in S_3$	$g \in N$	$\lvert ccl(g)\rvert$	CI_1	CI_2	$\chi(g)$
1	e	e	e	1	x_1^8	x_1^{15}	$\binom{15}{3}$
2	e	$(5,6,7)$	$(5,6,7)$	2	$x_1^5x_3$	x_3^5	5
3	$(0,1)$	$(5,6)$	$(0,1)(5,6)$	30	$x_1^4x_2^2$	$x_1^3x_2^6$	19
4	$(0,1)(2,3)$	e	$(0,1)(2,3)$	15	$x_1^4x_2^2$	$x_1^3x_2^6$	19
5	$(0,1)(2,3)$	$(5,6,7)$	$(0,1)(2,3)(5,6,7)$	30	$x_1x_2^2x_3$	$x_3x_6^2$	1
6	$(0,1,2)(3,4)$	$(5,6)$	$(0,1,2)(3,4)(5,6)$	60	$x_1x_2^2x_3$	$x_3x_6^2$	1
7	$(0,1,2)$	e	$(0,1,2)$	20	$x_1^5x_3$	x_3^5	5
8	$(0,1,2)$	$(5,6,7)$	$(0,1,2)(5,6,7)$	40	$x_1^2x_3^2$	x_3^5	5
9	$(0,1,2,3)$	$(5,6)$	$(0,1,2,3)(5,6)$	90	$x_1^2x_2x_4$	$x_1x_2x_4^3$	1
10	$(0,1,2,3,4)$	e	$(0,1,2,3,4)$	24	$x_1^3x_5$	x_5^3	0
11	$(0,1,2,3,4)$	$(5,6,7)$	$(0,1,2,3,4)(5,6,7)$	48	x_3x_5	x_{15}	0

with the aid of Table 3. (In it, e denotes the identity group element, $|ccl(g)|$ the size of the N-conjugacy class containing g, CI_1 the contribution of g to the cycle index polynomial $Z(N, \{0,\ldots,7\})$, CI_2 the contribution of g to the cycle index polynomial $Z(N, \Omega)$, and χ the permutation character corresponding to the action $(N, \{{\Omega \atop 3}\})$.)

From the data in Table 3, we now get that $t_3(N, \Omega)$ is equal to

$$\frac{1}{360}(5 \cdot 7 \cdot 13 + 2 \cdot 5 + (45) \cdot 19 + (90) \cdot 1 + (60) \cdot 5 + 90 \cdot 1) = 5.$$

Thus, using only the orbit-counting lemma we are able to determine that $(N, \{{\Omega \atop 3}\})$ has precisely 5 orbits.

7.7 Constructive Enumeration of Orbits of $(N, \{{\Omega \atop 3}\})$

We would like now to physically construct the orbits enumerated in Sect. 7.6, by which we mean exhibit at least one representative for each orbit, determine each orbit's length and, ideally speaking, explain its structure. The results of such enumeration are presented below, in terms of the set $\Omega' = \mathfrak{X}^N$ (cf. Sect. 7.5). To eliminate mistakes, this activity was performed with the aid of a computer; however, in principle one can avoid such usage during the explanation stage of enumeration.

From this information we see that the only available option to form 35 blocks of $\mathfrak{S}$ is to take the union of orbits 1 and 5 (see Table 4). Although it is not requested for our goals, perhaps the easiest way to explain these orbits is to perform a constructive enumeration of the 9 orbits of the group S_5 (see Sect. 7.1.2), and then merge these to form the orbits of N. One explanation for this is that the enumeration of the 9 orbits of S_5 is a rather simple task,

Table 4. Orbit lengths and representatives

Orbit#	Length	Representative
1	30	$\{0, 9, 13\}$
2	180	$\{0, 13, 14\}$
3	180	$\{0, 10, 13\}$
4	60	$\{0, 5, 13\}$
5	5	$\{0, 1, 3\}$

due to the fact that these orbits are in bijective correspondence with the isomorphism classes of 3-edge subgraphs of the Petersen graph.

The following facts may also be useful in this endeavor:

- S_5 is normal in N with quotient group $\mathbb{Z}_3$. Thus, there are only two possibilities for a given orbit of N: either it is an orbit of S_5, or it is fused from three S_5-orbits of equal length.
- A cleverly arranged bijection between Ω and Ω' could greatly simplify the job.

We stress that in our eyes this last activity has very high methodological significance, in fact much more so than the practical significance of the resulting explanation (or even interpretation) of the analytically enumerated orbits of $(N, \{{\Omega \atop 3}\})$.

7.8 Summary of Results About N

For the reader's convenience, we have compiled below all objective results from Sects. 6 and 7 about the group (N, Ω) and its related structures. We will not burden the reader with a further discussion of our reasonings, other than to say that a rigorous proof depends strongly on the "rules of the game," as set forth in the preamble to Sect. 7. Finally, we again mention that an alternate proof (more literate for a group theorist), based on the exceptional isomorphism between A_8 and $PSL(4, 2)$, will be presented in [40] and [41].

Proposition 3.

(a) $\mathcal{M}$ *is a Schurian Siamese association scheme.*
(b) $\mathfrak{S}$ *is a coherent Siamese* $STS(15)$.
(c) $Aut(\mathfrak{S}) = A_8$.
(d) $N = N_{S_{15}}(A_5) \cong (S_5 \times S_3)^+$.
(e) N *is the automorphism group of a Siamese partition of* $\mathfrak{S}$.
(f) $Aut(\mathcal{M}) = A_5$.
(g) N *acts transitively on the point set of* $\mathfrak{S}$*, and has two orbits on the block set of* $\mathfrak{S}$ *of respective lengths 30 and 5.*
(h) N *coincides with the stabilizer in* $Aut(\mathfrak{S})$ *of a spread in* $\mathfrak{S}$.

Table 5. Representatives of 2-orbits of A_4

Ψ_1	Ψ_2	Ψ_3	Ψ_4	Ψ_5	Ψ_6
(e, e)	$(e, (1, 3, 2))$	$(e, (1, 2, 3))$	$(e, (0, 3)(1, 2))$	$(e, (0, 2, 3))$	$(e, (0, 1, 3))$
Ψ_7	Ψ_8	Ψ_9	Ψ_{10}	Ψ_{11}	
$(e, (0, 1)(2, 3))$	$(e, (0, 2, 1))$	$(e, (0, 1, 2))$	$(e, (0, 3, 2))$	$(e, (0, 3, 1))$	
Ψ_{12}	Ψ_{13}	Ψ_{14}	Ψ_{15}	Ψ_{16}	
$(e, (0, 2)(1, 3))$	(e, H)	$(e, H(1, 2, 3))$	$(e, H(1, 3, 2))$	(H, e)	
Ψ_{17}	Ψ_{18}	Ψ_{19}	Ψ_{20}	Ψ_{21}	
$(H, (1, 3, 2))$	$(H, (1, 2, 3))$	(H, H)	$(H, H(1, 2, 3))$	$(H, H(1, 3, 2))$	

8 More About 15 Points

To illustrate the fact that the notion of a Siamese scheme is stronger than that of Siamese color graph, we give an example of a Siamese color graph which does not admit a Siamese association scheme.

8.1 Starting Group

We consider the group A_4, a one-point stabilizer in A_5, in its intransitive action on the set Ω as defined in Sect. 6.1. The group (A_4, Ω) has two orbits V_1, V_2 of respective lengths 3 and 12. Evidently, A_4 acts regularly on V_2; hence it is convenient to identify the elements of V_2 with those of A_4. To make our representation consistent, we also describe the elements of V_1 in group theoretic terms, namely as cosets of the unique Sylow 2-subgroup H in A_4.

Very simple reasoning shows that the action of A_4 on $V_1 \cup V_2$ has rank 21. We choose representatives of the 2-orbits of $(A_4, V_1 \cup V_2)$ as shown in Table 5 (where e denotes the identity element of A_4).

Now we define the following binary relations on $V = V_1 \cup V_2$:

$$\begin{aligned} Id_V &= \Psi_1 \cup \Psi_{19} \\ S &= \Psi_8 \cup \Psi_9 \cup \Psi_{20} \cup \Psi_{21} \\ \Phi_1 &= \Psi_5 \cup \Psi_{10} \cup \Psi_{12} \cup \Psi_{14} \cup \Psi_{18} \\ \Phi_2 &= \Psi_6 \cup \Psi_7 \cup \Psi_{11} \cup \Psi_{15} \cup \Psi_{17} \\ \Phi_3 &= \Psi_2 \cup \Psi_3 \cup \Psi_4 \cup \Psi_{13} \cup \Psi_{16} \end{aligned}$$

Finally, we consider the color graph $W = (V, \{Id_V, S, \Phi_1, \Phi_2, \Phi_3\})$.

Proposition 4.

(a) $Aut(W) = A_4$,

(b) W *is a geometric Siamese color graph of Siamese rank 3. However,* W *does not admit a Siamese association scheme.*

8.2 A Non-coherent Siamese Partition of $STS(15)\#7$

Because W is a geometric Siamese color graph, its existence implies that of a Siamese $STS(15)$. We constructed this Siamese Steiner system, and realized that it is isomorphic to $STS(15)\#7$, in the notation of [48].

According to [48] $STS(15)\#7$ has group of order 288 with orbits of length 3 and 12 as its points. Using GAP, we identified this group as $(S_4 \times S_4)^+$ in its natural action on 8 points. This allowed us to obtain an interesting model of $STS(15)\#7$ in which there are two types of points and three types of blocks. To better describe this model we start from the action of $S_4 \times S_4$ on $O_1 \cup O_2$, where the two evident copies of S_4 act naturally and independently on $O_1 = \{1, 2, 3, 4\}$ and $O_2 = \{5, 6, 7, 8\}$.

8.2.1 Description of the Model of $STS(15)\#7$

Points of the first type will be partitions of O_2 of the form $2+2$. There are three such points, and they form a single orbit under the action of $(S_4 \times S_4)^+$.

To describe points of the second type, we consider all partitions of $O_1 \cup O_2$ of the form $2+2+2+2$, where each pair in the partition contains exactly one letter from each of O_1 and O_2. There are $4! = 24$ such partitions; however we require only half of these. As these 24 partitions fall into two orbits under the action of $(S_4 \times S_4)^+$, we define points of the second type to be the partitions in one of these orbits, say the one which contains the partition $\{\{1,5\},\{2,6\},\{3,7\},\{4,8\}\}$. Thus in total we have $3 + 12 = 15$ points.

Blocks of the first type will consist of partitions of $O_1 \cup O_2$ of the form $4+4$ such that each 4-tuple of the partition contains two elements from each of O_1 and O_2. The number of such blocks is $\frac{1}{2}\binom{4}{2}^2 = 18$.

Blocks of the second type consist of pairs of elements, one element from O_1 the other from O_2. There are $4^2 = 16$ such blocks.

Finally, we assign one additional block b_∞. Altogether, we get $18+16+1 = 35$ blocks.

Incidence is defined as follows: A block of the first type is incident to exactly one point of the first type (the partition it induces on O_2) and to exactly two points of the second type (the partitions of type $2+2+2+2$ which are refinements of the partition of type $4+4$). A block of the second type is incident to exactly three points of the second type (the partitions which contain it), while block b_∞ is incident to all three points of first type.

Proposition 5. *The incidence structure defined above is an* $STS(15)$*. Moreover, it is isomorphic to* $STS(15)\#7$*.*

The second statement of the proposition may be proved by computer, however it also follows easily from theoretical considerations. Indeed, it is evident by construction that our model is invariant with respect to $(S_4 \times S_4)^+$, and consulting [48] one sees that $STS(15)\#1$ and $STS(15)\#7$ are the only $STS(15)$ which admit a group of automorphisms of that order. By using certain invariants provided in [48], the task is easily completed.

Now we wish to describe a Siamese partition of $STS(15)\#7$. For this purpose, let $\{x, y, z\}$ be an arbitrary 3-element subset of O_1. Starting from the pair $\{x, y\}$, let B_1 consist of all blocks of the first type in which these two elements appear in the same partition cell. Let B_2 consist of all blocks of the second type which contain either x or y. Finally, let Ω denote the set of all points of the Steiner system and set $B = B_1 \cup B_2 \cup \{b_\infty\}$. Of course, in a similar manner we could define a corresponding incidence structure starting from $\{x, z\}$ or $\{y, z\}$. Note also that there are $\binom{4}{3} = 4$ choices for the initial selection of a 3-element subset from O_1.

Proposition 6.

(a) Each of the incidence structures defined with respect to $\{x, y\}$, $\{x, z\}$ and $\{y, z\}$ is a $GQ(2)$.
(b) Any two of these three generalized quadrangles intersect in the same spread.
(c) The generalized quadrangles defined by $\{x, y, z\}$ form a Siamese partition of $STS(15)\#7$.
(d) This Siamese partition is non-coherent (i.e., it is implied solely by the existence of the Siamese color graph W).
(e) The automorphism group of the Siamese partition has order 72.

8.3 All Siamese Color Graphs on 15 Points are Obtained

Above we gave the construction of two geometric Siamese color graphs on 15 points. To prove that there are no others, the following steps have been taken:

- A geometric color Siamese graph on 15 points provides a Siamese partition of an $STS(15)$.
- There are 80 non-isomorphic $STS(15)$.
- For each $STS(15)$, a computer search was performed to enumerate all embedded $GQ(2)$.
- It was checked if three of these generalized quadrangles form a Siamese partition.

Thus, we get the following result:

Proposition 7 (Computer search). *The only $STS(15)$ admitting a Siamese partition are the designs $STS(15)\#1$ and $STS(15)\#7$. In both cases, the partition is unique up to isomorphism.*

Corollary 2. *There are exactly two non-isomorphic geometric Siamese color graphs on 15 vertices.*

It is well known that the triangular graph $T(n)$ is uniquely determined by its parameters, except for the case $n = 8$. In particular, $\overline{T(6)}$ is the only srg$(15, 6, 1, 3)$, and it is the point graph of the unique $GQ(2)$. Thus we get the following:

Corollary 3. *Every Siamese color graph on 15 vertices is necessarily geometric. Hence, there are exactly two non-isomorphic Siamese color graphs on 15 vertices, one of which admits a Siamese association scheme.*

Remark 2. Using a computer, one could pursue an alternative line of proof which involves constructing all color graphs on 15 points which are "pretending" to be Siamese, then checking these one-by-one for confirmation. We believe, however, that the approach we have outlined above is preferable for the following reasons:

- It is more consistent with the goals of our presentation.
- It is well controlled because the catalog [48] is a very reliable source of information.
- As a by-product, we get a nice opportunity to compare "Siamese properties" with other features of $STS(15)$ (see also the discussion in Sect. 10).
- In principle, ad hoc techniques may be used to eliminate most $STS(15)$ from consideration.

As a sequel to the above remark, we believe that finding a computer-free proof of Corollary 3 constitutes a worthwhile and challenging open problem. In this initiative, one could perhaps exploit methods from topological graph theory (in the sense of [26]) to elaborate a proof. In any case, we feel that some rather innovative ideas may be required to remove this computer dependence.

9 Objects on 40 Points

9.1 Classical Objects

According to Sect. 5, we are aware of the existence of an infinite series of Siamese objects which we choose to call "classical." Our interest here is to scrutinize the one on 40 points. As usual, we refer to [40, 41] and [55] for extra details.

Recall the following well known facts about generalized quadrangles of order 3 (e.g., see [54, 52, 9, 65, 18]):

- Up to isomorphism, there exist only two GQ of order 3: $W(3)$, and its dual $Q(4,3)$.
- $W(3)$ has a unique spread (up to isomorphism), however $Q(4,3)$ has no spreads.
- $Aut(W(3)) \cong P\Gamma U(4,2) \cong Sp(4,3).\mathbb{Z}_2$.
- A spread in $W(3)$ is invariant with respect to S_6, which is a maximal subgroup of $PSU(4,2)$.

With the aid of COCO, we constructed a Siamese association scheme on 40 points from the action of $A_6 \cong PSL(2,9)$ on the cosets of a fixed Sylow 3-subgroup. This action is not 2-closed; COCO returned its 2-closure as a group of order 720 which we subsequently identified as $A_6 \times S_2$. According to the information above, we knew in advance (at both the theoretical and computational levels) that the automorphism group of the unique antipodal geometric drg on 40 points has order 1440. Using a computer, we identified this latter group as $S_6 \times S_2$ and gave an *a posteriori* interpretation of it which we believe to be quite beautiful.

Let $\Omega_0 = \{1,2,3,4,5,6\}$. Let P be the set of directed cycles of length 3 in Ω_0. Let B_1 be the set of all partitions of Ω_0 into two triples. Let B_2 be the set of directed arcs from Ω_0. Let $B = B_1 \cup B_2$.

Define incidence as follows: Let a cycle be incident to a partition if its vertex set is one of the partition cells; let it be incident to an arc if it contains this arc. Then we have the following:

Proposition 8.

(a) The incidence structure (P,B) is a $GQ(3)$.
(b) The 10 blocks in B_1 provide a spread in this $GQ(3)$.
(c) The pair $(GQ(3),B_1)$ is invariant with respect to $S_6 \times S_2$.

Remark 3. One can identify a natural involutory automorphism of (P,B) which interchanges two oppositely directed 3-cycles, yet it is not induced from any permutation in S_6. In fact, one can define the direct factor S_2 as being generated by exactly this involution.

Corollary 4. *The automorphism group of the antipodal geometric drg on 40 points is isomorphic to $S_6 \times S_2$.*

Remark 4. We call attention to a beautiful model of $W(3)$ constructed by S. E. Payne in [53]. Though his model and ours are done in much the same spirit, we believe that ours has a certain advantage, in that it conveys in a very transparent manner the full symmetry of the considered object. We hope that such constructions, and their subsequent analyses, will shed light on higher order Siamese objects, and perhaps lead to a characterization of the classical ones.

9.2 Circulant Example

In this section and the next, we introduce two other interesting examples of Siamese color graphs on 40 points, neither of which admits a Siamese association scheme. The one given here may be described in a very transparent manner, since it admits a point-transitive regular cyclic group of automorphisms. Thus, we identify the points with the set $\mathbb{Z}_{40}$ of integers modulo 40. We take the following blocks as basic:

$$\begin{aligned} b_1 &= \{0,10,20,30\} \\ b_2 &= \{0,1,6,32\} \\ b_3 &= \{0,4,11,23\} \\ b_4 &= \{0,3,16,18\} \end{aligned}$$

and we form the block system $B = \bigcup B_i$, where B_i is the orbit of $\mathbb{Z}_{40}$ which contains b_i, $i = 1, 2, 3, 4$. Then we get:

Proposition 9.

(a) The incidence structure $(\mathbb{Z}_{40}, B)$ is a Steiner system $S(2, 4, 40)$ with a spread.
(b) $(\mathbb{Z}_{40}, B)$ is invariant with respect to a subgroup K of order 160 in the full affine group over $\mathbb{Z}_{40}$, namely $K = \{x \mapsto ax + b \mid a \in \{1, 3, 9, 27\}, b \in \mathbb{Z}_{40}\}$.
(c) Group K is the full automorphism group of $(\mathbb{Z}_{40}, B)$.

Note that the reader can easily check parts (a) and (b) of Proposition 9 by hand. We used a computer to verify part (c), though a computer-free proof can also be elaborated.

Next, we describe just one of the generalized quadrangles of this system. Consider the subgroup $H = \{x \mapsto ax + b \mid a \in \{1, 9\}, 4|b\}$. Take the spread B_1 together with the orbits $(b_2 - 1)^H$, $(b_3 + 2)^H$, and b_4^H under the action of H. Altogether we get 40 blocks which form a generalized quadrangle. The other three GQ are constructed by shifting the given one with the aid of the elements of $\mathbb{Z}_{40}$. Thus we get:

Proposition 10.

(a) $(\mathbb{Z}_{40}, B)$ is a (non-coherent) Siamese Steiner design.
(b) The stabilizer of a Siamese partition coincides with the group K of Proposition 9.

9.3 One More Point-Transitive Example

Consider the group $H = A_5 \times Z_4$. We assume that A_5 acts naturally on $\Omega_1 = \{1, 2, 3, 4, 5\}$, and that Z_4 is generated by the cycle $(6, 7, 8, 9)$ acting on $\Omega_2 = \{6, 7, 8, 9\}$. Thus the base set is $\Omega_0 = \Omega_1 \cup \Omega_2 = \{1, 2, \ldots, 9\}$.

We now describe points and blocks of the design. In what follows, we shall denote by α' the antipode of $\alpha \in \Omega_2$ with respect to the cycle $(6, 7, 8, 9)$; thus $6' = 8$ and $7' = 9$. Otherwise, a, b, c, d, e will always denote pairwise distinct elements of Ω_1, while α, β will denote distinct non-antipodal elements of Ω_2. Finally, we abbreviate by $a\alpha$ the pair $\{a, \alpha\}$.

Points of the design will be pairs $\{a\alpha, b\alpha'\}$. As there are $\binom{5}{2} = 10$ choices for $a, b \in \Omega_1$ and 4 choices for $\alpha \in \Omega_2$, there are altogether 40 points, which in fact comprise a single orbit of H. A typical point is $\{16, 28\}$.

We introduce four types of blocks in our design by way of indicating orbit representatives (see Table 6). The arguments we used to deduce their lengths will be omitted due to space limitations.

In all, we have 40 points and 130 blocks in our design. In fact, by way of computer we get the following result:

Table 6. Block types in terms of orbit representatives

Type	Length	Representative
I	60	$\{\{16,28\},\{27,59\},\{56,48\},\{47,19\}\}$
II	40	$\{\{16,28\},\{18,46\},\{26,48\},\{37,59\}\}$
III	10	$\{\{16,28\},\{18,26\},\{17,29\},\{19,27\}\}$
IV	20	$\{\{16,28\},\{28,36\},\{28,56\},\{28,46\}\}$

Proposition 11. *The structure defined above is a $S(2,4,40)$ containing exactly 4 generalized quadrangles which form a Siamese partition. Its automorphism group is generated by $A_5 \times Z_4$ and the additional automorphism $(1,2)(6,8)$; hence it is a group of index two in $S_5 \times D_4$.*

9.4 Other Siamese Objects

The reader is now acquainted with three Siamese designs on 40 points. One is the classical $PG(3,3)$, whose automorphism group A_8 is both point- and block-transitive; the other two designs have groups that act transitively on points, but intransitively on blocks.

In fact, 475 such designs were found by us with the aid of a computer, many of which have small automorphism group. Here we are interested in nine such designs which have relatively large group.

Table 7 contains information about the groups related to these nine Steiner systems. In it we denote by G the automorphism group of the Steiner system, and by H the automorphism group of the partition (hence, the color group of the corresponding graph). For each group, we also include the lengths of orbits in its actions on points and blocks.

Recall that there are exactly two geometric srg$(40,12,2,4)$, one of which is the point graph of $W(3)$. Only this latter srg has a spread, and it is unique up to isomorphism. Therefore, in a geometric Siamese color graph on 40 points there is only one option for the drg and the srg. Note that all Siamese color graphs described here are geometric.

Table 7. Some Siamese Steiner designs on 40 points

	$S(2,4,40)$			Partition		
#	$\|G\|$	P	B	$\|H\|$	P	B
1	24261120	40	130	11520	40	$10,120$
2	160	40	$10,40,80$	160	40	$10,40,80$
3	480	40	$10,20,40,60$	480	40	$10,20,40,60$
4	1296	$4,36$	$1,24^2,81$	144	$4,36$	$1,9,24^2,36^2$
5	648	$4,36$	$1,12^4,81$	72	$4,36$	$1,9,12^4,36^2$
6	288	$4,36$	$1,9,24^2,36^2$	288	$4,36$	$1,9,24^2,36^2$
7	10368	$4,36$	$1,48,81$	576	$4,36$	$1,9,48,72$
8	1296	$4,36$	$1,24^2,81$	144	$4,36$	$1,9,24^2,36^2$
9	144	$4,36$	$1,9,24^2,36^2$	144	$4,36$	$1,9,24^2,36^2$

At present, the designs in Table 7 (apart from the first three) have only strict computer dependent descriptions, hence we see no sense in disturbing the reader any further with their details. We do however mention a few reasons why we decided to include Table 7 in our presentation:

- It can be used in the future for the verification of further results on Siamese objects on 40 points, particularly serving as a baseline for the measurement of progress in the area.
- Some numerical data appearing in it may be a source of important observations, even conjectures, which may lead later to theoretical interpretations and explanations.
- A few objects in the table will be discussed below and again in Sect. 10.

Note that the interested reader will find a catalog of Siamese partitions for these nine designs in Appendix A.1 of [55].

9.5 Discussion on Methodology

At this point we are well positioned to discuss the methodology used by us for the computer-aided enumeration of all Siamese color graphs of a prescribed order. We mention that while we achieved a complete list of all desired objects on 15 points, we were only partially successful in the case of 40 points.

9.5.1 Strategy A: Combinatorial Analogue of Transitive Extension

This was the first heuristic approach used by us to construct Siamese objects on 15 and 40 points. Recall that all classical objects described to this point were obtained in a unified manner:

- Start from the action of $PSL(2,q^2)$ acting on the q^2+1 points of the projective line.
- Construct the corresponding induced transitive action of degree $\frac{q^4-1}{q-1}$.
- Describe all Siamese color graphs which are invariant with respect to this action.

Let us now consider a generalization of this procedure. Namely, at the first step we replace $PSL(2,q^2)$ by its stabilizer of a projective point (point at infinity). This stabilizer is a certain affine subgroup of $PSL(2,q^2)$. With this subgroup we now fulfill the second and third steps as described above. Clearly, this generalization will produce all results attainable from the original three steps, however it will hopefully generate other graphs as well, possibly with intransitive automorphism groups.

In fact this procedure gave us all Siamese objects on 15 points (here "all" turned out to be just two!), while on 40 points we were able to get a number of such objects (each with automorphism group of order a multiple of 36).

9.5.2 Strategy B: Construction and Investigation of Steiner Designs

The idea behind this second approach is quite simple: Construct as many Steiner designs as possible, and for each isomorphism class of such designs describe all Siamese partitions.

As previously mentioned, it was exactly this strategy that enabled us to complete the classification of Siamese color graphs on 15 points. In fact, for all but two classes we were able to prove non-existence of Siamese partitions, while for designs coming from each of the two remaining classes (known to be Siamese by Strategy A) we confirmed that a Siamese partition was unique, up to isomorphism.

For a given Steiner design, a search for Siamese partitions is achieved in four main steps:

- Enumerate all embedded GQ in the Steiner design.
- Find all spreads occurring as intersections of the line sets of pairs of such GQ.
- For a given such spread, find all GQ which contain it.
- Among the GQ which contain this spread, determine if the line sets of certain of them (together with spread) form a Siamese partition.

The problem of enumerating all Steiner designs on 40 points seems hopeless; thus only special classes of designs were constructed and investigated. The technique used for their construction was the Kramer-Mesner method (see Sect. 2.4). We enumerated all cyclic Steiner systems, as well as those invariant with respect to an intransitive group of order 36, with point orbits of respective lengths 4 and 36.

Altogether, using Strategies A and B we were able to construct nine Siamese partitions on 40 points, specifically the ones which are presented in Table 7.

9.5.3 Strategy C: Direct Enumeration of Siamese Color Graphs

This third strategy grew out of our attempts to enumerate all Siamese color graphs on 40 points, when we recognized the need to arrange our search objectives in a more sophisticated way. Currently, the algorithm is formulated strictly for *geometric* Siamese color graphs, however its extension to the general case should not be extremely difficult to achieve. We provide the following outline.

Let Δ be the point graph of $W(3)$, S a fixed spread in Δ, and $\Gamma = \Delta \setminus S$ the drg obtained by removing the spread S from graph Δ. We set $G = Aut(S)$ and $H = Aut(\Gamma)$. In our case, $G \cong S_{10} \wr S_4$ is a huge group of order $(4!)^{10} \cdot 10!$, while $H \cong S_6 \times S_2$ is a relatively small subgroup of G.

Now each geometric srg Δ' intersecting Δ in S may be described via an embedding of an isomorphic copy Γ' of Γ inside the complement graph $\overline{\Delta}$ of Δ. Therefore, if we are interested in enumerating all non-isomorphic Siamese pairs (Δ, Δ') (recall that these may be interpreted as triples (Γ, S, Γ')), then we need to enumerate all inequivalent embeddings of Γ' into $\overline{\Delta}$ which, in a sense, preserve S.

Such enumeration of Siamese pairs forms the first stage of our algorithm. Note that this task may be formulated entirely in group theoretic terms: Enumerate all double cosets in G of the form HgH. Indeed, in accordance with our general methodology for implementing COCO, together with the interpretation suggested in Sect. 2.5, we proceed as follows:

- Consider the action of G on Γ, where Γ is regarded as a combinatorial structure.
- Formulate an action of G on an appropriate coset space G/H which is similar to the action of G on the orbit Γ^G containing Γ.
- Enumerate all 2-orbits $(\Gamma, \Gamma')^G$ in terms of the action $(G, G/H)$ (equivalently, enumerate all double cosets HgH in G).

At the next stage of the algorithm, we need to extend each obtained pair (Γ, Γ') to a complete Siamese color graph. Let $g \in G$ be a permutation for which $\Gamma^g = \Gamma'$ is embedded into $\overline{\Delta}$. Then $B = \Gamma \cup \Gamma'$ is a regular graph of valency 18. We now remove Δ from the complement graph $\overline{B}$ of B, giving another regular graph B' of valency 18.

Now comes a crucial observation in the concrete case we are considering: B' admits a partition into two drg with common spread S if and only if B' is isomorphic to one of the graphs B (which are presumed to be known after successful fulfillment of the first stage of the algorithm). Thus each isomorphism between B' and B yields a Siamese color graph, and every such graph arises in exactly this manner.

Unfortunately, in principle COCO is not able to enumerate all 2-orbits of the action $(G, G/H)$. Equally fortunate, however, is the fact that we do not fully require this; indeed, we are only interested in those 2-orbits $(\Gamma, \Gamma')^G$ for which the graphs Γ and Γ' are disjoint. In any case, both stages of the algorithm were implemented with GAP. The specific feature of this implementation is that it allowed simultaneous manipulation of both group theoretic and combinatorial information.

We will discuss efficiency of this algorithm in the next section. At present, we wish only to mention that in order to describe all Siamese color graphs on 40 points we had to initially take into account all antipodal drg having identical spread (not just the geometric ones). In group theoretic terms, this corresponds to replacing the initial transitive action $(G, G/H)$ by a more sophisticated intransitive one. From our experience, implementation and efficient execution of a suitably modified algorithm would seem to be a quite difficult and time-consuming task.

Remark 5. In actuality, our implementation of Strategy C was fulfilled at the level of Siamese partitions rather than color graphs. Namely, in place of Δ and other corresponding objects, we considered instead the (unique) $GQ(3)$ having a spread. Taking into account that $Aut(GQ(3)) = Aut(\Delta)$, there are no changes at all in the group theoretical manipulations involved in the described process, only that the combinatorial criteria used in the selection of suitable double cosets are formulated in slightly different terms. In fact, Strategy C as we described it above has definite advantages over the strategy implemented by us. For one thing, it allows a more plausible generalization to all Siamese color graphs (as briefly discussed in the previous paragraph).

9.5.4 Review of Results on 40 Points

As we previously mentioned, a combination of Strategies A and B enabled us to construct nine Siamese color graphs, thus implying the nine Siamese Steiner systems which appear in Table 7 of Sect. 9.4. Further attempts to extend our scope within the confines of these strategies did not lead to any new Siamese objects.

This explains our motivation for creating Strategy C, which in principle turns out to be much more productive than the other ones. Although this algorithm is still in development, much recent progress has been made. For example, running two successive versions of the algorithm, each over a two-day period, gave roughly 130 and exactly 475 Siamese designs, respectively. We find this increase very encouraging.

Interestingly, for each Siamese Steiner design we investigated it turns out that there is a unique Siamese partition, up to isomorphism.

Further note that in [55], in addition to the geometric drg on 40 points with automorphism group $S_6 \times S_2$ (see Corollary 1), two more drg's on 40 points were discovered which, in fact, give the smallest possible answer to a question posed by Godsil–Hensel (see [22, 10] for more details). Finally, it was proved in [55] that these two non-geometric drg are the only ones possible. The proof is computer-dependent and is based on inspection of Ted Spence's catalog of strongly regular graphs (see [63], and also the recent paper [15]).

Currently we do not have any example of a Siamese color graph involving either of these two drg; in other words, we do not know an example of a non-geometric Siamese association scheme.

9.5.5 Further Perspectives

In principle, it should be possible to enumerate all Siamese color graphs on 40 points, or at least all geometrical ones.

Since we found Siamese color graphs related to $W(2)$ and $W(3)$, it is natural to consider $W(4)$ on 85 points. Several heuristic strategies were used to search for Siamese color graphs related to $W(4)$ but only one example was found; it is an association scheme which belongs to our infinite series of classical Siamese objects.

While it seems a worthwhile goal to arrange a systematic search for all geometric Siamese color graphs on 85 points, no efficient algorithm for this task has as yet been created. A reasonable guess is that there exist Siamese color graphs on 85 points which do not admit Siamese association schemes.

However, at present no evidence has surfaced to either support or refute this contention.

10 Additional Remarks

There are a number of topics which we regard as essentially important yet only implicitly related to the main line of our presentation. Not wishing to distract the reader's attention during the initial stages of our message, we instead decided to postpone their discussion to here.

10.1 Theory and Algorithms

10.1.1 Kramer–Mesner Method

Though the Kramer–Mesner method was originally presented in [43], it definitely has numerous predecessors. In particular, the methodology of so-called "tactical decompositions" can be traced to the end of the 19th century, especially to the classical paper [51]. Although such historical roots as these certainly deserve a detailed treatment, this task lies well beyond the scope of the present paper.

10.1.2 Computer Package Discreta

In the implementation of the Kramer–Mesner method there are three essential steps:

(a) Generate all orbits on t-sets and k-sets.
(b) Derive the corresponding Kramer–Mesner matrix.
(c) Find all solutions to the related system of Diophantine equations.

Since each of these steps requires extensive computations, the Kramer–Mesner method is greatly facilitated by machine. The computer package Discreta was conceived in Bayreuth (Germany) for this main purpose, and to this day it remains, in our judgment, the most powerful software tool for implementation of the Kramer–Mesner method. The project of the development of Discreta was initiated by R. Laue, and a first implementation elaborated by B. Schmalz [57]. At the next stage, crucial input was provided by A. Betten and A. Wasserman [6]. The reader is referred to [3, 5] for some beautiful examples of applications of Discreta.

We mention that the crucial role of Discreta in the fulfillment of step (c) assumes the form of the LLL-algorithm, see [67] for details.

10.1.3 Double Coset Enumeration

Another reason for the great success of Discreta is the existence of a special algorithm, called *Leiterspiel,* which operates at the level of subgroup ladders and double coset graphs. This algorithm goes back to [45], see [58] for an outline of its implementation.

10.1.4 Automorphism Group of a Design

Imagine a design $\mathfrak{S} = (P, B)$, obtained with the aid of the Kramer–Mesner method via a suitable merging of orbits of a permutation group (H, P). Clearly, $H \leq G$, where $G = Aut(\mathfrak{S})$.

The task of describing (G, P) is greatly facilitated by an *a priori* knowledge of the lattice of overgroups of (H, P) within the symmetric group defined over set P. This idea was touched upon a few times in the present paper, though in a naive and rudimentary way. In fact, in rigorous form it constitutes a quite deep and interesting topic which lies on the edge between group theory and combinatorics. We refer to [4] for an elementary discussion, and to [34] for a more advanced treatment.

10.1.5 Transitive Extension

A familiar concept from group theory, a transitive extension of a permutation group (H, Ω) is a transitive permutation group of the form $(G, \Omega \cup \{x\})$, where $x \notin \Omega$ and $G_x = H$. We also refer to *transitive extension* when describing the general procedure of determining all groups $(G, \Omega \cup \{x\})$ given (H, Ω) as above. See Sect. 3.7.6 of [18] for a brief discussion of this procedure.

The notion of transitive extension has a very nice combinatorial generalization. If, in particular, we make the restriction that the groups $(G, \Omega \cup \{x\})$ be 2-closed, then we are effectively investigating all regular color graphs with vertex set $\Omega \cup \{x\}$ which are invariant with respect to (H, Ω). Our Strategy A of Sect. 9.5.1.1 is nothing else but a special case of this "combinatorial" extension. See [33] for what is perhaps the earliest attempt to implement computers along this line of investigation.

10.1.6 Factorization of Graphs

Factorization of the complete graph K_v into a number of specified (regular) v-vertex graphs is a quite general form of our main problem: Search for Siamese color graphs. Here again we briefly discuss predecessors, particularly the paper [23]. Table 2 on page 83 of this paper depicts results of the enumeration of all amorphic association schemes on 16 points, and a portion of those on 25 points. Computer techniques used for this enumeration are quite close in spirit to our Strategy C of Sect. 9.5.3.3, however without evident use of double cosets.

We confess that in the case of this general class of problems there has been no essential progress made in the last two decades. At present, it would seem that such progress depends heavily on the emergence of some new and clever ideas in computer algebra.

10.2 Implementation and Logistics

10.2.1 Interplay Between Computer Packages

As we mentioned, our main tools in this project were COCO and GAP (together with its share packages). Oddly, we did not use Discreta but the reasons were purely logistical: We faced essential difficulties in attempting to install it on our existing machines. Ultimately, the author SR wrote his own version of the Kramer–Mesner method in GAP in order to search for Siamese Steiner designs on 40 points. Thus, in this extended sense, Discreta also had an indirect presence in our project.

One of the logistical difficulties we faced was the incompatibility of data formats used by GAP and COCO. Specifically, the output of one package had to be constantly adjusted to the format style of the other package prior to data input. Unfortunately, this problem seems to be of growing severity, particularly as computer algebra packages continue to become more diverse and prolific over time.

10.2.2 Future Incarnations of COCO

The package COCO was created more than a decade ago, however its philosophical principles go back to an even earlier time. Nowadays, there is strong motivation to replace it with a more modern version which is wider in scope and more powerful in utility. Preferably, such a version should also be available as a share package of GAP.

In fact, some efforts in this direction have already been undertaken by the authors MK, SR and a group of their colleagues, mostly at the level of conceptualization and design. Moreover, certain experimental programs have actually been written in GAP which perform functions of this future "COCO II." Hopefully, support and promotion of this project will one day result in the creation of a package which will have applicability to a wide audience of researchers in the field of algebraic combinatorics.

10.3 Siamese Reflections

10.3.1 Weighing Matrices

Here we discuss the methods used in [35] to construct the first infinite series of color graphs of order $q^3 + q^2 + q + 1$ and Siamese rank $q + 1$, q any prime. These methods are based on so-called balanced generalized weighing matrices, in the spirit of an approach developed by Y. Ionin [32]. (See also [21] for further information on weighing matrices.)

In [35], the authors start with a weighing matrix

$$W = \begin{pmatrix} 0 & I & I & I & I \\ I & 0 & I & U & U^2 \\ I & I & 0 & U^2 & U \\ I & U & U^2 & 0 & I \\ I & U^2 & U & I & 0 \end{pmatrix},$$

and interpret each entry of W as a certain 3×3 matrix. From this, they obtain the adjacency matrix of a color graph on 15 vertices. In some sense, W can be viewed as the 'DNA' of this color graph, incorporating all of its properties into a convenient, compact form. We checked by computer that, up to isomorphism, this is the same color graph as the one which admits the first member of our infinite series of classical Siamese association schemes. In fact, the members of the infinite series constructed by Kharaghani and Torabi have rank $q + 1$ (q a prime power), and parameters

$$v = q^3 + q^2 + q + 1, \quad k = q^2 + q, \quad \lambda = q - 1, \quad \mu = q + 1, \quad \sigma = q + 1.$$

Using a computer, we verified for $q = 2, 3, 4, 5, 7$ that members of this series are isomorphic to our respective classical Siamese color graphs on 15, 40, 85, 156 and 400 points. Indeed, it seems that our series and the one of Kharaghani and Torabi are in fact the same, up to isomorphism.

We strongly believe that further computer algebra experimentation with weighing matrices, followed by a subsequent theoretical comprehension, is a task of essential importance. Hopefully, such activity will shed new light on the origins of Siamese objects.

10.3.2 Vertex Transitivity

Recall that there exist two non-classical vertex-transitive Siamese color graphs on 40 vertices, one of which is circulant. Note also that every classical Siamese color graph is circulant. This leads to the following intriguing question: Are there any other circulant (or at least vertex-transitive) Siamese color graphs besides the classical ones and those on 40 vertices?

10.3.3 Group $N \cong (S_5 \times S_3)^+$

Group N, which was discussed by us in Sect. 7, deserves special attention for a variety of reasons. We mention just a couple.

Enumeration of transitive permutation groups has a very long history, see [61] and [14]. In particular, all transitive permutation groups of degree at most 14 were classified during the 19th century. The problem of classifying all imprimitive permutation groups of degree 15 was only considered in [46], and a bit later in [44]. According to [46] there are 70 such groups, while in [44] 98 groups were found.

In [11] G. Butler, who it seems may have been only partially aware of the results of his predecessors, described all transitive permutation groups

Table 8. Siamese property versus existence of resolutions and maximal arcs

Design	Group order	Spread orbits	Resolution orbits	Max arcs	Siamese
#1	12130560	2	Many	–	Yes
#6	288	45	7217	–	Yes
#7	10368	10	At least 91838	–	Yes
#8	1296	16	1232	–	Yes
#9	144	21	36	–	Yes
[25]	39	13	One resolution	3	No
[50]	13	One spread	–	3	No

of degree 15 with the aid of a computer. Among these, he found exactly 98 imprimitive ones, a result that was later confirmed in [14]. To our knowledge, no one has ever investigated the list of H. W. Kuhn [44] to make sure that the groups appearing there are pairwise dissimilar, and thus in agreement with the list of Butler. Nonetheless, we can affirm that group N was definitely known to Kuhn.

Group N, as a particular case ($q = 2$) of the automorphism groups of regular spreads in $PG(3, q)$, also appears in the context of the famous Kirkman Problem, see Corollary 2 on page 73 of [30].

10.3.4 Empirical Observations

Among the 80 Steiner designs on 15 points, there are four resolvable designs which altogether contain seven non-isomorphic Kirkman systems, see [48]. According to the numeration given in [48], these designs are $STS\#1$, $STS\#7$, $STS\#19$ and $STS\#61$, with respective automorphism groups of order 20160, 288, 12 and 21. Among these four, only #1 and #7 are Siamese, and these two are also "extremal" in the sense of having the greatest number of spreads (56 and 32, respectively).

A lower bound on the number of Steiner designs on 40 points is evaluated to be about one million. Nevertheless, for a long time only one such design was known to be resolvable, namely the classical $PG(3, 3)$. Recently, however, M. Greig and A. Rosa found a non-classical example [25] which has three maximal arcs. One more example, also with maximal arcs, appears in [50].

This prompted us to arrange a comparison between a few of our Siamese designs on 40 points and these two newly discovered ones, which we also regard as exceptional. Our first impression is that the Siamese property seems to be in direct correlation with "abundance of resolutions," however it is inversely correlated to the existence of maximal arcs, see Table 8 (the designations $\#n$ carry over from Table 7). We intend to investigate these speculations after the enumeration of all Siamese objects on 40 points has been completed.

Acknowledgments

An essential part of this research was conducted while M.K. was a visitor, and S.R. was a graduate student, at the University of Delaware. These authors express their gratitude to that institution. All authors are grateful to S.-Y. Song and his colleagues for comments which contributed to improving the quality of the paper. Finally, it is our pleasure to acknowledge Bruno Buchberger and the coordinators of the Special Semester on Groebner Bases (February 1 – July 31, 2006), organized by RICAM, Austrian Academy of Sciences, and RISC, Johannes Kepler University, Linz Austria. Their collective interest in tutorials which cover algorithmic aspects of algebra and combinatorics mirrors our own, and is most appreciated.

References

1. E. Bannai and T. Ito, *Algebraic Combinatorics I. Association Schemes*, Benjamin/Cummings, Menlo Park, 1984.
2. T. Beth, D. Jungnickel, and H. Lenz, *Design Theory*, Cambridge University Press, Cambridge, 1993.
3. A. Betten, A. Kerber, R. Laue, and A. Wasserman, Simple 8-designs with small parameters, *Designs Codes Cryptogr.*, **15** (1998), 5–27.
4. A. Betten, M. Klin, R. Laue, and C. Pech, *A Computer Approach to the Enumeration of Block Designs which are Invariant with Respect to a Prescribed Permutation Group*, Preprint MATH-AL-13-1997 Technische Univesitität Dresden, 1–51, supplement 52–74, 1997.
5. A. Betten, M. Klin, R. Laue, and A. Wasserman, Graphical t-designs via Kramer-Mesner matrices, *Discrete Math.*, **197/198** (1999), 83–109.
6. A. Betten, R. Laue, and A. Wasserman, DISCRETA: A program system for the construction of t-designs with a prescribed automorphism group, University of Bayreuth, 1998, http://www.mathe2.uni-bayreuth.de/betten/DISCRETA/index.html.
7. N. L. Biggs, Distance-regular graphs with diameter three, *Ann. Discrete Math.*, **15** (1982), 69–80.
8. A. E. Brouwer, http://www.win.tue.nl/~aeb/
9. A. E. Brouwer, A. M. Cohen, and A. Neumaier, *Distance Regular Graphs*, Springer-Verlag, Berlin, 1989.
10. A. E. Brouwer, J. H. Koolen, and M. H. Klin, A root graph that is locally the line graph of the Petersen graph, *Discrete Math.*, **264** (2003), 13–24.
11. G. Butler, The transitive groups of degree fourteen and fifteen, *J. Symbol. Comput.*, **16** (1993), 413–422.
12. P. J. Cameron, *Permutation Groups*, Cambridge University Press, Cambridge, 1999.
13. F. N. Cole, L. D. Cummings, and H. S. White, Complete classification of the triad systems of 15 elements, *Proc. Nat. Acad. Sci. USA*, **3** (1919), 197–199.

14. J. H. Conway, A. Hulpke, and J. McKay, On transitive permutation groups, *London Math. Soc. J. Comput. Math.*, **1** (1998), 1–8.
15. J. Degraer and K. Coolsaet, Classification of three-class association schemes using backtracking with dynamic variable ordering, *Discrete Math.*, **300** (2005), 71–81.
16. R. H. Dye, Maximal subgroups of symplectic groups stabilizing spreads II, *Journal of London Math. Society* (40), **2** (1989), 215–226.
17. I. A. Faradžev and M. H. Klin, Computer package for computations with coherent configurations, in S. M. Wat (ed.) *Proceedings ISSAC-91 Bonn*, pp. 219–223, Assoc. Comput. Math. Press, New York, 1991.
18. I. A. Faradžev, M. H. Klin, and M. E. Muzichuk, Cellular rings and groups of automorphisms of graphs, in I. A. Faradžev, A. A. Ivanov, M. H. Klin and A. J. Woldar (eds.) *Investigations in Algebraic Theory of Combinatorial Objects.* Mathematics and Its Applications, Vol. 84, pp. 1–152, Kluwer Academic, Dordrecht, 1994.
19. R. A. Fisher, An examination of the different possible solutions of a problem in incomplete blocks, *Ann. Eugenics*, **10** (1940), 52–75.
20. D. E. Flesner, The geometry of subgroups of $PSp_4(2^n)$, *Illinois J. Math.*, **19** (1975), 42–70.
21. P. B. Gibbons and R. Mathon, Construction methods for Bhaskar Rao and related designs, *J. Austral. Math. Soc., Ser. A*, **42** (1987), 5–30.
22. C. D. Godsil and A. D. Hensel, Distance regular covers of complete graphs, *J. Comb. Theory, Ser. B*, **36** (1992), 205–238.
23. Ja. Ju. Gol'fand, A. V. Ivanov, and M. H. Klin, Amorphic cellular rings, in I. A. Faradžev, A. A. Ivanov, M. H. Klin, and A. J. Woldar (eds.) *Investigations in Algebraic Theory of Combinatorial Objects.* Mathematics and Its Applications, Vol. 84, pp. 167–186, Kluwer Academic, Dordrecht, 1994. Translated from Russian.
24. M. J. Grannel, T. S. Griggs, and J. P. Murphy, Equivalence classes of Steiner triple systems, *Congressus Numerantum*, **86** (1992), 19–25.
25. M. Greig and A. Rosa, Maximal arcs in Steiner systems $S(2, 4, v)$. Combinatorics 2000 (Gaeta), *Discrete Math.* (1–3), **267** (2003), 143–151.
26. J. L. Gross and T. W. Tucker, *Topological Graph Theory*, Dover, New York, 1987.
27. M. Hall and J. D. Swift, Determination of Steiner triple systems of order 15, *Math. Tables Other Aids Comput.*, **9** (1955), 146–152.
28. F. Harary, *Graph Theory*, Addison–Wesley, Reading, 1969.
29. D. G. Higman, Coherent configurations, *I. Rend. Sem. Mat. Univ. Padova,* **44** (1970), 1–25.
30. J. W. P. Hirschfeld, *Finite Projective Spaces of Three Dimensions*, Oxford University Press, Oxford, 1985.
31. D. R. Hughes and F. C. Piper, *Design Theory*, Cambridge University Press, Cambridge, 1985.
32. Y. J. Ionin, A technique for constructing symmetric designs, *Designs Codes Cryptogr.*, **14** (1998), 147–158.

33. A. A. Ivanov, Construction of some new automorphic graphs using computers, in *Air Physics and Applied Mathematics (Moscow) MPhTI*, pp. 144–146, 1981 (in Russian).
34. A. Kerber and R. Laue, Group actions, double cosets and homomorphisms: Unifying concepts for the constructive theory of discrete structures, in M. Hazewinkel, A. A. Ivanov, and A. J. Woldar (eds.) *Algebra and Combinatorics: Interactions and Applications.* Acta Applicandae Mathematicae, Vol. 52 (1–3), pp. 63–90, Kluwer Academic, Dordrecht, 1998.
35. H. Kharaghani and R. Torabi, On a decomposition of complete graphs, *Graphs Comb.*, **19** (2003), 519–526.
36. M. H. Klin, M. Meszka, S. Reichard, A. Rosa, and E. Spence, The smallest non-rank 3 strongly regular graphs which satisfy the 4-vertex condition, *Bayreuther Mathematische Schriften*, **74** (2005), 145–205.
37. M. Ch. Klin, R. Pöschel, and K. Rosenbaum, *Angewandte Algebra*, Vieweg, Braunschweig, 1988.
38. M. H. Klin and S. Reichard, Siamese association schemes in miniature: A link between generalized quadrangles, graphs and Steiner designs. Abstracts of AMS Meeting (Bloomington), 2003, 985-05-232.
39. M. H. Klin, R. Reichard, and A. J. Woldar, Siamese objects and their relation to color graphs, association schemes and Steiner designs, *Bull. Belgian Math. Soc.* (5), **12** (2005), 845–857.
40. M. H. Klin, S. Reichard, and A. J. Woldar, Siamese association schemes and Siamese Steiner designs. I. Introduction (in preparation).
41. M. H. Klin, S. Reichard, and A. J. Woldar, Siamese association schemes and Siamese Steiner designs. II. Objects on 15 points (in preparation).
42. M. Klin, C. Rücker, G. Rücker, and G. Tinhofer, Algebraic combinatorics in mathematical chemistry. Methods and algorithms. I. Permutation groups and coherent (cellular) algebras. *MATCH*, **40** (1999), 7–138.
43. E. S. Kramer and D. M. Mesner, t-Designs on hypergraphs, *Discrete Math.*, **15** (1976), 263–296.
44. H. W. Kuhn, On imprimitive substitution groups, *Am. J. Math.*, **26** (1904), 45–102.
45. R. Laue, Computing double coset representatives for the generation of solvable groups, *Lect. Notes Comput. Sci.*, **144** (1982), 65–70.
46. E. N. Martin, On the imprimitive substitution groups of degree fifteen and the primitive substitution groups of degree eighteen, *Am. J. Math.*, **23** (1901), 259–286.
47. R. Mathon, Lower bounds for Ramsey numbers and association schemes, *J. Comb. Theory, Ser. B*, **42** (1987), 122–127.
48. R. A. Mathon, K. T. Phelps, and A. Rosa, Small Steiner triple systems and their properties, *Ars Combinatoria*, **15** (1983), 3–110.
49. B. D. McKay, nauty User's Guide (Version 1.5), Technical Report TR-CS-90-02, Computer Science Department, Australian National University, 1990.

50. S. Milici, A. Rosa, and V. Voloshin, Colouring Steiner systems with specified block colour patterns, *Discrete Math.* (1–3), **240** (2001), 145–160.
51. E. H. Moore, Tactical memoranda I–III, *Am. J. Math.*, **18** (1896), 264–303.
52. S. E. Payne, All generalized quadrangles of order 3 are known, *J. Comb. Theory*, **18** (1975), 203–206.
53. S. E. Payne, Tight point sets in finite generalized quadrangles, *Congressus Numerantium*, **60** (1987), 243–260.
54. S. E. Payne, and J. A. Thas, *Finite Generalized Quadrangles*, Research Notes in Mathematics, Pitman, London, 1984.
55. S. Reichard, Computational and Theoretical Analysis of Coherent Configurations and Related Incidence Structures, Thesis, University of Delaware, Newark, 2003.
56. R. Remak, Über Untergruppen direkter Produkte von drei Faktoren, *Crelle's J. Reine Angewandte Math.*, **166** (1931), 65–100.
57. B. Schmalz, t-Designs zu vorgegebener Automorphismengruppe, Thesis, Universität Bayreuth, *Bayreuther Mathematische Schriften*, **41** (1992), 1–164.
58. B. Schmalz, The t-designs with prescribed automorphism group, new simple 6-designs, *J. Comb. Des.*, **1** (1993), 126–170.
59. M. Schönert et al., GAP – Groups, Algorithms, and Programming, Lehrstuhl D für Mathematik, Rheinisch-Westfälische Technische Hochschule, Aachen, 1995.
60. W. R. Scott, *Group Theory*, Dover, New York, 1987.
61. M. W. Short, *The Primitive Soluble Permutation Groups of Degree Less than 256*, Lecture Notes in Mathematics, Vol. 1519, Springer, Berlin, 1992.
62. L. H. Soicher, GRAPE: A system for computing with graphs and groups, in L. Finkelstein and W. M. Kantor (eds.) *Groups and Computation*, DIMACS Series in Discrete Mathematics and Theoretical Computer Science, Vol. 11, pp. 287–291, Amer. Math. Soc., Providence, 1993.
63. E. Spence, http://www.maths.gla.ac.uk/~es/srgraphs.html.
64. J. J. Sylvester, Elementary researches in the analysis of combinatorial aggregation, *Philos. Mag.*, **24** (1844), 285–296.
65. J. A. Thas and S. E. Payne, Spreads and ovoids in finite generalized quadrangles, *Geometrica Dedicata*, **52** (1994), 227–253.
66. J. A. Todd, As it might have been, *Bull. London Math. Soc.*, **2** (1970), 1–4.
67. A. Wasserman, Finding simple t-designs with enumeration techniques, *J. Comb. Des.* (2), **6** (1998), 79–90.

Using Gröbner Bases to Investigate Flag Algebras and Association Scheme Fusion

Douglas A. Leonard

Department of Mathematics and Statistics, Auburn University, Auburn, AL, USA.
leonada@auburn.edu

Summary. This paper is meant primarily as a *tutorial* on how to phrase problems in association schemes in the language of Gröbner bases and use the computational results provided by those bases, though it does contain fusion scheme computations not previously found in the literature.

Key words: Gröbner bases, Flag algebras, Association schemes

1 Introduction

It is instructive to see how the use of *Gröbner bases* [2] can clarify certain computational and theoretic aspects of various topics. Here the choice of topics come from *association schemes* and *fusion of flag algebras* arising from *generalized n-gons* and 2-$(v, k, 1)$ *designs* [6]. Three different ways of using concepts related to Gröbner bases are given as a guide for using such concepts in similar types of problems.

Often it is the case that combinatorialists have to sort out some form of truth about a given object based on its parameters. This can translate into pages and pages of manipulations of intermediate results involving polynomial equations, with the attendant derivations and justifications. But these problems can usually be easily rephrased in terms of finding common zeros of a system of polynomial equations, and knowing for what values of the parameters this happens. This is exactly what Gröbner bases are meant to accomplish. So Gröbner basis theory can provide direct computational answers without the need for further justification, once problems are phrased in terms of *generators* (polynomials that should be zero) for an *ideal* in a *multivariate polynomial ring*, and the need to know the corresponding *varieties* (set of common zeros). This should be useful either to those wishing to use Gröbner bases to clarify and/or simplify computations or to those interested in seeing how Gröbner bases can be used in "applied theoretical" settings.

M. Klin et al. (eds.), Algorithmic Algebraic Combinatorics and Gröbner Bases,
DOI 10.1007/978-3-642-01960-9_3, © Springer-Verlag Berlin Heidelberg 2009

And it could conceivably use the same system of equations to produce results overlooked in some ad hoc approach to solving them.

Now consider the concrete problem of investigating a parameterized series of *association schemes*, describing the *structure constants*, determining fusion partitions, and investigating those fusions found. Much of this work was done by the Soviet school for *metric* (one parameter) schemes (see, for example, the survey [4] and http://www.ricam.oeaw.ac.at/specsem/SRS/groeb/download/Muzychuk.pdf). See also [9].

The next natural stage is to investigate *dihedral* (two parameter) schemes in the terminology of Zieschang, coming mainly from flag algebras, which were introduced by a number of authors [5, 10, 7, 11, 13].

The paper [6], which started this investigation, used ad hoc methods to do the computations. So here those methods have been replaced by the systematic Gröbner basis methods, which make the actual computations invisible to the reader. There are new computations relative to *fusion* expanding on this previous work, with the MAGMA [8] code that generated them, to whet the appetite of the audience.

2 Gröbner Basis Preliminaries

There are many good books covering this material, with [3] being the first author's favorite. The following is a brief introduction to ideals and varieties in polynomial rings for those unfamiliar with the topic. Given some *variables* $x_n, \ldots, x_1$ and a *coefficient ring* R, $R[x_n, \ldots, x_1]$ denotes the ring of (finite) R-linear combinations of (finite) products in these variables, and seems to be called either a *multivariate polynomial ring* or a *free(associative) algebra*, depending on whether the variables *commute* with each other or not. Even to be able to write down polynomials in a canonical way, it is necessary to have a *monomial order* (a *total order* with obvious extra properties). The (default) *lexicogrphical monomial order* is based on comparing products by considering their *indices* (that is, *exponents*) lexicographically. So, for instance, in the multivariate polynomial ring $R[x_3, x_2, x_1]$ the order looks like

$$1 \prec x_1 \prec x_1^2 \prec \cdots \prec x_2 \prec x_2x_1 \prec \cdots \prec x_2^2 \prec \cdots \prec x_3 \prec \cdots$$

which can be described by $x_3^{i_3}x_2^{i_2}x_1^{i_1} \succ x_3^{j_3}x_2^{j_2}x_1^{j_1}$ iff ($i_3 > j_3$) or ($i_3 = j_3$ and $i_2 > j_2$) or ($i_3 = j_3$ and $i_2 = j_2$ and $i_1 > j_1$). The generic *total degree orders* are the *grevlex* and *glex* orders (short for *graded reverse lexicographical* and *graded lexicographical*), in which total degree is the first concern. These give orders that look like

$$1 \prec x_1 \prec x_2 \prec x_3 \prec x_1^2 \prec x_2x_1 \prec x_3x_1 \prec x_2^2 \prec \cdots$$

and

$$1 \prec x_1 \prec x_2 \prec x_3 \prec x_1^2 \prec x_2x_1 \prec x_2^2 \prec x_3x_1 \prec \cdots$$

respectively. These can be defined in ways similar to those above; but the non-singular matrices

$$A_{grevlex} := \begin{pmatrix} 1 & 1 & 1 \\ 1 & 1 & 0 \\ 1 & 0 & 0 \end{pmatrix} \qquad A_{glex} := \begin{pmatrix} 1 & 1 & 1 \\ 1 & 0 & 0 \\ 0 & 1 & 0 \end{pmatrix}$$

can be used to reduce these to the lexicographical case by converting the column vector of exponents

$$\begin{pmatrix} i_3 \\ i_2 \\ i_1 \end{pmatrix} \quad \text{to} \quad \begin{pmatrix} i_3 + i_2 + i_1 \\ i_3 + i_2 \\ i_3 \end{pmatrix} \quad \text{or} \quad \begin{pmatrix} i_3 + i_2 + i_1 \\ i_3 \\ i_2 \end{pmatrix}$$

respectively.

The *leading monomial* of f, denoted here by $LM(f)$ is the largest monomial occurring in f, relative to the given order. Its coefficient is called the *leading coefficient*, denoted by $LC(f)$, and the product $LC(f)LM(f) =: LT(f)$ is called the *leading term*. (*Warning*: Some authors interchange the words term and monomial, and there is no uniformity of notation in the literature.)

An *ideal* I (two-sided in the case of free algebras) is a subset of a ring closed under addition of its elements and under multiplication by ring elements. Ideals describing *finitely-presented algebras* (that is, free associative algebras modulo said ideals) are commonly given in terms of a (finite) set of *generators* (referred to as *relations* when the variables are called generators!) $I := \langle g_1, g_2, \ldots, g_m \rangle$, so that I consists of all (finite) R-linear combinations of $g_1, \ldots, g_m$.

As with vector spaces, where arbitrary generating sets are not as useful as maximal linearly independent sets, called (*vector space*) *bases*, it should be no surprise that arbitrary generating sets for ideals (which are unfortunately called *bases* in some of the literature) are not as important or useful as *Gröbner bases*, those generating sets B relative to which a "canonical" remainder $NormalForm(f, B)$ for f after division by the elements of B (in any order) can be defined.

A more useful computational definition of Gröbner bases is in terms of *SPolynomials*. If h is a *least common multiple* of $LM(f)$ and $LM(g)$, then $SPoly(f, g) := \frac{h}{LT(f)} f - g \frac{h}{LT(g)}$ (with suitable generalization in the non-commutative case). Then B is a Gröbner basis if $SPoly(f, g)$ reduces to 0 after division by the elements of B for all choices of $f, g \in B$. In fact this is the foundation for all forms of the *Buchberger algorithm*; namely that given any generating set, it is possible to produce a Gröbner basis by computing SPolynomials, reducing them by division relative to the current generating set, and appending the results to the generating set if necessary, possibly reducing the other elements along the way, until the set is closed under this combined operation of computing SPolynomials and reducing them.

If one has a collection of equations of the form $f_i = h_i$ that can be written in polynomial form in terms of some (not necessarily polynomial) variables, then it is possible to use $g_i := f_i - h_i$ as generators for an ideal in a polynomial ring. Common solutions to the collection of equations then become common zeros of the elements of the ideal I or equivalently of the elements of the Gröbner basis B. These common zeros are called *points* of the *variety* $\mathcal{V}(I) := \{v : g(v) = 0 \text{ for all } g \in I\}$. This variety is generally computed by finding a Gröbner basis relative to some easy ordering, changing it to a (factored) Gröbner basis relative to a lex order, and recursively computing coordinates of the elements of the variety.

The importance of all this to a field such as *algebraic combinatorics* is that in the study of a collection of objects relative to a set of parameters, it is typical that questions may be phrased in terms of what parameter values are necessary or sufficient to guarantee the feasibility or existence of such objects. Then equations satisfied by the parameters can be turned into generators for an ideal for which the variety gives such parameter values. That means that rather than haphazardly sorting through a collection of equations, looking for results about the parameters, it is possible to merely compute a Gröbner basis and a variety, letting the Gröbner basis theory replace all the intermediate results, derivations, and justifications.

Sometimes there are unexpected added benefits beyond this, in that proper phrasing of problems in terms of Gröbner bases may give insight into the combinatorial structures, definitions, or other aspects of a problem. This will be exemplified below by considering *generalized n-gons* and 2-$(v, k, 1)$ *designs.*

3 Algebraic Combinatorics Preliminaries

A matrix algebra W over the field $\mathbf{C}$ of complex numbers is called a *coherent algebra* iff the identity matrix, I, is in W, the all-ones matrix, J is in W, the transpose, A^T of A is in W, if A is, and the entry-wise products of elements of W are in W. It is well known that each coherent algebra W contains an unique basis of $(0,1)$ matrices $\{A_0, A_2, \ldots, A_{r-1}\}$ such that $A_0 + A_1 + \cdots + A_{r-1} = J$ and $A_i^T = A_{i'}$. The use of this standard basis can be used to reformulate the notion of a coherent algebra in terms of relations and their graphs, called *coherent configurations.* And if $A_0 = I$, these are called *association schemes.*

Many significant examples of classes of combinatorial objects may be reformulated in terms of coherent configurations or association schemes, a few being discussed below.

The fact that W is an algebra implies the existence of (non-negative integer) *structure constants* $p_{i,j,k}$ such that $A_i A_j = \sum_{k=1}^{r} p_{i,j,k} A_k$. If the A_i's are treated as variables instead of $(0,1)$ matrices, the algebras are called *table algebras* [1]. While in many cases it is crucially important to keep track of all combinatorial information about a coherent algebra W, many significant feasible conditions are obtained by viewing it merely as a table algebra. In particular, a question about the existence of coherent subalgebras of a given coherent algebra (*fusions*) may be solvable based only on the knowledge of the structure constants. Classes of table algebras may be defined in terms of parameters and relations among those parameters, which suggests applying Gröbner basis techniques to produce feasibility conditions.

4 Definitions

Start with finite sets $\mathcal{V}$, of v *vertices*, and $\mathcal{B}$, of b *blocks*. The basic combinatorial structure called a *1-design* is merely a subset $\mathcal{D}$ of $\mathcal{V} \times \mathcal{B}$ such that $r := \#\{l \in \mathcal{B} : (p,l) \in \mathcal{D}\}$ is independent of $p \in \mathcal{V}$ and $k := \#\{p \in \mathcal{B} : (p,l) \in \mathcal{D}\}$ is independent of $l \in \mathcal{V}$. Elements $(p,l) \in \mathcal{D}$ are called *flags*; and there are $vr = bk$ such.

A 1-design such that any two vertices are incident with *at most one* block and any two blocks are incident with at most one vertex is called a *partial linear space*. (Geometers tend to use the word *point* in place of vertex, the word *line* in place of block, and the parameters $s := k-1$ and $t := r-1$ instead of k and r respectively.) Normally the study of vertices is linked to the study of the (symmetric) $(0,1)$ *adjacency matrix*, A, with

$$A(p_i,p_j) = 1 \quad \text{if and only if} \quad (p_i,l),(p_j,l) \in \mathcal{D} \quad \text{for some } l \in B.$$

But here the focus is initially on flags; so let L denote the (symmetric) $(0,1)$ *collinearity matrix* indexed by the flags, with

$$L((p_1,l_1),(p_2,l_2)) = 1 \quad \text{if and only if} \quad l_1 = l_2,\ p_1 \neq p_2;$$

and let N denote the (symmetric) $(0,1)$ *concurrence matrix* indexed by the flags, with

$$N((p_1,l_1),(p_2,l_2)) = 1 \quad \text{if and only if} \quad p_1 = p_2,\ l_1 \neq l_2.$$

These matrices already satisfy the conditions

$$L^2 = (s-1)L + sI, \qquad N^2 = (t-1)N + tI$$

(with I denoting the $vr \times vr$ identity matrix).

Generalized n-gons were first introduced by Tits [12] in 1959. Standard definitions are in terms of *distance* and *adjacency*, and are geometric in flavor, descriptive of the fact that each point should be contained in an n-gon and in no smaller polygon. In terms of flags and the matrices L and N above this translates into the following matrix-theoretic definition, which obscures the geometric origin. Let

$$\begin{aligned} A_{4i} &:= (NL)^i, & A_{4i+1} &:= L(NL)^i, \\ A_{4i+2} &:= N(LN)^i, & A_{4i+3} &:= (LN)^{i+1}, \end{aligned}$$

denote the various *flag adjacency matrices* (a possibly unexpected benefit of this non-geometric approach). Then $\mathcal{D}$ will be called a *generalized n-gon* if $(A_j,\ 0 \leq j \leq 2n-1)$ is a linearly independent set over $\mathbf{Z}$, and hence a *basis* for the $\mathbf{Z}$-module it generates; but $A_{2n} = A_{2n-1}$. (It is a straightforward, instructive exercise to see that this really corresponds to the geometric concept above.)

Since $A_0 = I$, $A_{4i+1}^T = A_{4i+1}$, $A_{4i+2}^T = A_{4i+2}$, $A_{4i+3}^T = A_{4i+4}$, and $\sum_{j=0}^{2n-1} A_j = J$, the all 1's matrix; the ordered set $(A_j : 0 \leq j \leq 2n-1)$ is a basis for the $\mathbf{Z}$-module it generates, lacking only a multiplication to be the *adjacency algebra* of an *association scheme.*

If the two matrices L and N are *not* known, and the parameters t and s are *not* known either, then it is possible to study generalized n-gons by starting with the *coefficient ring* $\mathbf{Z}[t,s]$ (with a *lex* monomial order $t \succ s$), and then a *free (associative) algebra* $\mathbf{Z}[t,s][y,x]$ in two *non-commutative* variables (with a *total degree* monomial order with $y \succ x$). The former means that when writing polynomials in the variables t and s, $t^i s^j$ should be treated as larger than $t^k s^l$ if either $(i > k)$ or $(i = k$ and $j > l)$ (and that one shouldn't write $s^j t^i$). The latter means that when writing polynomials in the (non-commuting) variables y and x monomials, a string such as $yxxyx$ should be treated as larger than either yxy or $yxxxy$, the first because the total degree is greater, the second because the total degree is the same but at the leftmost difference in the strings $y \succ x$. (One of the first important lessons one learns when working with multivariate polynomial rings is that a proper choice of monomial order is perhaps the most critical step in any problem, and the step most easily ignored completely by those not used to this area.)

Then this should be made into a *finitely-presented algebra* (quotient ring in two non-commuting variables, representing the two non-commuting matrices N and L respectively)

$$\mathbf{Z}[t,s][y,x]/\langle f_1, f_2, f_3\rangle$$

with $f_1 := x^2 - (s-1)x - s$, $f_2 := y^2 - (t-1)y - t$, and $f_3 := (yx)^l - (xy)^l$ if $n = 2l$ or $f_3 := y(xy)^l - (xy)^l x$ if $n = 2l+1$. The import here is that the *ideal* $I := \langle f_1, f_2, f_3\rangle$ consists of all those polynomials that are supposed to be equal to zero under the given assumptions. (Such a (table) algebra, depending only on the structure constants and not on some $(0,1)$ matrices, may exist whether or not a corresponding n-gon does. At this stage it is not even necessary to assume knowledge of classical necessary conditions for the existence of such n-gons.)

5 An Application of SPolynomials and Reduction

Buchberger's algorithm for computing *Gröbner bases* for ideals (such as $\langle f_1, f_2, f_3\rangle$ above) is based on the central observation above that *bases* (as opposed to *generating sets*) for ideals should be *closed* under the operation of *reduction of SPolynomials.* Consider an example of this type of computation for the n-gon case with $n = 2l+1$. Since $LM(yf_3) = y^2(xy)^l = LM(f_2(xy)^l)$, the corresponding (non-commutative) *SPolynomial* would be

$$SPolynomial(f_3, f_2) := yf_3 - f_2(xy)^l = -y(xy)^l x + (t-1)y(xy)^l + t(xy)^l,$$

which would then be *reduced* to a *remainder*

$$(yf_3 - f_2(xy)^l) + f_3 x + (xy)^l f_1 - (t-1)f_3 = (t-s)\left((xy)^l x + (xy)^l\right)$$

using the explicit division by the set $\{f_1, f_2, f_3\}$ given.

Thus a single SPolynomial computation, and the use of the assumption that $(xy)^l x$ and $(xy)^l$ are linearly independent, already forces $t - s = 0$. So it follows that:

Proposition 1. *Generalized* $(2l+1)$*-gons with parameter pairs* (t, s) *can only exist if* $t = s$.

Remark. Again, this can be viewed as a result about table algebras.

6 Structure Constants of Association Schemes

In light of the proposition above, consider only generalized $2l$-gons in their finitely-presented algebra form. If a *total degree* monomial order (with $a_i \succ a_j$ for $i > j$) is used, then a *minimal, reduced* Gröbner basis (that is, one with a minimal number of elements, and no leading monomial of one dividing any monomial of another) for the ideal of relations, $\mathcal{I}$, will have *all* its elements of the form

$$a_i a_j - NormalForm(a_i a_j, \mathcal{I})$$

with

$$NormalForm(a_i a_j, \mathcal{I}) = \sum_{k=0}^{2n-1} p_{i,j,k} a_k,$$

describing the multiplication in the algebra, as well as determining the *structure constants* $p_{i,j,k}$ that make this the *adjacency algebra* of an *association scheme.* (If this is not immediately obvious, see the proof below.)

Thus a simple Gröbner basis computation relative to an appropriate total degree monomial order gives constructively the following result (with concrete examples below):

Proposition 2.

$$\mathbf{Z}[t, s][a_{4l-1}, \ldots, a_1]/\mathcal{I}$$

with $\mathcal{I}$ *the* ideal of relations *with generating set containing the basis relations from the definition of a generalized 2l-gon:*

$$f_1 := a_1^2 - (s-1)a_1 - s,$$
$$f_2 := a_2^2 - (t-1)a_2 - t,$$
$$f_3 := (a_2 a_1)^l - (a_1 a_2)^l;$$

together with the relations gotten from the definitions of the a_j's:

$$(a_2a_1)^i - a_{4i}, \qquad a_1(a_2a_1)^i - a_{4i+1}, \qquad a_2(a_1a_2)^i - a_{4i+2},$$
$$(a_1a_2)^{i+1} - a_{4i+3}, \qquad 0 \le i < l,$$

corresponds to the adjacency algebra of an association scheme. (Note that $a_0 = 1$ is implicit in the calculations, and sometimes explicit in the theory.)

Moreover, if a total degree *monomial order (with $a_i \succ a_j$ for $i > j$) is used, then a* minimal, reduced *Gröbner basis for $\mathcal{I}$ will have* all *its elements of the form*

$$a_ia_j - NormalForm(a_ia_j, \mathcal{I})$$

with

$$NormalForm(a_ia_j, \mathcal{I}) = \sum_{k=0}^{4l-1} p_{i,j,k}a_k,$$

which corresponds to a complete description of the algebra multiplication and a determination of the structure constants $p_{i,j,k}$.

Proof. Given that the a_i are linearly independent, there can be no relations with leading monomial of degree 1. Given that the a_i's form an algebra, a_ia_j must be expressible as a linear combination $\sum_k p_{i,j,k}a_k$, since all elements of the algebra are of this form. Hence $a_ia_j - \sum_{k=0}^{4l-1} p_{i,j,k}a_k$ must be basis elements in any total degree monomial order Gröbner basis. And any monomial of total degree greater than 2 is divisible by one of degree 2, so can't be a leading monomial in any *minimal* total degree Gröbner basis. □

7 Fusion

Now consider a *partition Π* of the set $\{0, \ldots, 4l-1\}$, and the corresponding *fusion* of classes (sum of their respective matrices)

$$B_\gamma := \sum\{A_k : k \in \gamma\}, \quad \gamma \in \Pi.$$

This could conceivably determine a *fusion scheme*; that is, be an association scheme with fewer classes than the original, if $\{0\}$ is a part, $\gamma' := \{k' : k \in \gamma, A_k^T = A_{k'}\}$ is a part for each $\gamma \in \Pi$, and the structure constant $P_{\alpha,\beta,\gamma} := \sum_{i\in\alpha}\sum_{j\in\beta} p_{i,j,k}$ is independent of the choice of $k \in \gamma$, for all parts α, β, and γ in Π.

This can be viewed as another Gröbner basis problem, but this time in the *multivariate polynomial coefficient ring* $\mathbf{Z}[t, s]$ (with two *commuting* variables t and s, the *integer* parameters). The ideal in question is generated by the relations (forced by the third item above):

$$\sum_{i\in\alpha}\sum_{j\in\beta} p_{i,j,k_1} - \sum_{i\in\alpha}\sum_{j\in\beta} p_{i,j,k_2}$$

for all α, β, and γ in Π and all $k_1, k_2 \in \gamma$, provided Π is a good partition (that is, satisfying the first two items above). (This is a slightly different condition than that of *good sets* in [6].) Then the *variety* of this ideal determines all possible parameter pairs (t, s) for which the good partition Π produces a fusion scheme. Of course, it is necessary to restrict the variety to pairs of *positive integers*, and probably to ignore the *trivial* case $t = 1 = s$ in which the original generalized $2l$-gon is merely a $2l$-gon, and fusion corresponds to *Schur rings* over the dihedral group of order $2l$. The varieties involved are computed most easily using a *lexicographical* monomial order on $\mathbf{Z}[t, s]$ and a *factored, minimal, reduced* Gröbner basis.

8 2-$(v, k, 1)$ Designs

Consider a similar problem of *flag adjacency matrices* for 2-$(v, k, 1)$ *designs* (also called Steiner systems $S(2, k, v)$ or just 2-designs); that is, 1-designs for which every 2 vertices determine an unique block. The matrices for this are similar in flavor to those for the generalized n-gons:

$$A_0 := I, \quad A_1 := L, \quad A_2 := N, \quad A_3 := LN,$$
$$A_4 := NL, \quad A_5 := LNL, \quad A_6 := NLN - LNL$$

with

$$LNLN = sA_6 + (s-1)A_5 + sA_4, \quad NLNL = sA_6 + (s-1)A_5 + sA_3.$$

Since $A_0 = I$, $A_1^T = A_1$, $A_2^T = A_2$, $A_3^T = A_4$, $A_5^T = A_5$, $A_6^T = A_6$, and $\sum_{j=0}^{6} A_j = J$, the all 1's matrix; the ordered set $(A_j : 0 \leq j \leq 6)$ is a basis for the $\mathbf{Z}$-module it generates, lacking only a multiplication to be the *adjacency algebra* of an *association scheme*.

Then this should be made into a *finitely-presented algebra* as before

$$\mathbf{Z}[t, s][y, x]/\langle f_1, f_2, f_3, f_4 \rangle$$

with $f_1 := x^2 - (s-1)x - s$, $f_2 := y^2 - (t-1)y - t$, and $f_3 := (yx)^2 - s(yxy - xyx) - (s-1)xyx) - sxy)$, $f_4 := (xy)^2 - s(yxy - xyx) - (s-1)xyx - syx$.

Remark. When $s < t$, $A_6 \neq 0$, so this is a rank 7 association scheme. However, if $s = t$, then $A_6 = 0$ and the 2-$(v, k, 1)$ design is a *projective plane* or equivalently a *generalized 3-gon*, covered earlier.

9 Code and Output

There are also technical difficulties to be considered when using existing computer algebra packages. For instance, in MAGMA, the coefficient ring of a finitely-presented algebra needs to be a field. So it is necessary to use $\mathbf{Q}(t, s)$, the function field in two variables over the rational field $\mathbf{Q}$, with a total degree monomial order, and then map results to $\mathbf{Q}[t, s]$ with the desired *lex* monomial order, where factorization can be done.

The MAGMA code below has been automated so that the only parameter input necessary is l (meaning $n/2$). The output is a list of partitions giving rise to fusion schemes, together with a factored Gröbner basis from which it is relatively easy to read the corresponding parameter pairs (t, s), if such positive integer pairs exist.

Although the computations are (for better or worse) no longer visible to the reader, it is instructive to have some idea of what is happening. As an example from the generalized quadrangle case, the partition $\Pi := [[1, 2, 3, 4, 7][5, 6]]$ could only give a fusion scheme if $ts^2 - 2ts$, $t^2 - 2t - s^2 + 2s$, $t^2s - ts^2$, and $t^2 - ts^2 + 3ts - 4t + s^2 - 4s + 4$ (gotten from the fusion condition 3 above) are all zero. An *interreduction* of these generators already gives a Gröbner basis $(t^2 - 2t, ts - 2t, s^2 - 2s)$ for the ideal they generate. And a recursive calculation of s and t gives $s = 0, t = 0$, $s = 2, t = 0$, or $s = 2, t = 2$. Of these 3 rational points in the variety, clearly only $(t, s) = (2, 2)$ is useful in this fusion context. (Currently the code actually produces a modified Gröbner basis $(t-2, s-2)$ by removing factors such as t and s which can't ever lead to positive integer solutions.) There are examples for which the solutions must be done by hand. For instance, in partition 48 in the hexagon output below, one factor is $T^2 - TS + 2T + 1$ which gives a one dimensional variety $s = (t+1)^2/t$ over the rationals, but only the single positive integer solution $s = 4, t = 1$.

The code is written to search for a good partition, compute differences of coefficients that should be equal for fusion to occur, find a Gröbner basis for ideal of all such differences, and output same, at least in the cases in which there might conceivably be an element in the variety with all coordinates positive. In most cases, it is relatively easy to read off any parameter pairs with both entries positive integers, and to ignore those with only the trivial solution $t = 1 = s$.

Note that the code produces a Gröbner basis describing the original adjacency algebra before tackling the fusion problem (an expected benefit); so it is possible to see the structure constants, not in a table or even in the form $a_i a_j = \sum_k p_{i,j,k} a_k$, but in the form $a_i a_j - \sum_k p_{i,j,k} a_k$.

10 Concluding Remarks

There are further computations available at the author's home page: http://www.dms.auburn.edu/~leonada. Interpretation of the results obtained is of independent interest. Such interpretation for 4-gons and Steiner designs is contained in [6]. But as a striking example of the significance of this interpretation step, note that 3-gons with $s = t = 4$, [4] correspond to a sporadic strongly regular graph with $v = 105$, $k = 32$, $\lambda = 4$, $\mu = 12$. This graph is a subgraph of the famous sporadic McLaughlin graph on 275 vertices.

Note also that even though a scheme itself may not exist, some fusion of the table algebra may have a combinatorial interpretation.

Acknowledgments

It is important to recognize RICAM (Austrian Academy of Sciences), RISC (Johannes Kepler University), and Bruno Buchberger personally for the opportunities provided by the D1 Workshop during the Special Semester on Gröbner bases at Linz, Austria in May, 2006, without which this paper would not have been written. The desirability of such a tutorial paper on the use of Gröbner bases in algebraic combinatorics was suggested by Mikhail Klin, who is also responsible for the underlying topics and much of the historical combinatorial context as well.

Appendix A

```
//common code to search for fusion
R<T,S>:=PolynomialRing(Q,2);
co:=function(f,mon) return MonomialCoefficient(f,mon);
end function;
fusion:=function(J) return &+[A.(N+1-j): j in J];
end function;
prod:=function(I,J) return NormalForm(fusion(I)*fusion(J),ID);
end function;
relations:=function(I,J,K)
   W:=prod(I,J);
   return [(co(W,A.(N+1-K[1]))-co(W,A.(N+1-k)))
   @hom<FF->R|T,S>: k in K];
end function;
RELATIONS:=function(part)
   rel:=[];
   for i in [1..#part] do for j in [1..#part]
   do for k in [1..#part] do
      rel:=rel cat relations(part[i],part[j],part[k]);
   end for; end for; end for;
   return rel;
end function;
max:=function(PV,j)
   if j gt 1 then
      return Maximum({PV[i]: i in [1..j-1]});
   else
      return -1;
   end if;
```

```
end function;
partno:=0; goodpartno:=0; vector:=[1:i in [1..N]]; v1:=N;
while v1 gt 1 do
   Bound:=max(vector,N+1);
   B:=[[]: i in [1..Bound]];
   for i in [1..N] do Append(~B[vector[i]],i); end for;
   symmetric:=true;
   for i in [1..Bound] do
      if #{vector[AT[j]]: j in B[i]} ne 1
      then symmetric:=false; break; end if;
   end for;
   if symmetric then
      partno+:=1;
      id:=ideal<R|RELATIONS(B)>;
      gb:=GroebnerBasis(id);
      pos_sol:=true;
      if #gb ne 0 then
         for i in [1..#gb] do
            m:=Minimum(Coefficients(gb[i]));
            if m gt 0 then pos_sol:=false; break; end if;
         end for;
      end if;
      if pos_sol then
         goodpartno+:=1;
         partno, "partition" cat IntegerToString(goodpartno)
         cat "=",B;
         if #gb ne 0 then
            for i in [1..#gb] do Factorization(gb[i]);
            end for;
         end if;
      end if;
   end if;
   v2:=v1;
   bound:=1+max(vector,v1);
   if vector[v1] ge bound then
      while vector[v1] ge bound and v1 gt 1 do
         vector[v1]:=1;
         v1-:=1;
         v2:=N;
         if v1 gt 1 then bound:=1+max(vector,v1); end if;
      end while;
      if v1 gt 1 then
         vector[v1]+:=1;
         v1:=v2;
      end if;
```

```
   else
      if v1 gt 1 then
         vector[v1]+:=1;
      end if;
   end if;
end while;
//preamble code for generalized n-gons
l:=2;//the only parameter that needs to be changed from
//one run to the next n-gons for n=2l, as rank N+1
//association schemes of flags
N:=4*l-1;
AT:=[i : i in [1..N]];
for i in [1..l-1] do AT[4*i]:=4*i-1; AT[4*i-1]:=4*i; end for;
Q:=RationalField();
FF<t,s>:=FunctionField(Q,2);
A:=FreeAlgebra(FF,N);
AssignNames(~A,["a" cat IntegerToString(N+1-i): i in [1..N]]);
a1:=A.N;a2:=A.(N-1);
rel1:=a1^2-s-(s-1)*a1;
rel2:=a2^2-t-(t-1)*a2;
rel3:=(a2*a1)^l-(a1*a2)^l;
def1:=[(a1*a2)^i-A.(N-4*i+2): i in [1..l]];
def2:=[(a2*a1)^i-A.(N-4*i+1): i in [1..l-1]];
def3:=[a1*(a2*a1)^i-A.(N-4*i): i in [1..l-1]];
def4:=[a2*(a1*a2)^i-A.(N-4*i-1): i in [1..l-1]];
ID:=ideal<A|rel1,rel2,rel3,def1,def2,def3,def4>;
G:=GroebnerBasis(ID);G;#G;
//Gr\"obner basis for the association scheme of a generalized
//quadrangle
a7^2 -(t^2s^2-t^2s+t^2-ts^2-t+s^2-s+1)*a7
     -(t^2s^2-t^2s-ts^2+ts+s^2-s)*a6
     -(t^2s^2-t^2s+t^2-ts^2+ts-t)*a5-(t^2s^2-t^2s-ts^2+ts)*a4
     -(t^2s^2-t^2s-ts^2+ts)*a3-(t^2s^2-ts^2)*a2
     -(t^2s^2-t^2s)*a1-(t^2s^2)*a0,
a7*a6-(t^2s-t^2-ts+t+s-1)*a7-(t^2s-2ts+s)*a6
     -(t^2s-t^2-ts+t)*a5-(t^2s-ts)*a4
     -(t^2s-ts)*a3-(t^2s)*a1,
a7*a5-(ts^2-ts+t-s^2+s-1)*a7-(ts^2-ts-s^2+s)*a6
     -(ts^2-2ts+t)*a5-(ts^2-ts)*a4
     -(ts^2-ts)*a3-(ts^2)*a2,
a7*a4-(ts-t-s+1)*a7-(ts-s)*a6-(ts-t)*a5-(ts)*a3,
a7*a3-(ts-t-s+1)*a7-(ts-s)*a6-(ts-t)*a5-(ts)*a4,
a7*a2-(t-1)*a7-(t)*a5,
a7*a1-(s-1)*a7-(s)*a6,
a6*a7-(t^2s-t^2-ts+t+s-1)*a7-(t^2s-2ts+s)*a6
```

```
     -(t^2s-t^2-ts+t)*a5
     -(t^2s-ts)*a4-(t^2s-ts)*a3-(t^2s)*a1,
a6^2 -(t^2-2t+1)*a7-(ts-t)*a6-(t^2-t)*a5-(t^2s-ts)*a2
     -(t^2s)*a0,
a6*a5-(ts-t-s+1)*a7-(ts-s)*a6-(ts-t)*a5-(ts)*a3,
a6*a4-(t-1)*a7-(ts-t)*a4-(ts)*a2,
a6*a3-(t-1)*a7-(t)*a5,
a6*a2-(t-1)*a6-(t)*a4,
a6*a1-(1)*a7,
a5*a7-(ts^2-ts+t-s^2+s-1)*a7-(ts^2-ts-s^2+s)*a6
     -(ts^2-2ts+t)*a5-(ts^2-ts)*a4-(ts^2-ts)*a3-(ts^2)*a2,
a5*a6-(ts-t-s+1)*a7-(ts-s)*a6-(ts-t)*a5-(ts)*a4,
a5^2 -(s^2-2s+1)*a7-(s^2-s)*a6-(ts-s)*a5-(ts^2-ts)*a1-(ts^2),
a5*a4-(s-1)*a7-(s)*a6,
a5*a3-(s-1)*a7-(ts-s)*a3-(ts)*a1,
a5*a2-(1)*a7,
a5*a1-(s-1)*a5-(s)*a3,
a4*a7-(ts-t-s+1)*a7-(ts-s)*a6-(ts-t)*a5-(ts)*a3,
a4*a6-(t-1)*a7-(t)*a5,
a4*a5-(s-1)*a7-(ts-s)*a4-(ts)*a1,
a4^2 -(1)*a7,
a4*a3-(s-1)*a6-(ts-s)*a2-(ts)*a0,
a4*a2-(1)*a6,
a4*a1-(s-1)*a4-(s)*a2,
a3*a7-(ts-t-s+1)*a7-(ts-s)*a6-(ts-t)*a5-(ts)*a4,
a3*a6-(t-1)*a7-(ts-t)*a3-(ts)*a2,
a3*a5-(s-1)*a7-(s)*a6,
a3*a4-(t-1)*a5-(ts-t)*a1-(ts)*a0,
a3^2 -(1)*a7,
a3*a2-(t-1)*a3-(t)*a1,
a3*a1-(1)*a5,
a2*a7-(t-1)*a7-(t)*a5,
a2*a6-(t-1)*a6-(t)*a3,
a2*a5-(1)*a7,
a2*a4-(t-1)*a4-(t)*a1,
a2*a3-(1)*a6,
a2^2 -(t-1)*a2-(t)*a0,
a2*a1-(1)*a4,
a1*a7-(s-1)*a7-(s)*a6,
a1*a6-(1)*a7,
a1*a5-(s-1)*a5-(s)*a4,
a1*a4-(1)*a5,
a1*a3-(s-1)*a3-(s)*a2,
a1*a2-(1)*a3,
a1^2 -(s-1)*a1-(s)*a0
```

```
//edited fusion partitions for generalized 4-gons,
//eliminating those with only t=1=s
//with the Gr\"obner basis for each to describe those pairs
//(t,s) for which fusion occurs
partition1=[[1,2,3,4,5,6,7]]
[]
partition3=[[1,2,3,4,5,7],[6]]
[T-1]
partition4=[[1,2,3,4,6,7],[5]]
[S-1]
partition5=[[1,2,3,4,7],[5,6]]
[T-2,S-2]
partition10=[[1,2,7],[3,4],[5,6]]
[T-2,S-2]
partition11=[[1,2],[3,4],[5,6],[7]]
[T-S]
partition13=[[1,3,4,5,6,7],[2]]
[]
partition14=[[1,3,4,6],[2,5,7]]
[S-1]
partition15=[[1,3,4,6],[2,5],[7]]
[S-1]
partition17=[[1,3,4,6],[2,7],[5]]
[S-1]
partition18=[[1,3,4,6],[2],[5,7]]
[]
partition19=[[1,3,4,6],[2],[5],[7]]
[S-1]
partition20=[[1,6,7],[2,3,4,5]]
[T-1]
partition21=[[1,6],[2,3,4,5],[7]]
[T-1]
partition22=[[1],[2,3,4,5,6,7]]
[]
partition23=[[1,7],[2,3,4,5],[6]]
[T-1]
partition24=[[1],[2,3,4,5],[6,7]]
[]
partition25=[[1],[2,3,4,5],[6],[7]]
[T-1]
partition27=[[1,5,6],[2],[3,4,7]]
[T-1,S^2-4*S+3]=[T-1,(S-3)(S-1)]
partition29=[[1,6,7],[2,5],[3,4]]
[T-1,S-3]
partition30=[[1,6],[2,5,7],[3,4]]
```

```
[T-3,S-1]
partition32=[[1],[2,5,6],[3,4,7]]
[T^2-4*T+3,S-1]=[(T-3)(T-1),S-1]
partition33=[[1],[2,5,7],[3,4,6]]
[S-1]
partition34=[[1],[2,5],[3,4],[6],[7]]
[S-1]
partition35=[[1,6,7],[2],[3,4,5]]
[T-1]
partition36=[[1,6],[2],[3,4],[5],[7]]
[T-1]
partition38=[[1],[2],[3],[4],[5],[6],[7]]
[]

//edited fusion partitions for generalized 6-gons,
//eliminating those with only t=1=s
//with the Gr\"obner basis for each to describe those pairs
//(t,s) for which fusion occurs
partition1=[[1,2,3,4,5,6,7,8,9,10,11]]
[]
partition7=[[1,2,3,4,5,7,8,9,10,11],[6]]
[T-1]
partition8=[[1,2,3,4,6,7,8,9,10,11],[5]]
[S-1]
partition26=[[1,2],[3,4],[5,6],[7,8],[9,10],[11]]
[T-S]
partition30=[[1,3,4,5,6,7,8,9,10,11],[2]]
[]
partition34=[[1,3,4,5,6,7,8,10],[2],[9,11]]
[T-S]
partition37=[[1,3,4,6,9,11],[2,5,7,8,10]]
[S-1]
partition41=[[1,3,4,6,9,11],[2,5,10],[7,8 ]]
[T^2-5*T+4,S-1]=[(T-1)(T-4),S-1]
partition42=[[1,3,4,6,9,11],[2,5],[7,8],[10]]
[S-1]
partition45=[[1,3,4,6,9,11],[2,7,8,10],[5]]
[S-1]
partition46=[[1,3,4,6,9,11],[2,7,8],[5,10]]
[T-2,S-1]
partition47=[[1,3,4,6,9,11],[2,10],[5,7,8]]
[T-2,S-1]
partition48=[[1,3,4,6,9,11],[2],[5,7,8,10]]
[T^2*S-T^2-T*S^2+3*T*S-2*T+S-1]=[(S-1)(T^2-TS+2T+1)] so S=1
or (T=1,S=4)
```

```
partition49=[[1,3,4,6],[2],[5,7,8,10],[9,11]]
[]
partition54=[[1,3,4,6,9,11],[2],[5],[7],[8],[10]]
[S-1]
partition55=[[1,3,4,9],[2,7,8,10],[5],[6,11]]
[S-1]
partition58=[[1,6,7,8,9],[2,3,4,5,10,11]]
[T-1]
partition62=[[1,6,9],[2,3,4,5,10,11],[7,8]]
[T-1,S^2-5*S+4]=[T-1,(S-4)(S-1)]
partition63=[[1,6],[2,3,4,5,10,11],[7,8],[9]]
[T-1]
partition67=[[1],[2,3,4,5,6,7,8,9,10,11]]
[]
partition69= [[1],[2,3,4,5,6,7,8,9],[10,11]]
[T-S]
partition72=[[1,7,8,9],[2,3,4,5,10,11],[6]]
[T-1]
partition73=[[1,7,8],[2,3,4,5,10,11],[6,9]]
[T-1,S-2]
partition74=[[1,9],[2,3,4,5,10,11],[6,7,8]]
[T-1,S-2]
partition75=[[1],[2,3,4,5,10,11],[6,7,8,9]]
[T^2*S-T*S^2-3*T*S-T+S^2+2*S+1]=[(T-1)(TS-S^2-2S-1)] so T=1
or (S=1,T=4)
partition76=[[1],[2,3,4,5],[6,7,8,9],[10,11]]
[]
partition81=[[1],[2,3,4,5,10,11],[6],[7],[8],[9]]
[T-1]
partition83=[[1,7,8,9],[2,3,4,10],[5,11],[6]]
[T-1]
partition100=[[1],[2,5,7,8,10],[3,4,6,9,11]]
[S-1]
partition105=[[1],[2,5,10],[3,4,11],[6,9],[7,8]]
[T^2-5*T+4,S-1]=[(T-4)(T-1),S-1]
partition106=[[1],[2,5],[3,4],[6,9],[7,8],[10],[11]]
[S-1]
partition108=[[1,6,7,8,9],[2],[3,4,5,10,11]]
[T-1]
partition113=[[1,6,9],[2],[3,4,11],[5,10],[7,8]]
[T-1,S^2-5*S+4]=[T-1,(S-4)(S-1)]
partition114=[[1,6],[2],[3,4],[5,10],[7,8],[9],[11]]
[T-1]
partition124=[[1],[2],[3],[4],[5],[6],[7],[8],[9],[10],[11]]
[]
```

```
//edited fusion partitions for generalized 6-gons,
//eliminating those with only t=1=s
//with the Gr\"obner basis for each to describe those pairs
//(t,s) for which fusion occurs
partition1=[[1,2,3,4,5,6,7,8,9,10,11,12,13,14,15]]
[]
partition10=[[1,2,3,4,5,7,8,9,10,11,12,13,14,15],[6]]
[T-1]
partition16=[[1,2,3,4,6,7,8,9,10,11,12,13,14,15],[5]]
[S-1]
partition40=[[1,2],[3,4],[5,6],[7,8],[9,10],[11,12],[13,14],
[15]]
[T-S]
partition49=[[1,3,4,5,6,7,8,9,10,11,12,13,14,15],[2]]
[]
partition53=[[1,3,4,5,6,7,8,9,10,11,12,14],[2],[13,15]]
[T-S]
partition55=[[1,3,4,5,6,7,8,10,13,15],[2],[9,11,12,14]]
[T-1,S-2]
partition56=[[1,3,4,6,9,11,12,14],[2,5,7,8,10,13,15]]
[S-1]
partition61=[[1,3,4,6,9,11,12,14],[2,5,7,8,15],[10,13]]
[T-2,S-1]
partition67=[[1,3,4,6,9,11,12,14],[2,5,15],[7,8],[10,13]]
[T-2,S-1]
partition68=[[1,3,4,6,9,11,12,14],[2,5],[7,8],[10,13],
[15]]
[S-1]
partition73=[[1,3,4,6,9,11,12,14],[2,7,8,10,13,15],[5]]
[S-1]
partition78=[[1,3,4,6,9,11,12,14],[2,7,8,13],[5],[10,15]]
[S-1]
partition82=[[1,3,4,6,9,11,12,14],[2],[5,7,8,10,13,15]]
[S-1]
partition84=[[1,3,4,6,9,11,12,14],[2],[5,7,8,10],[13,15]]
[S-1]
partition87=[[1,3,4,6,13,15],[2],[5,7,8,10],[9,11,12,14]]
[T-1,S-2]
partition88=[[1,3,4,6],[2],[5,7,8,10],[9,11,12,14],[13,15]]
[]
partition100=[[1,3,4,6,9,11,12,14],[2],[5],[7],[8],[10],[13],[15]]
[S-1]
partition101=[[1,3,4,9],[2,7,8,13],[5],[6,11,12,14],[10,15]]
[S-1]
partition104=[[1,6,7,8,9,14,15],[2,3,4,5,10,11,12,13]]
```

```
[T-1]
partition110=[[1,6,7,8,15],[2,3,4,5,10,11,12,13],[9,14]]
[T-1,S-2]
partition115=[[1,6,15],[2,3,4,5,10,11,12,13],[7,8],[9,14]]
[T-1,S-2]
partition116=[[1,6],[2,3,4,5,10,11,12,13],[7,8],[9,14],
[15]]
[T-1]
partition120=[[1],[2,3,4,5,6,7,8,9,10,11,12,13,14,15]]
[]
partition122=[[1],[2,3,4,5,6,7,8,9,10,11,12,13],[14,15]]
[T-S]
partition124=[[1],[2,3,4,5,6,7,8,9,14,15],[10,11,12,13]]
[T-2,S-1]
partition125=[[1,7,8,9,14,15],[2,3,4,5,10,11,12,13],[6]]
[T-1]
partition130=[[1,7,8,14],[2,3,4,5,10,11,12,13],[6],[9,15]]
[T-1]
partition134=[[1],[2,3,4,5,10,11,12,13],[6,7,8,9,14,15]]
[T-1]
partition136=[[1],[2,3,4,5,10,11,12,13],[6,7,8,9],[14,15]]
[T-1]
partition138=[[1],[2,3,4,5,14,15],[6,7,8,9],[10,11,12,13]]
[T-2,S-1]
partition139=[[1],[2,3,4,5],[6,7,8,9],[10,11,12,13],[14,15]]
[]
partition152=[[1],[2,3,4,5,10,11,12,13],[6],[7],[8],[9],
[14],[15]]
[T-1]
partition156=[[1,7,8,14],[2,3,4,10],[5,11,12,13],[6],[9,15]]
[T-1]
partition183=[[1],[2,5,7,8,10,13,15],[3,4,6,9,11,12,14]]
[S-1]
partition185=[[1],[2,5,7,8,15],[3,4,6,9,14],[10,13],[11,12]]
[T-2,S-1]
partition197=[[1],[2,5,15],[3,4,14],[6,9],[7,8],[10,13],
[11,12]]
[T-2,S-1]
partition198=[[1],[2,5],[3,4],[6,9],[7,8],[10,13],[11,12],
[14],[15]]
[S-1]
partition203=[[1,6,7,8,9,14,15],[2],[3,4,5,10,11,12,13]]
[T-1]
partition205=[[1,6,7,8,15],[2],[3,4,5,10,13],[9,14],[11,12]]
[T-1,S-2]
```

```
partition221=[[1,6,15],[2],[3,4,13],[5,10],[7,8],[9,14],
[11,12]]
[T-1,S-2]
partition222=[[1,6],[2],[3,4],[5,10],[7,8],[9,14],[11,12],
[13],[15]]
[T-1]
partition259=[[1],[2],[3],[4],[5],[6],[7],[8],[9],[10],[11],
[12],[13],[14],[15]]
[]

//preamble code for 2-(v,k,1) designs
N:=6;
AT:=[1,2,4,3,5,6];
Q:=RationalField();
FF<t,s>:=FunctionField(Q,2);
A:=FreeAlgebra(FF,N);
AssignNames(~A,["a" cat IntegerToString(N+1-i): i in
[1..N]]);
a1:=A.6;
a2:=A.5;
a3:=A.4;
a4:=A.3;
a5:=A.2;
a6:=A.1;
rel1:=a1^2-s-(s-1)*a1;rel1;
rel2:=a2^2-t-(t-1)*a2;rel2;
rel3:=(a2*a1)^2-s*a6-(s-1)*a5-s*a3;rel3;
rel4:=(a1*a2)^2-s*a6-(s-1)*a5-s*a4;rel4;
def1:=a1*a2-A.4;def1;
def2:=a2*a1-A.3;def2;
def3:=a1*a2*a1-A.2;def3;
def4:=(a2*a1*a2)-(a1*a2*a1)-A.1;def4;
ID:=ideal<A|rel1,rel2,rel3,rel4,def1,def2,def3,def4>;
G:=GroebnerBasis(ID);G;#G;
//Gr\"obner basis for the association scheme of a 2-(v,k,1)
//design with parameters (t,s)
a6^2 -(t^2s-2ts^2-2ts-t+s^3+3s^2+2s)*a6
     -(t^2s-2ts^2-ts+s^3+s^2)*a5-(t^2s-2ts^2-ts+s^3+s^2)*a4
     -(t^2s-2ts^2-ts+s^3+s^2)*a3-(t^2s-2ts^2-ts+s^3+s^2)*a2
     -(t^2s-ts^2)*a1-(t^2s-ts^2)*a0,
a6*a5-(ts^2-s^3-s^2)*a6-(ts^2-ts-s^3+s^2)*a5
     -(ts^2-ts-s^3+s^2)*a4
     -(ts^2-s^3)*a3-(ts^2-s^3)*a2,
a6*a4-(ts-s^2-s)*a6-(ts-t-s^2+s)*a5-(ts-t-s^2+s)*a4
     -(ts-s^2)*a3-(ts-s^2)*a2,
```

```
a6*a3-(ts-s^2-s)*a6-(ts-s^2)*a5-(ts-s^2)*a4,
a6*a2-(t-s-1)*a6-(t-s)*a5-(t-s)*a4,
a6*a1-(s)*a6,
a5*a6-(ts^2-s^3-s^2)*a6-(ts^2-ts-s^3+s^2)*a5-(ts^2-s^3)*a4
     -(ts^2-ts-s^3+s^2)*a3
     -(ts^2-s^3)*a2,
a5^2 -(s^3-s^2)*a6-(ts+s^3-3s^2+2s-1)*a5-(s^3-2s^2+s)*a4
     -(s^3-2s^2+s)*a3
     -(s^3-s^2)*a2-(ts^2-ts)*a1-(ts^2)*a0,
a5*a4-(s^2)*a6-(s^2-2s+1)*a5-(s^2-s)*a4-(s^2-s)*a3-(s^2)*a2,
a5*a3-(s^2-s)*a6-(s^2-2s+1)*a5-(s^2-s)*a4-(ts- s)*a3-(ts)*a1,
a5*a2-(s)*a6-(s-1)*a5-(s)*a4,
a5*a1-(s-1)*a5-(s)*a3,
a4*a6-(ts-s^2-s)*a6-(ts-s^2)*a5-(ts-s^2)*a3,
a4*a5-(s^2-s)*a6-(s^2-2s+1)*a5-(ts-s)*a4-(s^2-s)*a3-(ts)*a1,
a4^2 -(s)*a6-(s-1)*a5-(s)*a3,
a4*a3-(s-1)*a6-(s-1)*a5-(ts-s)*a2-(ts)*a0,
a4*a2-(1)*a6-(1)*a5,
a4*a1-(s-1)*a4-(s)*a2,
a3*a6-(ts-s^2-s)*a6-(ts-t-s^2+s)*a5-(ts-s^2)*a4-(ts-t-s^2+s)
     *a3-(ts-s^2)*a2,
a3*a5-(s^2)*a6-(s^2-2s+1)*a5-(s^2-s)*a4-(s^2-s)*a3-(s^2)*a2,
a3*a4-(t-1)*a5-(ts-t)*a1-(ts)*a0,
a3^2 -(s)*a6-(s-1)*a5-(s)*a4,
a3*a2-(t-1)*a3-(t)*a1,
a3*a1-(1)*a5,
a2*a6-(t-s-1)*a6-(t-s)*a5-(t-s)*a3,
a2*a5-(s)*a6-(s-1)*a5-(s)*a3,
a2*a4-(t-1)*a4-(t)*a1,
a2*a3-(1)*a6-(1)*a5,
a2^2 -(t-1)*a2-(t)*a0,
a2*a1-(1)*a4,
a1*a6-(s)*a6,
a1*a5-(s-1)*a5-(s)*a4,
a1*a4-(1)*a5,
a1*a3-(s-1)*a3-(s)*a2,
a1*a2-(1)*a3,
a1^2 -(s-1)*a1-(s)*a0
//fusion partitions for 2-(v,k,1) designs (with t>s),
//except for those partitions with only t=1=s
partition1=[[1,2,3,4,5,6]]
[]
%partition2=[[1,2,3,4,5],[6]]
%[T-S]
partition3=[[1,2,3,4,6],[5]]
```

```
[S-1]
%partition4=[[1,2,3,4],[5,6]]
%[T-1]
%partition6=[[1,2,5],[3,4],[6]]
%[T-S,S^2-5*S+4]=[T-1,(S-4)(S-1)]
%partition7=[[1,2],[3,4],[5],[6]]
%[T-S]
partition9=[[1,3,4,5,6],[2]]
[]
%partition11=[[1,3,4,5],[2],[6]]
%[T-S]
%partition12= [[1,3,4],[2,5],[6]]
%[T-2,S-2]
partition13=[[1,5],[2,3,4,6]]
[S-2]
%partition14=[[1,5],[2,3,4],[6]]
%[T-2,S-2]
partition15=[[1,6],[2,3,4,5]]
[T-S-1]
partition16=[[1],[2,3,4,5,6]]
[]
partition17=[[1],[2,3,4,5],[6]]
[]
partition18=[[1,6],[2,3,4],[5]]
[T-2,S-1]
partition20=[[1,6],[2,5],[3,4]]
[T-4,S-3]
partition21=[[1],[2,5],[3,4],[6]]
[S-1]
partition22=[[1,6],[2],[3,4,5]]
[T-S-1]
partition23= [[1],[2],[3],[4],[5],[6]]
[]
```

References

1. Z. Arad and M. E. Muzychuk (eds.) *Standard Integral Table Algebras Generated by a Non-real Element of Small Degree.* Lecture Notes in Math., Vol. 1773, Springer, Berlin, 2002.
2. B. Buchberger, Gröbner bases, an algorithmic method in polynomial ideal theory, in N. K. Bose (ed.) *Multidimensional System Theory*, pp. 184–232, Reidel, Dordrecht, 1985.

3. D. Cox, J. Little, and D. O'Shea, *Ideals, Varieties, and Algorithms, An Introduction to Computational Algebraic Geometry and Commutative Algebra*, 3rd edn. Springer, Berlin, 2007.
4. I. A. Faradžev, M. H. Klin, and M. E. Muzichuk, Cellular rings and groups of automorphisms of graphs, in *Investigations in Algebraic Theory of Combinatorial Objects*. Math. Appl. (Soviet Ser.), Vol. 84, pp. 1–152, Kluwer Academic, Dordrecht, 1994.
5. R. Kilmoyer and L. Solomon, On the theorem of Feit–Higman, *J. Comb. Theory, Ser. A*, **15** (1973), 310–322.
6. M. Klin, C. Pech, and P.-H. Zieschang, Flag Algebras of Block Designs, I. Initial Notions. Steiner 2-designs and Generalized Quadrangles, *Math-AL-10-1998*, Technische Universität, Dresden, November 1998
7. R. A. Liebler, Tactical configurations and their generic ring, *Eur. J. Comb.*, **9** (1988), 581–592.
8. The MAGMA computational algebra system for algebra, number theory and geometry, The University of Sydney Computational Algebra Group, http://magma.maths.usyd.edu.au/magma.
9. E. Martinez-Moro, Computations on character tables of association schemes, in *Computer Algebra in Scientific Computing*, CASC'99, Munich, pp. 293–307, Springer, Berlin, 1999.
10. U. Ott, Some remarks on representation theory in finite geometries, in *Geometries and Groups*. Lecture Notes in Math., Vol. 893, pp. 68–110, Springer, Berlin, 1981.
11. K. W. Smith, Flag algebras of a symmetric design, *J. Comb. Theory, Ser. A*, **48** (1988), 209–228.
12. J. Tits, Sur la trialité et certains groupes qui s'en déduisent, *Inst. Hautes Etudes Sci. Publ. Math.*, **2** (1959), 14–60.
13. P. -H. Zieschang, On dihedral configurations and their Coxeter geometries, *Eur. J. Comb.*, **18** (1997), 341–354.

Enumerating Set Orbits

Christian Pech[1*] and Sven Reichard[2**]

[1] Department of Mathematics, Ben-Gurion University of the Negev, Beer Sheva, Israel. pech@cs.bgu.ac.il
[2] University of Western Australia, Crawley 6009, Western Australia. reichard@maths.uwa.edu.au

Summary. We describe a practically efficient canonicity test for orbit representatives of permutation group acting on sets. This allows us to perform a wide range of combinatorial searches. We describe an implementation of the algorithm in GAP. We give a few applications, one of which answers a question by De Wispelaere regarding the classification of all two-ovoids in the classical generalized hexagon of order 4.

Key words: Set-orbits, Orderly generation, Computer algebra package, Spreads, Incidence structures, Fusions of coherent configuration, Steiner triple systems, Generalized quadrangles, Generalized hexagons, Schur rings

1 Introduction

Many combinatorial problems can be stated in the framework of *subset problems*: Given a finite set Ω find all subsets $S \subseteq \Omega$ satisfying a given property. We will call these subsets *solutions*. Some examples:

1. Independent sets of a graph Γ: Here, Ω is the set of all vertices of Γ. Solutions are subsets S of Ω such that no two elements of S are adjacent.
2. Proper graph colorings: Here, Ω consists of all independent sets. Solutions are collections of sets which partition the set of vertices.
3. Arcs in projective planes: Ω is the set of points; an arc is a set of points such that no three are collinear.

Often the problem exhibits some degree of symmetry. This can be described as a group G acting on Ω leaving the set of solutions of the problem invariant. In the examples above, the automorphism groups of the given

* C. Pech was supported by the Skirball postdoctoral fellowship of the Center of Advanced Studies in Mathematics at the Mathematics Department of Ben-Gurion University.
** S. Reichard was funded by ARC Discovery Projects Grant No. 35400300.

M. Klin et al. (eds.), Algorithmic Algebraic Combinatorics and Gröbner Bases, DOI 10.1007/978-3-642-01960-9_4, © Springer-Verlag Berlin Heidelberg 2009

graphs or projective planes serve as such groups of symmetries. The action of G on Ω induces an action of G on the solutions. Usually, one is interested in *non-isomorphic* solutions, i.e., in solutions from different orbits under this action.

This raises the question of how to enumerate orbits of G acting on subsets of Ω. The most straightforward method is to store all subsets of Ω, and compute all orbits as sets of subsets. This can be improved a bit by noting that all sets in the same orbit have the same number of elements; thus we can generate the orbits of sets of each size independently. This approach is used in Computer Algebra Systems like GAP [9]. The drawback is the memory requirement, which is exponential in the size of Ω, restricting $|\Omega|$ to currently around 30.

Another approach is to search for representatives of set orbits incrementally, adding one element at a time. Here we note that once we encounter two equivalent sets S_1, S_2 we may immediately discard one of them, since any solution containing S_2 will be equivalent to a solution containing S_1.

A depth-first approach was first described in by Zaichenko [21], based on the general idea of "orderly generation" [13]. This has the advantage that only a single set has to be kept in memory at any given time, making it very memory efficient, at the cost of some computational efficiency. It relies on the concept of canonical orbit representatives and their algorithmical recognition.

A general theoretical description of similar algorithms was given by Laue [11]; a breadth-first implementation was used to construct t-designs by Schmalz [16].

In this article we describe another depth-first approach which uses techniques of dynamic programming for an efficient implementation. In Sect. 2 we give some combinatorial and algebraic preliminaries. In Sect. 3 we formulate the general problem. In Sect. 4 we give a straightforward implementation of a canonicity test; improvements that make it efficient in practice are described in Sect. 5. In Sect. 6 we describe how to use this algorithm to enumerate set orbits. In Sect. 7 we give some details of an implementation in GAP. Finally, in Sect. 8, we give some applications to combinatorial problems.

2 Preliminaries

2.1 Finite Permutation Groups

Let Ω be a finite set. A bijection $g : \Omega \to \Omega$ is called a *permutation* of Ω. We use an exponential notation for permutations, so for $x \in \Omega$ and a permutation g we denote the image of x under g by x^g.

The composition of two permutations g, h of Ω is again a permutation, denoted by gh. We follow the convention of writing compositions left-to-right, such that for $x \in \Omega$, $x^{gh} = (x^g)^h$.

Given a permutation g its inverse mapping g^{-1} is again a permutation. For any g we have that $gg^{-1} = g^{-1}g = e$, where e denotes the identity mapping.

The set of all permutations of Ω forms a group under these operations, the symmetric group $S(\Omega)$. Any subgroup G of $S(\Omega)$ is a *permutation group* acting on Ω. We call the elements of Ω *points*, and the number of points the *degree* of G.

Given a set $\{g_1, \ldots, g_k\}$ of permutations in $S(\Omega)$ we may consider the group *generated* by the permutations, i.e., the smallest subgroup of $S(\Omega)$ containing all of the given permutations. We denote this group by $\langle g_1, \ldots, g_k \rangle$.

Two points $x, y \in \Omega$ are said to be *equivalent* under a permutation group G if there is an element $g \in G$ which maps x to y. Since G is a group this is in fact an equivalence relation. The equivalence classes are the *orbits* of G; we denote the orbit containing x as x^G.

Given a point $x \in \Omega$, let G_x be the set of all elements g of G mapping x to itself. G_x is a subgroup of G; we call it the *stabilizer* of x under G. Similarly, given several points $x_1, \ldots, x_k$, we define the *pointwise stabilizer* $G_{x_1,\ldots,x_k}$ to consist of those elements fixing each x_i.

Given a set $S = \{x_1, \ldots, x_k\} \subseteq \Omega$, we denote by G_S the set of all elements of G mapping S to itself, thus $x_i^g \in S$ for all i. G_S is a group, the *setwise stabilizer* or simply the stabilizer of S. It contains the pointwise stabilizer as a subgroup.

A sequence $x_1, \ldots, x_k$ is a *base* of G if its pointwise stabilizer is trivial. Given a base we define a chain of subgroups $G^{(i)}$, $i = 0 \ldots, k$, by $G^{(0)} = G$, $G^{(i)} = G^{(i-1)}_{x_i}$, such that $G^{(i)}$ is the pointwise stabilizer of the first i points. We call this the *stabilizer chain* for the given base. A set of generators of G which also contains generators for all the $G^{(i)}$ is known as a *strong generating set* (relative to the given base).

Stabilizer chains can be computed and are a basic tool for computations in permutation groups. For details, see [17, 2].

2.2 Combinatorial Search

Many problems in Combinatorics can be formulated in the following manner: Given a set of basic "components", construct from them larger structures which satisfy certain properties. In many cases the larger structure is simply formed by taking an appropriate subset of the given components, however at times the construction is more involved.

As an example, consider the construction of a projective plane of a given order n. Here, we are given a point set P, and we have to find a set B of lines, i.e., point sets of size $n + 1$, such that each pair of distinct points appears together in exactly one line. Here, we can consider all subsets of size $n + 1$ as given (further on called "lines"), and we need to select several of those that will fulfill the axioms of projective planes.

The naive way would be to look at all possible subsets and check whether the wanted properties are satisfied. However, this is feasible only in the smallest cases since the number of subsets grows exponentially with the number of points.

The standard way to slow down this "combinatorial explosion" has been described under various names such as "backtrack search" and "branch and bound". The idea is to construct solutions incrementally, checking after each step whether it is possible in theory to extend the current, *partial solution* to a complete solution. In the example of the projective plane, assume that in the beginning we pick two lines that intersect in two points. It is clear that no solution can contain these two lines simultaneously, so we may immediately discard this partial solution and search elsewhere.

More formally, we start with a search space (a set) of candidate solutions on which is defined a predicate SOL such that SOL(x) is true if and only if x is a solution. Moreover, we define a predicate PSOL for which we require that for $y \subseteq x$, SOL(x) implies PSOL(y). Thus if we find that for a partial solution y, PSOL(y) does not hold we may discard all candidate solutions containing y. Candidate solutions that fulfill PSOL are called partial solutions.

Returning to the projective planes we may require that for a partial solution y any two lines intersect in a single point. This will reduce the search space significantly.

It is always possible to find such a *partial predicate* PSOL, by setting PSOL(y) to true for all y. In practice there is a trade-off between the cost of computing PSOL and its power to discard large parts of the search space.

3 Search and Symmetry

Another important way to speed up a combinatorial search is by using symmetry. If the original configuration of given components is symmetrical this means that the set of solutions reflects that symmetry. Usually one is interested only in solutions that are not equivalent under this symmetry.

More formally, suppose that there is a group G acting on the given set which leaves the predicate SOL invariant, thus it maps solutions to solutions. We would like to find inequivalent solutions, i.e., representatives of the G-orbits acting on all solutions.

If we have an *isomorphism test* for solutions, i.e., a test whether two solutions are equivalent, then we can construct all solutions and discard those that are isomorphic to those found earlier. This approach has two drawbacks. Firstly we need to store all solutions (or at least all representatives). Secondly we still need to search equivalent portions of the search tree multiple times.

We can get around the second problem by not only storing the full solutions, but also the partial solutions, and compare each partial solution to everything previously encountered. If we find that the current partial solution is already known, then we can again discard the remaining part of the search tree. However, this requires even more storage, and a quadratically growing number of isomorphism tests.

A way around both problems is the use of *canonical representatives*. Suppose we have another predicate CSOL that is true for exactly one element of each orbit of solutions; we call this particular element canonical. Then we may reject a solution if it is not canonical, without the need to compare it to other solutions.

Extending this idea to partial solutions requires some care in order not to lose any solution. In particular we need to make sure that for each canonical solution c there is a chain of canonical partial solutions leading up to c. If we can achieve this then we only have to consider canonical partial solutions, and we will obtain all canonical solutions. This approach is known as *orderly generation*, as described by Read [13] and Faradžev [6].

Canonicity tests have been developed for several combinatorial structures, such as graphs [12] and designs [10]. Here, we will describe such a test for subsets in general. This will allow to perform orderly generation for any problem that may be formulated in terms of subsets.

4 The Basic Algorithm

Let G be a permutation group acting on a set Ω, and let $T \subset \Omega$ be a set of points. Assume that we have a linear order '$<$' on Ω and as a consequence a natural lexicographical order on the subsets of Ω. We say that T is canonical (under G) if it is lexicographically smallest in its orbit (by definition, the empty set is canonical in its orbit because it is the only element of its orbit).

The following lemma shows that this concept of canonicity is well-behaved:

Lemma 1. *Let G be group acting on Ω, $\emptyset \neq T \subset \Omega$, $t = \max T$, and $T' = T - t$. If T is canonical, then so is T'.*

Proof. By way of contradiction assume that T' is not canonical. Then there is an element $g \in G$ such that $S' = T'^g$ is lexicographically smaller that T'. Let $S = T^g$, then $S = S' \cup \{t^g\}$. This is lexicographically smaller than T, contradicting the canonicity of T. □

Thus the problem to be solved is to check whether a given set is lexicographically smallest in its orbit. This problem has been previously studied in [1].

Instead of solving the original question whether under the action of a permutation group G a given set T is mapped to any smaller set we look at the slightly more general problem of mapping a set S to any set smaller than another given set T. If $S = T$ this solves the original question; however in the implementation we use the more general approach recursively.

Now a set S is mapped to a set smaller than T if either some element of S is mapped to an element smaller than anything in T, or if some element of S is mapped to the minimal element of T, and the remainder of S is mapped to

Algorithm 1: IsReducibleSet1(G, T)

```
return IsMappedToSmallerSet1(G,T,T)
```

Algorithm 2: IsMappedToSmallerSet1(G, S, T)

Check if any permutation in G maps the set S to a set which is lexicographically smaller than the set T

Data: A permutation group G, two sets S and T

Result: A boolean value

```
if T is empty then
    return false;
if S is empty then
    return true;
t0 ← min T;
if ∃g ∈ G, ∃s ∈ S : s^g < t0 then
    return true;
for s ∈ S do
    if ∃g ∈ G : s^g = t0 then
        S' ← (S \ {s})^g;
        T' ← T \ {t0};
        G' ← G_{t0}; // point stabilizer
        if IsMappedToSmallerSet1(G', S', T') then
            return true
return false;
```

something smaller than the remainder of T. This leads to the straightforward Algorithm 2.

The double existence quantifiers in the middle of the algorithm amount to computing the orbits under G of the elements of S. Similarly, the final loop has to cover only those elements of S which lie in the orbit of t_0 under G.

5 Improvements

5.1 Recycling Information

If we look at Algorithm 2, and in particular at its parameter T, we note except for removing the first element between one level and the next, this set is never changed. This allows us to precompute a lot of the data needed in the algorithm.

First, we can compute the various stabilizers occurring in the algorithm. If we assume that $T = \{t_1, \ldots, t_k\}$, then let $G^{(i)}$ denote the stabilizer of the first i points in G (in other words, the $G^{(i)}$ form a stabilizer chain).

Together with the stabilizers we can compute *Schreier vectors* for the various levels. (Schreier vectors are data structures that efficiently encode the coset representatives of a point stabilizer in the complete group. For a detailed description, see [4].) This allows us to efficiently compute an element of $G^{(i-1)}$ mapping t_i to a given point.

Finally, the fact that $G^{(i-1)}$ maps a given point x to something smaller than t_i also depends only on i and x. Thus, we may either save this information as it comes up, or we can precompute it before starting the recursion.

5.2 Computing the Stabilizer

In the algorithms above the termination condition is that the set T is empty. In fact, since in each invocation the sizes of S and T are equal, this also means that S is empty. This means that we have mapped the points of the initial S one by one to the points of T.

In the case that we invoked the top level with $S = T$ this means that we have found an element in the setwise stabilizer of T under G. Using a common *first-in-orbit* argument we can break off the search at that point, tracking back to the first point of T that was not moved.

This allows us to reduce the search space by using symmetry. In fact, we may use the permutations found in that way, and construct a strong generating set of the setwise stabilizer of T under G step by step. This is given as Algorithm 3 and Algorithm 4.

Algorithm 3: `IsReducibleSet2`(G, T)

Check whether T is canonical under G; compute the stabilizer H of T in G in the process. Note that if T is reducible, H will be a possibly proper subgroup of the stabilizer.

Let H be the trivial group;
for $i = |T|, |T| - 1, \ldots, 1$ **do**
  Let G' be the stabilizer of the first $i - 1$ elements of T;
  Let T' be T with the first $i - 1$ elements removed;
  Let e be the trivial permutation;
  if `IsMappedToSmallerSet2`(G', T', T', e) **then**
    return true;
Now, $G_T = H$;
return false;

5.3 Using the Stabilizer

Assume that during the search we know some group H that fixes the set S, hence a subgroup of the setwise stabilizer G_S. We select an element $s \in S$, map it to t_0, and find that this doesn't lead to S being mapped to something smaller than T. If s' is in the orbit s^H, then there is an element $h \in H$ which maps s to s'. Then, mapping s' to t will not lead to a reduction neither.

Algorithm 4: `IsMappedToSmallerSet2`(G, S, T, g_0)

Check if any permutation in G maps the set S to a set which is lexicographically smaller than the set T; compute the stabilizer of the original set in the process.

Data: A permutation group G, two sets S and T, a permutation g_0
Result: A boolean value

```
if T is empty then
    Add g0 as a new generator to the known stabilizer H;
    return to Algorithm 3;
if S is empty then
    return true;
t0 ← min T;
if ∃g ∈ G, ∃s ∈ S : s^g < t0 then
    return true;
for s ∈ S do
    if ∃g ∈ G : s^g = t0 then
        S' ← (S \ {s})^g;
        T' ← T \ {t0};
        G' ← G_t0; // point stabilizer
        if IsMappedToSmallerSet2(G', S', T', g0 · g) then
            return true
return false;
```

Thus, we use the information about the stabilizer of T in G as described in the previous section to compute part of the stabilizer of S as follows: Removing a point from S amounts to finding the stabilizer of that point; moving S under a permutation g requires conjugating the stabilizer by that permutation.

This method gives us the required subgroup of G_S to work with. This allows to prune the search tree at the expense of conjugating stabilizer chains, and finding point stabilizers.

It turns out that conjugation is quite expensive, and it can actually be avoided.

5.4 Avoiding Conjugation

Above we described how the stabilizer G_S of the set S is used to prune the search tree. When S is mapped under a permutation g, the stabilizer G_S needs to be conjugated. This involves the conjugation of all generators and hence it is computationally expensive.

However, the stabilizer is only used to test whether two elements of S are equivalent. If for each element of S we keep track of its origin, i.e., of its preimage in the original set, then we can check the equivalence of the preimages by using a stabilizer of the original group, without the need of any conjugation.

This gives the final version of the algorithm, Algorithm 5.

Algorithm 5: IsMappedToSmallerSet3(G, S, S_0, T, g_0, H)

Check if any permutation in G *maps the set* S *to a set which is lexicographically smaller than the set* T*; compute the stabilizer of the original set in the process.*

Data: A permutation group G, sets S, S_0, and T, a permutation g_0, and a subgroup H of G (a subgroup of the stabilizer of the original set)

Result: A boolean value

if T *is empty* **then**
  Add g_0 as a new generator to the known stabilizer H;
  return to Algorithm 3;
if S *is empty* **then**
  return true;
$t_0 \leftarrow \min T$;
if $\exists g \in G, \exists s \in S : s^g < t_0$ **then**
  return true;
for $i \in 1, \ldots, |S|$ **do**
  Let s be the i-th element of S;
  Let s_0 be the i-th element of S_0;
  if s_0 *is marked as used* **then**
    continue;
  if $\exists g \in G : s^g = t_0$ **then**
    $S' \leftarrow (S \setminus \{s\})^g$;
    $S'_0 \leftarrow S_0 \setminus \{s_0\}$;
    $T' \leftarrow T \setminus \{t_0\}$;
    $G' \leftarrow G_{t_0}$ // point stabilizer
    $H' \leftarrow H_{s_0}$;
    if IsMappedToSmallerSet3($G', S', S'_0, T', g_0 \cdot g, H'$) **then**
      return true
    Add all new generators of H' to H;
    Mark all elements of s_0^H as used;
return false;

6 Enumerating Set Orbits

Given the algorithm described above, we can now enumerate the orbits of (G, Ω) acting on the power set 2^Ω, by using Lemma 1 and the following:

Lemma 2. *If* S *is canonical, then* $\max S$ *is minimal in its orbit under* $G_{S'}$.

Proof. Suppose that $x = \max S$ is not minimal. Then there is a $g \in G_{S'}$ such that $y = x^g < x$. But then $S^g = (S' \cup \{x\})^g = S'^g \cup \{x^g\} = S' \cup \{y\} < S$, and since $g \in G$, this contradicts the canonicity of S. □

We get the following algorithm (see Algorithm 6).

There are a couple of things to note here:

- The stabilizer H needed here is obtained as a byproduct in the canonicity test.

Algorithm 6: EnumerateSubsets($G, \Omega, S = \emptyset$)

```
H ← G_S;
orbitReps ← {x ∈ Ω : max S < x ≤ x^h ∀h ∈ H};
for x ∈ orbitReps do
    S̃ ← S ∪ {x};
    if not IsReducibleSet(G, S̃) then
        output S̃;
        EnumerateSubsets(G, Ω, S̃);
```

- Much of the precomputed data used in the canonicity test for S, such as the stabilizer chain and the reducible orbits for all but the final level, can be reused in the test for the set $\tilde{S}$, which is obtained by adding a single element to S.
- The point stabilizer H_x is a large subgroup of the stabilizer $G_{\tilde{S}}$; in fact it is the stabilizer of x in that group. So, injecting H_x in the test for $\tilde{S}$ will speed up that procedure.

7 Implementation

The algorithm as described above has been implemented and tested in GAP [9]. The user has access to it on three different levels, providing a trade-off of control and ease of use.

The most straightforward access to the algorithm is a function `IsCanonicalSetOrbitRepresentative`, which takes as arguments a permutation group G and a set T, and returns a boolean value indicating whether T is smallest in its orbit.

```
gap> G := Group((1,2,3,4,5), (2,5)(3,4));
Group([ (1,2,3,4,5), (2,5)(3,4) ])
gap> IsCanonicalSetOrbitRepresentative(G, [1,2]);
true
gap> IsCanonicalSetOrbitRepresentative(G, [1,3]);
true
gap> IsCanonicalSetOrbitRepresentative(G, [1,5]);
false
```

Any additional information obtained during the execution of the algorithm, such as the stabilizer of T, is lost. If this information is needed, a more involved invocation is needed:

```
can := CanonicityTest(G, [1,2], 5);;
gap> result := RunTest(can);
true
```

```
gap> stabilizer := GetStabilizer( can );
Group( [ (1,2)(3,5) ] )
```

Finally, for the common case that some action is to be performed for each representative it is possible to define an object representing the domain of all such representatives. For this we need to provide the group G, the set on which G acts, and the desired sizes of the sets to consider. Hence for example we can find all orbits of sets of size 2 or 3 by the following:

```
gap> domain := [1..5];
[ 1 .. 5 ]
gap> sizes := [2,3];
[ 2, 3 ]
gap> representatives
    := SetOrbitRepresentatives(G, domain, sizes);;
gap> Size(representatives);
4
gap> for set in representatives do
> Print( set, "\n");
> od;
[ 1, 2 ]
[ 1, 2, 3 ]
[ 1, 2, 4 ]
[ 1, 3 ]
```

8 Applications

We tested the algorithm mainly with two types of problems: Enumeration of spreads in incidence structures, and enumeration of fusions in coherent configurations.

An incidence structure can be thought of as a generalized geometry. It consists of two types of objects, "points" and "lines", and some incidence relation between the two types. In other words, each point is either incident to a given line, or it is not incident to it. The notions of point and line suggest using language from planar geometry, thus, if a point P and a line L are incident, we say that P lies on L, and that L goes through P.

Given an incidence structure, its dual is obtained by interchanging the roles of points and lines, and reverting the incidence relation accordingly. Thus, a point of the original becomes a line of the dual, and vice versa.

A *spread* in an incidence structure is a subset of lines which covers each point exactly once. In particular, any two lines in a given spread are disjoint. The horizontal lines form a spread of the euclidean plane $\mathbb{R}^2$.

8.1 STS(15)

For the spreads we ran several test cases with known results. One of them was the classification of Steiner triple systems on 15 points. It is a well known result [20] that there are exactly 80 such structures up to isomorphism.

This can be translated to a spread problem as follows: Take as points all unordered pairs of a 15-element set; as lines take triples. Define incidence as inclusion. This gives an incidence structure with $\binom{15}{2}$ "points" and $\binom{15}{3}$ "lines". A Steiner triple system on the original set is a set of triples such that each pair of elements is contained in exactly one triple. In other words, each "point" is incident to exactly one "line", thus we have a spread.

As a group we have the symmetric group of degree 15 acting on the original elements. The algorithm had to deal with the group acting on triples, which gives a degree of 455. The construction of all non-isomorphic solutions took 6 minutes of CPU time.[3]

8.2 GQ(3, 9)

We also determined all spreads in the unique generalized quadrangle $GQ(3,9)$. We found a total of 26 orbits of solutions, which confirms a result by A. E. Brouwer [3]. The computation took 2 s.. The group in question has degree 112.

These spreads are interesting also from another point of view. The point graph of the generalized quadrangle is the first subconstituent of the McLaughlin graph of order 275, which corresponds to a rank 3 representation of the McL simple group. It is still an open question whether this bigger graph is the point graph of a suitable partial geometry $pg(5,28,2)$ (see [19, 15, 18]). Supposing that such a geometry exists, we call a point P together with all lines through it a *bundle*.

If we take a bundle and remove the point P we obtain a set of lines which covers each neighbor of P exactly once. In other words, we get a spread in the first subconstituent, which we know is isomorphic to $GQ(3,9)$. However, the group acting here is the stabilizer of a point in McL, which is smaller than the full automorphism group of the generalized quadrangle. In fact, some of the orbits of spreads split, and we get a total of 36 orbits of possible bundles.

8.3 Generalized Hexagon

In her thesis, A. De Wispelaere [5, p. 159] noted that there are two known two-ovoids in the classical generalized hexagon of order 4. The question was raised whether there are any others.

Two-ovoids can be regarded as a spread of the dual structure, hence they fit in this framework. The number of points in the generalized hexagon is

[3] The computation was performed on an Intel Celeron processor at 1.6 GHz.

$1365 = \frac{4^6-1}{4-1}$, same as the number of lines. Each line contains 5 points, and each point lies on 5 lines. A two-ovoid is a set of points meeting each line exactly once; thus it contains $1365/5 = 273$ points. In the dual hexagon this corresponds to a set of lines containing each point exactly once, i.e., a spread.

We were able to give a negative answer to this question; all two-ovoids are known. The search took approximately 2 weeks of CPU time.

8.4 Schur Rings Over Small Groups

Schur rings or S-rings are structures related to (abstract) groups (see, e.g., [7, p. 56]). They form a special case of association schemes which are central to the area of Algebraic Combinatorics. The construction of S-rings over a given group G relies on a search of certain subsets of G. Hence, here we consider the action of G on itself.

In [8], all S-rings of order up to 31 were determined, as well as those of order 33 through 35. The S-rings of order 32 were computed by the first author using an earlier version of the above described algorithm. Recently the second author, using the algorithm described above, was able to extend this to all orders up to 47. Details will follow elsewhere [14].

9 Conclusion

We have described a fast canonicity test for sets under the action of permutation groups. This test will be made available as a GAP share package. It allows to treat a wide range of combinatorial problems in a uniform manner.

The authors wish to thank M. Klin, A. Niemeyer, and C. Praeger for helpful input.

The authors are pleased to acknowledge Prof. Bruno Buchberger and the coordinators of the Special Semester on Gröbner Bases (February 1 – July 31, 2006), organized by RICAM, Austrian Academy of Sciences, and RISC, Johannes Kepler University, Linz Austria for partial support of this project. The work of the second author has been funded by the Australian Research Council.

Last but not least, the authors would like thank the anonymous referees for their helpful remarks and suggestions.

References

1. L. Babai and E. M. Luks, Canonical labeling of graphs, in *Proc. 15th ACM STOC*, pp. 171–183, 1983.
2. L. Babai, E. M. Luks, and A. Seress, Fast management of permutation groups, in *29th Annual Symposium on the Foundations of Computer Science*, pp. 272–282, IEEE Press, New York, 1988.

3. A. E. Brouwer, Homepage, http://www.win.tue.nl/~aeb.
4. G. Butler, *Fundamental Algorithms for Permutation Groups*, Lect. Notes Comp. Sci., Vol. 559, Springer, Berlin/Heidelberg, 1991.
5. A. De Wispelaere, *Ovoids and Spreads of Finite Classical Generalized Hexagons and Applications*, Ph.D. Thesis, Universiteit Gent, 2006.
6. I. A. Faradžev, Constructive enumeration of combinatorial objects, in *Problèmes combinatoires et théorie des graphes*, Univ. Orsay, Orsay, 1976. Colloq. Internat. CNRS, Vol. 260, pp. 131–135, CNRS, Paris, 1978.
7. I. A. Faradžev, M. H. Klin, and M. E. Muzichuk, Cellular rings and groups of automorphisms of graphs, in I. A. Faradžev, et al. (eds.) *Investigations in Algebraic Theory of Combinatorial Objects*. Math. Appl. (Soviet Ser.), Vol. 84, pp. 1–152, Kluwer Academic, Dordrecht, 1994.
8. F. Fiedler, *Enumeration of Cellular Algebras Applied to Graphs with Prescribed Symmetry*, Master's thesis, Technische Universität Dresden, 1998.
9. The GAP Group, *GAP – Groups, Algorithms, and Programming, Version 4.4.10*; 2007, http://www.gap-system.org.
10. A. V. Ivanov and I. A. Faradžev, Constructive enumeration of incidence systems. I, II, III, *Rostock. Math. Kolloq.*, **24** (1983), 4–62 (in Russian).
11. R. Laue, Construction of combinatorial objects – a tutorial. *Bayreuther Math. Schr.*, **43** (1993), 53–96.
12. B. D. McKay, The nauty package, http://cs.anu.edu.au/people/bdm/nauty/.
13. R. C. Read, Every one a winner, or how to avoid isomorphism search when cataloguing combinatorial configurations, *Ann. Discrete Math.*, **2** (1978), 107–120.
14. S. Reichard, *S-rings of order up to 47*, in preparation.
15. S. Reichard, *An Algorithm for the Construction of Partial Geometries with Given Point Graphs*, Technical Report, Technische Universität Dresden, MATH-AL-12-1997, October 1997.
16. B. Schmalz, *t-Designs zu vorgegebener Automorphismengruppe*, Dissertation, Universität Bayreuth, 1992.
17. C. C. Sims, Computational methods in the study of permutation groups, in *Computational Problems in Abstract Algebra*, pp. 169–183, Pergamon, Oxford, 1970.
18. L. H. Soicher, Is there a McLaughlin geometry? *J. Algebra*, **300** (2006), 248–255.
19. J. H. van Lint, On ovals in $PG(2,4)$ and the McLaughlin graph, in *Papers Dedicated to J.J. Seidel*, pp. 234–255, Eindhoven, 1984.
20. H. S. White, F. N. Cole, and L. D. Cummings, Complete classification of the triad systems on fifteen elements, *Mem. Nat. Acad. Sci. U.S.A.* (2nd memoir), **14** (1919), 1–89.
21. V. A. Zaichenko, *An Algorithmic Approach to the Construction of Combinatorial Objects and Computations in Permutation Groups Based on the Method of Invariant Relations*, PhD thesis, Moscow, 1981 (in Russian).

The 2-dimensional Jacobian Conjecture: A Computational Approach

Ronen Peretz

Department of Mathematics, Ben-Gurion University, Beer Sheva, Israel.
ronenp@math.bgu.ac.il

Summary. The Jacobian Conjecture is one of the most important open problems in algebraic geometry. It was included among the millennium open problems in mathematics by Steve Smale in his address to the Millennium ICM (in the year 2000). Given a polynomial mapping $F : \mathbb{C}^n \to \mathbb{C}^n$ which has a nonzero constant determinant of its Jacobian matrix (**The Jacobian Condition**), the conjecture is that F is an invertible morphism. This means that it is injective, surjective and that its inverse mapping F^{-1} is also polynomial. Even in dimension $n = 2$ this is still open. This article is about the 2-dimensional Jacobian Conjecture. The degree of a polynomial mapping F is the maximum degree of its polynomial coordinate functions. A striking fact that we prove is that given a degree d, the 2-dimensional Jacobian Conjecture can be settled for **all** the polynomial mappings of degree d or less. If it is disproved then a counterexample is constructed. The magic tool used is the machinery of Gröbner bases. We use the powerful tools of computational algebra and in particular the theory of Gröbner bases in order to solve a certain ideal membership problem in an algebra of many variables polynomials over the complex field. The Jacobian condition satisfied by the mapping F (of degree d or less) is used in a canonical way to construct an ideal (the Jacobian ideal) within this algebra of polynomials. We consider the two relative resultants of the mapping F (one with respect to X and the second with respect to Y). The key theorem we prove is that the Jacobian Conjecture is valid for F if and only if the leading coefficients of these two resultants belong to the Jacobian ideal. We call this result: **The resultant reformulation of the Jacobian Conjecture**. Now the Gröbner bases machinery comes in to help to decide on the ideal membership problem we have. The algorithm was programmed and was used to prove the 2-dimensional Jacobian Conjecture up to degree 15. The theoretical importance of this algorithm is that it shows that the conjecture is a decidable problem.

The article contains many other important results, both new and old that are relevant to this particular computational approach to this famous open problem. We assume that the reader has some background in the theory and the folklore of the Jacobian Conjecture. The classical paper [H. Bass, E. Connell, and D. Wright, The Jacobian conjecture: reduction of degree and formal expansion of the inverse, *Bull. Amer. Math. Soc.* (2), **7** (1982), 287–330] and the only comprehensive book in this area [A. van den Essen, Polynomial Automorphisms and the Jacobian Conjecture,

M. Klin et al. (eds.), Algorithmic Algebraic Combinatorics and Gröbner Bases,
DOI 10.1007/978-3-642-01960-9_5, © Springer-Verlag Berlin Heidelberg 2009

Progress in Mathematics, Vol. 190, Birkhäuser, Basel, 2000] are excellent sources. We use the notation introduced in those sources whenever possible.

Key words: The Jacobian Conjecture, Étale morphisms, Inversion formulas, Polynomial automorphisms, Asymptotic values

1 Introduction

A polynomial map $F : \mathbb{C}^n \to \mathbb{C}^n$ is a map $F = (P_1, \ldots, P_n)$ whose coordinate functions are polynomials over the complex field $\mathbb{C}$ in the n variables $(X_1, \ldots, X_n)$. We write $(P_1, \ldots, P_n) \in \mathbb{C}[X_1, \ldots, X_n]^n$, where $\mathbb{C}[X_1, \ldots, X_n]$ is the algebra of polynomials in $(X_1, \ldots, X_n)$ over $\mathbb{C}$.

The polynomial map F is called invertible, or an automorphism of $\mathbb{C}^n$ if it is injective, surjective and its inverse mapping $G = F^{-1}$ is also polynomial. This means that $F \circ G = G \circ F = \mathrm{id}_{\mathbb{C}^n}$ where G is polynomial.

Such an automorphism preserves the algebro-geometric structure of the important affine variety $\mathbb{C}^n$ and hence its importance in algebraic geometry. It is natural to look for convenient criteria for a polynomial mapping to be invertible. Application of the chain rule to the functional equations $F \circ G = G \circ F = \mathrm{id}_{\mathbb{C}^n}$ gives the following equations on the Jacobian matrices of F and G, $J_F(G)J_G(X) = J_G(F)J_F(X) = I_n$. Here we use the notation $X = (X_1, \ldots, X_n)$. Taking determinants we get $\det J_G(F) \cdot \det J_F(X) = 1$. In particular the polynomial $\det J_F(X_1, \ldots, X_n)$ cannot have a zero and the Fundamental Theorem of the Algebra implies that $\det J_F(X)$ is a nonzero constant. We write $\det J_F(X) \in \mathbb{C}^*$ and call this the Jacobian condition. So far we obtained the implication

$$F \text{ is an automorphism} \quad \Rightarrow \quad \det J_F(X) \in \mathbb{C}^*.$$

This naturally leads to consider the reverse implication

$$\det J_F(X) \in \mathbb{C}^* \text{ and } F \text{ polynomial} \quad \Rightarrow \quad F \text{ is an automorphism.}$$

In dimension $n = 1$ this is true and is easy to prove. For $n \geq 2$ this is one of the famous open problems in mathematics, known as the Jacobian Conjecture. It appeared in the paper [11] by Ott-Heinrich Keller. An excellent survey of the problem appears in [2]. Also, a comprehensive book on the subject (as well as on related topics) is the one by Arno van den Essen [6].

A wealth of results were discovered over the years by many mathematicians that tried to resolve the Jacobian Conjecture. It is known to be true in dimension $n = 2$ up to about degree $d = 100$, [15]. Here the degree of a polynomial mapping is the maximum of the degrees of its coordinates. It is known to be true for quadratic polynomial mappings, i.e. $d \leq 2$, in any dimension n, [2, 6]. Amazingly enough it was proved that if the Jacobian Conjecture is valid for polynomial mappings of degree $d \leq 3$ in any dimension n, then it is valid for all degrees. The technique that was used to prove that result is called degree reduction. It is based on a K-theoretic principle, [2, 6]. This gives the impression that "we are almost there". After all only maps of degree $d = 3$ should be considered (and even here, only very special subclass of polynomial mappings will suffice). In spite of this feeling the Jacobian Conjecture is still open. This is a peculiar property of that riddle in mathematics. It gives one the impression of a not so difficult problem, and attracts mathematicians from various fields of research. Algebraic geometers as well as researchers in commutative algebra were involved. But also analysts that come from several complex variables, geometric function theory and geometers that do differential geometry tried to solve the problem. Topological methods were tried as well. Starting from manifold theory, varieties and complex spaces. Algebro-topological methods were used too. Representation theory and even theories on infinite dimensional varieties were applied. The list is still not complete! But it turns out to be a difficult problem. The by product of that, is the existence of many faulty "proofs" of this conjecture that appeared over the years. Some were even formally sent for publication and appeared in good mathematics journals. It is rather clear that the Jacobian Conjecture is a problem about polynomial mappings and over fields of characteristic zero. Thus even though it is tempting to think about a broader class of mappings (such as entire mappings on $\mathbb{C}^n$) the conjecture is known not to be true there. Thus there must be something very algebraic involved in the solution of this open problem. This might help us understand the origin of some faulty proofs. The author heard serious mathematicians claiming that "there is no obvious reason why such a conjecture should be true" and hence they believe that it is false. However, it might be that this is one of these stubborn problems that have a counterexample only in a very large dimension or degree, way beyond our computer technology, and hence it might be extremely difficult to construct one. One needs to look at the counterexample of S. Pinchuck, [23], for the so called "Real Jacobian Conjecture" to have a feeling of what might be involved.

The current paper describes an effort to deal with the Jacobian Conjecture in dimension $n = 2$. It is a follow up of a previous attempt that was geometric in nature. This geometric approach is described in [20, 21] and in particular in [22]. We outline the geometric ideas that are involved in this former attempt. A polynomial map F that satisfies the Jacobian condition $\det J_F(X) \in \mathbb{C}^*$ is called an étale mapping. From the point of view of differential geometry an étale mapping is a polynomial local diffeomorphism of $\mathbb{C}^n$. Hence we can restate the Jacobian Conjecture in this new language as follows,

$$F \text{ étale} \quad \Rightarrow \quad F \text{ is an automorphism.}$$

It is useful to recall that injective polynomial mappings $\mathbb{C}^n \to \mathbb{C}^n$ are surjective and their set theoretic inverse mapping is polynomial. So it is really only the injectivity property that we are after. The above restatement means that a local condition (being étale) implies for polynomial mappings over $\mathbb{C}$ the corresponding global property (of being a global diffeomorphism). A beautiful theory developed by J. Hadamard, [9], tells us that in order to prove that a local diffeomorphism is in fact a global one, it suffice to exclude the existence of asymptotic values (these are finite limits at infinity) of the mapping. This is the starting point of the geometric approach mentioned above. When one specializes the general Hadamard's theory to polynomial mappings, one immediately comes across fractional power series expansions of the asymptotic curves of the mapping. A very elegant set of results along these lines were discovered by Nguyen Van Chau.

As opposed to this approach the methods of the current paper are very algebraic and in some cases are even algorithmic. Thus computer algebra methods such as the computation of Gröbner bases of ideals are naturally used here. The connection to the previous (geometric) approach is via the study of asymptotic values of polynomial maps as suggested by J. Hadamard. Section 2 elaborates this algebraic version of Hadamard general theory. We restrict our attention to the 2-dimensional case of polynomial mappings $F = (P(X,Y), Q(X,Y)) : \mathbb{C}^2 \to \mathbb{C}^2$.

We translate the geometric study of asymptotic values into an algebraic study of the pair of resultants that are induced by F. If α, β are two new indeterminates then we can form the following two (so called) relative resultants,

$$\begin{aligned} R(X) &= \mathbf{resultant}(P(X,Y) - \alpha, Q(X,Y) - \beta, Y) \\ &= R_N(\alpha,\beta)X^N + \cdots + R_0(\alpha,\beta), \\ S(Y) &= \mathbf{resultant}(P(X,Y) - \alpha, Q(X,Y) - \beta, X) \\ &= S_M(\alpha,\beta)Y^M + \cdots + S_0(\alpha,\beta). \end{aligned}$$

A main result that we prove in Sect. 5 (Theorem 5) is called The Resultant Formulation of the Jacobian Conjecture. It shows that (under some standard normalization on F) the Jacobian conjecture is true iff the Jacobian condition on F implies that the coefficients $R_N(\alpha,\beta), S_M(\alpha,\beta)$ of the above resultants are nonzero constants. For related results see [6].

This suggests that there should be some kind of algebraic relation between the Jacobian of F, $\partial(P,Q)/\partial(X,Y)$ and the above resultants $R(X)$, $S(Y)$ (provided that the conjecture is true!). In spite of many efforts the author was unable to find such a relation.

However, the above method was fruitful and led to theorems that are reported in this paper. We now proceed to describe in some details the results that appear in the various sections of this paper.

In Sect. 3 it is proved that if $F : \mathbb{C}^2 \to \mathbb{C}^2$ is polynomial then any sequential asymptotic value of F is also an asymptotic value of F (Theorem 1). The proof relies on the technique of maximal domains [18, 19].

In Sect. 4 we prove some relations between the asymptotic values of a polynomial $F : \mathbb{C}^2 \to \mathbb{C}^2$ and the zeros of the highest coefficients of the above pair of resultants. In fact we show that the set of all the asymptotic values of F is the union of the two algebraic curves $R_N(\alpha,\beta) = 0, S_M(\alpha,\beta) = 0$ (Theorem 2). We also deduce that the set of X-infinite asymptotic values of F form the curve $R_N(\alpha,\beta) = 0$ and the set of Y-infinite asymptotic values of F is $S_M(\alpha,\beta) = 0$ (Theorem 3). This should be compared with the results in [20] for the real case.

Thus we see that the existence of asymptotic values of a polynomial map $F = (P,Q)$ is controlled by the structure of the highest coefficients of the induced pair of resultants. Once again we mention the related results of Nguyen Van Chau [7, 25].

If F satisfies the Jacobian condition $\partial(P,Q)/\partial(X,Y) \in \mathbb{C}^*$ then according to Sect. 2 F is invertible provided that it has no asymptotic values. Hence there should be a reformulation of the Jacobian conjecture in terms of the induced resultants.

Indeed that Resultant Reformulation is proved in Sect. 5, Theorem 5:

$$\partial(P,Q)/\partial(X,Y) \in \mathbb{C}^* \quad \Leftrightarrow \quad R_N(\alpha,\beta), S_M(\alpha,\beta) \in \mathbb{C}^*.$$

Probably the core of the Jacobian conjecture (if true) is to explore the algebraic relation that should exist between the Jacobian of a map and it's induced resultants.

As a first application of the Resultant Reformulation we prove in Sect. 6, Theorem 6 that the Jacobian conjecture is decidable. Namely, we describe an algorithm which gets as an input a positive integer d and decides (after finitely many steps) if the Jacobian conjecture holds true for all polynomial maps $\mathbb{C}^2 \to \mathbb{C}^2$ of degree at most d. Moreover, if the answer is negative, then the algorithm generates a counterexample.

The algorithm utilizes Buchberger's algorithm for calculating a Gröbner basis for the Jacobian variety of degree d [3, 8]. Naturally the complexity of our algorithm is high because the Buchberger-type algorithms are of high complexity [13]. The algorithm was implemented by the author (while visiting the University of Michigan) on a Sun platform. He used the standard **Gröbner package in Maple**. The input to the program was a positive integer d and the output was a **true/false** flag. If **true** then the Jacobian conjecture was found to be valid for all the complex polynomial mappings in dimension two and of degree d or less. A **false** flag meant the other alternative and a counterexample of degree d or less would have been given. The package was able to automatically prove the conjecture for mappings of degree 15 or less. The running time was about 40 h.

The Resultant Reformulation suggests natural inductive approach to prove the Jacobian conjecture. In Sect. 7 we discuss several such possibilities but these fail. This might indicate that the Jacobian conjecture is not true.

In Sect. 8 we explore some covering properties of a polynomial $F : \mathbb{C}^2 \to \mathbb{C}^2$ in terms of it's induced resultants. For example in Theorem 7 we prove what might be called “a polynomial Picard's Little Theorem”, namely, that there exist finitely many polynomials

$$R_N(\alpha, \beta), \ldots, R_1(\alpha, \beta)$$

(the highest N coefficients of $R(X)$) such that

$$\mathbb{C}^2 - F(\mathbb{C}^2) = V_C(R_N, \ldots, R_1).$$

So generically $F(\mathbb{C}^2)$ is the whole of $\mathbb{C}^2$ except, maybe, for a finite set whose size is controlled by the Bezout bound.

Motivated by the Resultant Reformulation we pose the following **conjecture:**

$\partial(P, Q)/\partial(X, Y)$ has a zero on $F^{-1}(R_N(\alpha, \beta) = 0)$.

This conjecture implies the Jacobian conjecture. We show that a slightly stronger version of this conjecture fails (Example 5) which shows how delicate the situation is.

For some results of Sect. 7 we use properties of certain well known derivations that are related to the Jacobian of a polynomial map. For that purpose a small theory on grading an algebra with a derivation is developed in Sects. 9, 10, 11 and 12. See related results in the book [6]. If k is a field of characteristic 0 and A is a k-algebra with 1 and D a derivation of A, then there is a natural gradation of A with respect to D: the elements of class n in the gradation are those that are annihilated by n iterations of D.

If we assume, further, the existence of an element $q \in A$ for which $D(q) = 1$ (called a slice) then in "reasonable" algebras the multiplication of an element in class n by q is an element of class $n+1$ and conversely (Theorem 8). As a consequence we obtain a structure theorem for the classes of this D-gradation in terms of $\ker(D, A)$ and of q (Theorem 9).

This leads immediately to the well known fact that $Nil(D) = \ker(D, A)[q]$ which was noted by several people [23, 25] and [17] (Corollary 1).

It also gives a necessary and sufficient condition for D to be a locally nilpotent derivation of A (Corollary 6).

One application of the above is for the polynomial ring over k in n variables, $A = k[x_1, \ldots, x_n]$. Given n elements $f_1, \ldots, f_n \in A$ that satisfy the Jacobian condition, namely, that $\partial(f_1, \ldots, f_n)/\partial(x_1, \ldots, x_n) = 1$ one can take $D = \partial(f_1, \ldots, f_{n-1}, \cdot)/\partial(x_1, \ldots, x_n)$, $q = f_n$ and apply the above. This is closely related to the results on the derivations $d_1, \ldots, d_n$ that were introduced in [17], however, the methods of proofs are different.

Section 13 is dedicated to invertible morphisms $F : \mathbb{C}^2 \to \mathbb{C}^2$ and the structure of their resultants. The main result is that

$$\textbf{resultant}(P(X, Y) - \alpha, Q(X, Y) - \beta, Y) = R_1 X + R_0(\alpha, \beta)$$

where $R_1 \in \mathbb{C}^*$ (Theorem 11).

A wealth of conclusions (many of which are known by other methods) are then derived. We are able to write down a simple inversion formula for F^{-1} (Theorem 12). Later on in Sect. 14 we note that the inverse map F^{-1} depends only on very few of the coefficients of F, namely only on the coefficients of the

4 face polynomials of F. We call that phenomenon The Rigidity of Morphisms (Theorem 16).

A second inversion formula is given using the inverse of the logarithmic derivatives of the pair of the induced resultants of F at the origin (Theorem 13).

Exact bounds on $\deg F^{-1}$ are given in terms of $\deg F$. These are known even in dimensions higher than 2 and are mentioned in [2, p. 292]. The result is attributed to O. Gaber and others, however, our method of the proof is different and yields a little sharpening of the known inequality. It gives the exact bounds on the degrees of the coordinates of F^{-1} in terms of the degrees of the coordinates of F (Theorem 14).

Other conclusions concern the integrality of the coefficients of F^{-1} with respect to those of F. Thus for example if the coefficients of F are rational then so are the coefficients of F^{-1} and so $F(\mathbb{Q}^2) = \mathbb{Q}^2$ (Theorem 15, Corollary 8).

Section 15 gives a proof of the famous "Fibre Theorem" for maps $\mathbb{C}^2 \to \mathbb{C}^2$ that satisfy the Jacobian condition. See Theorem 18 and compare to [14, 5].

In Sect. 16 we extract one more inversion formula and prove two more equivalent formulations to the Jacobian conjecture. This is done as follows:

If $F : \mathbb{C}^2 \to \mathbb{C}^2$ is an invertible morphism we write

$$P(X,Y) - \alpha = a_n Y^n + \cdots + a_0 - \alpha = a_n \prod_{i=1}^{n} (Y - \alpha_i),$$

$$Q(X,Y) - \beta = b_n Y^n + \cdots + b_0 - \beta = b_n \prod_{j=1}^{n} (Y - \beta_j)$$

and similarly with respect to X. We than utilize the inversion formula we found in terms of the logarithmic derivatives of the induced resultants in order to deduce the new inversion formula in terms of the zeros α_i, β_j (Theorem 22). This in turn leads to the new reformulations of the Jacobian conjecture (Theorem 23, Theorem 24).

In Sect. 17 we address the problem of parametrizing the Jacobian variety of degree n and to calculate its dimension. This variety underlies the algorithm that solves the Jacobian conjecture for degree n or less. We give linear algebraic condition on the coefficients of a polynomial $P(X,Y) \in \mathbb{C}[X,Y]$ in order for it to have a Jacobian mate (Theorem 25, Theorem 26).

We hope that this report will motivate further related research. Hopefully, if the Jacobian conjecture is true somebody will be able to find the algebraic relation between the Jacobian on one hand and the induced resultants on the other hand.

Finally, many results given here are known in that form or another. Our bibliography is not intended to be complete and we apologize for not giving the full account and credit. To mention just a few names that we skipped: J. Stein, M. Chamberlain, G. Meisters, P. Cassou-Nogues and S. Yu. Orevkov.

2 A Theorem of J. Hadamard

We refer to the paper by J. Hadamard [9]. A copy of this paper appears also in J. Hadamard Selecta (pp. 145–159).

Let $f : \mathbb{R}^n \to \mathbb{R}^n$ be a $\mathbb{C}'$ map such that for any $x \in \mathbb{R}^n$, $\det J(f)(x) \neq 0$. For each $x \in \mathbb{R}^n$ we let

$$N(x) = 1/\|D_f^{-1}(x)\|.$$

D_f is the differential of f and D_f^{-1} is its inverse operator. For any $R > 0$ we denote

$$\mu(R) = \min_{\|x\|=R} N(x).$$

Theorem (J. Hadamard [9]). *If f satisfies*

$$\int_0^\infty \mu(R)\, dR = \infty \tag{1}$$

then f is a global diffeomorphism of $\mathbb{R}^n$ onto $\mathbb{R}^n$.

We can restate condition (1) as follows

The image of any curve $\sigma(t)$, $0 \le t < \infty$, that extends to ∞ under f, i.e.

$$f(\sigma(t)), \quad 0 \le t < \infty, \quad \text{is a non-rectifiable curve.} \tag{2}$$

Definition 1. *Let $f : \mathbb{R}^n \to \mathbb{R}^n$ be any map. A point $x_0 \in \mathbb{R}^n$ is called an asymptotic value of f if there exists a curve $\sigma(t)$, $0 \le t < \infty$, that extends to ∞ such that*

$$\lim_{t\to\infty} f(\sigma(t)) = x_0,$$

where $\sigma(t)$ is called an asymptotic curve of f.

If we use the coordinate notations $f = (f_1, \ldots, f_n)$, $x_0 = (x_{01}, \ldots, x_{0n})$ then we call x_{0j} an asymptotic value of the function f_j, $j = 1, \ldots, n$, and $\sigma(t)$ is called an asymptotic curve of f_j.

Proposition 1. *Let $f : \mathbb{R}^n \to \mathbb{R}^n$ be a $\mathbb{C}'$ map. If f has no asymptotic values then f satisfies condition (2).*

Proof. We assume in order to get a contradiction that f does not satisfy condition (2). Let $\sigma(t)$, $0 \le t < \infty$, be a curve such that $\lim_{t\to\infty} \|\sigma(t)\|_2 = \infty$ but the length of $f(\sigma(t))$, $0 \le t < \infty$, is $L < \infty$.

We verify the Cauchy condition for $f(\sigma(t))$

$$\forall \epsilon > 0 \ \exists t(\epsilon) \text{ such that } \forall t_1, t_2 > t(\epsilon), \|f(\sigma(t_1)) - f(\sigma(t_2))\|_2 < \epsilon.$$

If not, then there is an $\epsilon_0 > 0$ and a sequence $0 < t_1 < t_2 < \cdots < t_m \to \infty$ such that

$$\|f(\sigma(t_{2j})) - f(\sigma(t_{2j-1}))\|_2 \ge \epsilon_0, \quad j = 1, 2, 3, \ldots.$$

But this implies the contradiction

$$L \geq \sum_{j=1}^{\infty} \|f(\sigma(t_{2j})) - f(\sigma(t_{2j-1}))\|_2 = \infty.$$

Hence $\lim_{t\to\infty} f(\sigma(t)) = x_0$ exists. But this contradicts the assumption on f. □

Thus a corollary of Hadamard's result is that in order to show that f is a global diffeomorphism it suffices to exclude the existence of asymptotic values for f. We will call that "Hadamard's condition".

Remark 1. It is clear that the above definition of asymptotic values and asymptotic curves can be formulated for maps over $\mathbb{C}$, i.e., $F : \mathbb{C}^n \to \mathbb{C}^n$. Likewise we conclude from Hadamard's theorem that also for the complex case it suffices to exclude the existence of asymptotic values for a local diffeomorphism $F : \mathbb{C}^n \to \mathbb{C}^n$ in order to prove that it is a global diffeomorphism.

3 Asymptotic and Sequential Asymptotic Values of Polynomial Maps

Definition 2. *Let $F : \mathbb{C}^n \to \mathbb{C}^n$ be a map (we may formulate the definition also for maps $f : \mathbb{R}^n \to \mathbb{R}^n$). A point $X_0 \in \mathbb{C}^n$ is called a sequential asymptotic value of F if there exists a sequence $\{X_j\}_{j=1}^{\infty}$ in $\mathbb{C}^n$ such that $\lim_{j\to\infty} \|X_j\|_2 = \infty$ and so that*

$$\lim_{j\to\infty} F(X_j) = X_0.$$

Remark 2. Clearly, any asymptotic value of F is a sequential asymptotic value of F. The opposite implication, however, is false.

Example 1. We shall demonstrate the remark over $\mathbb{R}^2$. We consider the map

$$f : \mathbb{R}^2 \to \mathbb{R}^2,$$
$$f(X,Y) = (e^X \cos Y, e^X \sin Y).$$

The only asymptotic value of f is $(0,0)$ as is easy to check. However, any point of $\mathbb{R}^2$ is a sequential asymptotic value of f, for if the point $(a,b) = (0,0)$ this follows by the fact that $(0,0)$ is an asymptotic value of f while if $(a,b) \neq (0,0)$ then there is a solution Z_0 to the complex equation $e^Z = a + ib$ (by Picard's Little Theorem). Hence the following

$$Z_j = Z_0 + 2\pi j i, \quad j = \ldots, -2, -1, 0, 1, 2, 3, \ldots$$

are also solutions by periodicity. Since $\lim_{j\to\infty} \|Z_j\|_2 = \infty$ it follows that (a,b) is a sequential asymptotic value of f.

Notation. We will denote by $A(F)$ the set of all the asymptotic values of F. We will denote by $A_S(F)$ the set of all the sequential asymptotic values of F.

Remark 3. By the above we always have $A(F) \subseteq A_S(F)$ but in general the two sets are not the same.

For polynomial maps the situation is simpler. For the proof of the next theorem we refer to the paper [19] and the preprint [18].

Theorem 1. *If $F : \mathbb{C}^2 \to \mathbb{C}^2$ is polynomial, then $A(F) = A_S(F)$.*

Proof. As in the analytic case we define a maximal domain Ω for F to be a domain in $\mathbb{C}^2$, i.e., an open connected subset of $\mathbb{C}^2$, with an almost smooth boundary such that F is a one to one map in Ω and Ω is maximal for inclusion with respect to these properties. Here almost smooth means smooth, except along finitely many algebraic curves. These curves are the singular locus of the mapping on the boundary.

Since F is polynomial, the fibre of F over any point in $\mathbb{C}^2$ is generically finite (unless F is constant in which case the theorem is obvious). Hence there is a tiling of $\mathbb{C}^2$ with finitely many maximal domains of F, $\{\Omega_1, \ldots, \Omega_N\}$.

Let $(a,b) \in A_S(F)$. Then there exists a sequence $(X_n, Y_n) \in \mathbb{C}^2$ such that $\lim \|(X_n, Y_n)\|_2 = \infty$ and so that $\lim F(X_n, Y_n) = (a,b)$. We can assume that for any n, the point (X_n, Y_n) lies outside the boundaries $\{\partial\Omega_1, \ldots, \partial\Omega_N\}$ (by shifting a little the point if necessary and using the continuity of F). Since the collection $\{\Omega_1, \ldots, \Omega_N\}$ is finite there exists an index j, $1 \le j \le N$, so that Ω_j contains infinitely many of the points $\{(X_n, Y_n)\}_{n=1}^{\infty}$. By restricting to a subsequence and re-indexing the tiles we may assume that $\{(X_n, Y_n)\}_{n=1}^{\infty} \subseteq \Omega_1$. Hence $(a,b) \in \partial F(\Omega_1)$. Since the boundary of Ω_1 is almost smooth it follows that so is the boundary of $F(\Omega_1)$. Let $\sigma(t)$, $0 \le t \le 1$, be a compact curve such that $\sigma(0) = (a,b)$, $\sigma(t) \in F(\Omega_1)$ for $0 < t \le 1$ and

$$\{F(X_n, Y_n)\}_{n=1}^{\infty} \subseteq \{\sigma(t) | 0 < t \le 1\}.$$

Let us pull back $\sigma(t)$, $0 < t \le 1$ via $F|_{\Omega_1}$. We will obtain a curve $\gamma(t)$ in Ω_1 which realizes (a,b) as an asymptotic value of F. Hence we obtain $(a,b) \in A(F)$. □

4 The Asymptotic Values of a Polynomial Map $\mathbb{C}^2 \to \mathbb{C}^2$ form a Variety Which is the Union of Two Distinguished Algebraic Curves in $\mathbb{C}^2$

Definition 3. *Let $F : \mathbb{C}^2 \to \mathbb{C}^2$, $F(X,Y) = (P(X,Y), Q(X,Y))$ be a polynomial map, i.e., $P(X,Y), Q(X,Y) \in \mathbb{C}[X,Y]$. Let α, β be two indeterminates. We will denote by*

$$R(\alpha, \beta, X) = \textbf{resultant}(P(X,Y) - \alpha, Q(X,Y) - \beta, Y)$$

the resultant of $P(X,Y)-\alpha, Q(X,Y)-\beta$ *considered as polynomials in* Y. *We have*

$$R(\alpha,\beta,X) \in \mathbb{C}[\alpha,\beta,X].$$

We suppose that we have the following standard representation of this resultant as a polynomial in X

$$R(\alpha,\beta,X) = R_N(\alpha,\beta)X^N + \cdots + R_0(\alpha,\beta)$$

where for each $j = 0,\ldots,N$ *we have* $R_j(\alpha,\beta) \in \mathbb{C}[\alpha,\beta]$ *and where* $R_N(\alpha,\beta) \not\equiv 0$, *i.e.,* $\deg_X R = N$ *generically in* α,β. *Similarly we denote the resultant of* $P(X,Y)-\alpha, Q(X,Y)-\beta$ *relative to* X *by*

$$S(\alpha,\beta,Y) = \mathbf{resultant}(P(X,Y)-\alpha, Q(X,Y)-\beta, X)$$

and assume the following standard representation

$$S(\alpha,\beta,Y) = S_M(\alpha,\beta)Y^M + \cdots + S_0(\alpha,\beta)$$

where $S_M(\alpha,\beta)$ *is not the zero polynomial.*

The purpose of this section is to prove the following representation of the variety of the asymptotic values of a polynomial map $\mathbb{C}^2 \to \mathbb{C}^2$.

Theorem 2. *Let* $F : \mathbb{C}^2 \to \mathbb{C}^2$, $F(X,Y) = (P(X,Y), Q(X,Y))$ *be a polynomial map, then*

$$A_S(F) = \{(\alpha,\beta) | R_N(\alpha,\beta)S_M(\alpha,\beta) = 0\}.$$

Proof. Let $(a,b) \in A_S(F)$. We will prove that $R_N(a,b)S_M(a,b) = 0$. There exists a sequence $(X_n,Y_n) \in \mathbb{C}^2$ such that $\lim(|X_n|^2 + |Y_n|^2) = \infty$ and so that $\lim F(X_n,Y_n) = (a,b)$. At least one of the coordinate sequences X_n or Y_n is unbounded. Let us assume that X_n is unbounded. By restricting to a subsequence we may thus assume that $\lim |X_n| = \infty$. We intend to show that if this is the case then

$$R_N(a,b) = 0$$

(similarly if the case is that $\lim |Y_n| = \infty$ then it will follow that $S_M(a,b) = 0$). We consider $P(X,Y)-\alpha, Q(X,Y)-\beta$ as polynomials in Y and apply the Euclidean algorithm:

$$\begin{aligned}
(P(X,Y)-\alpha) &= q_1(Q(X,Y)-\beta) + r_1 \\
(Q(X,Y)-\beta) &= q_2 r_1 + r_2 \\
r_1 &= q_3 r_2 + r_3 \\
&\vdots \\
r_{p-2} &= q_p r_{p-1} + r_p \\
r_{p-1} &= q_{p+1} r_p
\end{aligned}$$

where $q_j(\alpha,\beta,X,Y)$, $r_j(\alpha,\beta,X,Y)$ are polynomials in Y and rational in (α,β,X) and where

$$\deg_Y Q > \deg_Y r_1 > \deg_Y r_2 > \cdots .$$

Since clearly $GCD(P(X,Y)-\alpha, Q(X,Y)-\beta)=1$ (when we consider these as polynomials in Y, because α,β are independent) we have

$$r_p \in \mathbb{C}(\alpha,\beta,X).$$

Let us denote $(a_n,b_n)=F(X_n,Y_n)$. Then we have the following

$$\lim a_n = a, \ \lim b_n = b,$$
$$\forall n \ r_p(a_n,b_n,X_n)=0.$$

Hence we have

$$\forall n \ R(a_n,b_n,X_n)=0.$$

However, we are in the case where $\lim|X_n|=\infty$ and hence we must have

$$R_N(a,b)=0$$

as desired.

We now have to show the inverse containment, namely, if $(a,b)\in\mathbb{C}^2$ satisfies

$$R_N(a,b)S_M(a,b)=0$$

then $(a,b)\in A_S(F)$. In fact we will show more:

As in the definition we let

$$R(\alpha,\beta,X)=R_N(\alpha,\beta)X^N+\cdots$$

then we will show

(i) If $R(a,b,X_0)=0$ then $(a,b)\in F(\mathbb{C}^2)$ and $\{X_0\}\subseteq \pi_X F^{-1}(a,b)$.
(ii) If $R_N(a,b)\neq 0$ then $|\pi_X F^{-1}(a,b)|=N$.
(iii) If $R_N(a,b)=0$ then $(a,b)\in A_S(F)$.

We have

$$R(\alpha,\beta,X)=A(\alpha,\beta,X,Y)(P(X,Y)-\alpha)+B(\alpha,\beta,X,Y)(Q(X,Y)-\beta)$$

where $A(\alpha,\beta,X,Y), B(\alpha,\beta,X,Y)\in\mathbb{C}[\alpha,\beta,X,Y]$ satisfy

$$\deg_Y A(\alpha,\beta,X,Y) < \deg_Y Q(X,Y),$$
$$\deg_Y B(\alpha,\beta,X,Y) < \deg_Y P(X,Y).$$

(i) If $R(a,b,X_0)=0$ then it follows that $P(X_0,Y)-a, Q(X_0,Y)-b$ have a common factor of a positive degree in Y and so (i) follows.

(ii) If $R_N(a,b) \neq 0$ then $\deg_X R(a,b,X) = N$ and by (i) for each one of the N zeros of $R(a,b,X)$ there is an X-fibre of $F^{-1}(a,b)$ and so $|\pi_X F^{-1}(a,b)| \geq N$. On the other hand if $X_0 \in \pi_X F^{-1}(a,b)$ then there is a Y_0 such that $F(X_0,Y_0) = (a,b)$ so that $R(a,b,X_0) = 0$ which proves $|\pi_X F^{-1}(a,b)| \leq N$.

(iii) Let us pick a sequence $\{(a_n,b_n)\}$ such that

$$\lim_{n\to\infty}(a_n,b_n) = (a,b), \ \forall n \ R_N(a_n,b_n) \neq 0.$$

Then for each n there are exactly N base points for the X-fibres of $F^{-1}(a_n,b_n)$. Let us denote these by

$$\pi_X F^{-1}(a_n,b_n) = \{X_{n1},\ldots,X_{nN}\}.$$

Since $R_N(a,b) = 0$ it follows that $|\pi_X F^{-1}(a,b)| < N$ and so by standard compactness argument there is a j, $1 \leq j \leq N$ such that $\lim_{n\to\infty}|X_{nj}| = \infty$. Hence $(a,b) \in A_S(F)$. □

We can now classify the asymptotic values of a polynomial map $\mathbb{C}^2 \to \mathbb{C}^2$ into the types X-infinite and Y-infinite. This should be compared to [20] where the notions of X-finite and Y-finite asymptotic values were defined.

Definition 4. *Let $F : \mathbb{C}^2 \to \mathbb{C}^2$ be a map (we may formulate the definition also for maps $f : \mathbb{R}^2 \to \mathbb{R}^2$). A sequential asymptotic value (X_0,Y_0) of F is called X-infinite (Y-infinite) if there exists a sequence $\{(X_j,Y_j)\}_{j=1}^{\infty}$ in $\mathbb{C}^2$ such that $\lim_{j\to\infty}|X_j| = \infty$ (such that $\lim_{j\to\infty}|Y_j| = \infty$) and such that*

$$\lim_{j\to\infty} F(X_j,Y_j) = (X_0,Y_0).$$

An immediate consequence of the proof of the last theorem is the following

Theorem 3. *Let $F : \mathbb{C}^2 \to \mathbb{C}^2$, $F(X,Y) = (P(X,Y),Q(X,Y))$ be a polynomial map, then the set of all the X-infinite asymptotic values of F is given by*

$$\{(\alpha,\beta)|R_N(\alpha,\beta) = 0\},$$

and the set of all the Y-infinite asymptotic values of F is given by

$$\{(\alpha,\beta)|S_M(\alpha,\beta) = 0\}.$$

By the Bezout theorem it follows that

Theorem 4. *Let $F : \mathbb{C}^2 \to \mathbb{C}^2$, $F(X,Y) = (P(X,Y),Q(X,Y))$ be a polynomial map, then the set of asymptotic values of F which are both X-infinite and Y-infinite is either an infinite algebraic curve or contains at most $\deg R_N \times \deg S_M$ points.*

5 The Resultant Formulation of the Jacobian Conjecture

Before stating as a theorem this equivalent formulation of the Jacobian conjecture, it will be convenient to formally define few notions.

Definition 5. *Let n be a positive integer. A generic polynomial of degree n in the two indeterminates X, Y is a polynomial of the form*

$$P(X, Y, \{a_{ij}\}_{i+j\le n}) = \sum_{i+j\le n} a_{ij} X^i Y^j.$$

Thus $P(X, Y, \{a_{ij}\}_{i+j\le n}) \in \mathbb{C}[X, Y, \{a_{ij}\}_{i+j\le n}]$ and it has the following properties:

(1) All the coefficients of P are 1.
(2) The number of indeterminates of P is $\binom{n+2}{2} + 2$, for there are $\binom{n+2}{2}$ indeterminates of the form a_{ij}, $i + j \le n$ plus X, Y.
(3) P is a linear form in $\{a_{ij}\}_{i+j\le n}$.
(4) $\deg_X P = \deg_Y P = \deg P = n$.

Definition 6. *Let n be a positive integer. A generic polynomial map of degree n in the two indeterminates X, Y is an ordered pair of generic polynomials of degree n in X, Y, i.e.*

$$F = (P(X, Y, \{a_{ij}\}_{i+j\le n}), Q(X, Y, \{b_{ij}\}_{i+j\le n})),$$

thus F involves $2\binom{n+2}{2}+2$ indeterminates $\{a_{ij}\}_{i+j\le n}$, $\{b_{ij}\}_{i+j\le n}$, X, Y. The Jacobian of F is

$$J_F(X, Y, \{a_{ij}\}_{i+j\le n}, \{b_{ij}\}_{i+j\le n}) = \partial(P, Q)/\partial(X, Y),$$

thus $J_F(X, Y, \{a_{ij}\}_{i+j\le n}, \{b_{ij}\}_{i+j\le n}) \in \mathbb{C}[X, Y, \{a_{ij}\}_{i+j\le n}, \{b_{ij}\}_{i+j\le n}]$ and it has the following properties:

(1) The coefficients of J_F are integers m such that $|m| \le n^2$.
(2) The number of indeterminates of J_F is $2\binom{n+2}{2} + 2$.
(3) J_F is a bilinear form in $(\{a_{ij}\}_{i+j\le n}, \{b_{ij}\}_{i+j\le n})$.
(4) $\deg_X J_F = \deg_Y J_F = \deg J_F = 2(n-1)$.

Definition 7. *Let F be a generic map of degree n with the polynomial coordinates $P(X, Y, \{a_{ij}\}_{i+j\le n})$ and $Q(X, Y, \{b_{ij}\}_{i+j\le n})$. Let α, β be two new indeterminates and consider*

$$P(X, Y, \{a_{ij}\}_{i+j\le n}) - \alpha,\ \ Q(X, Y, \{b_{ij}\}_{i+j\le n}) - \beta$$

as polynomials in Y. As such, let us compute the resultant of these two polynomials. It will be denoted by

$$\textbf{resultant}(P - \alpha, Q - \beta, Y).$$

Thus **resultant**$(P-\alpha, Q-\beta, Y) \in \mathbb{C}[X, \alpha, \beta, \{a_{ij}\}_{i+j\le n}, \{b_{ij}\}_{i+j\le n}]$. *Similarly, let us consider $P-\alpha, Q-\beta$ as polynomials in X and as such let us compute the resultant of these two polynomials. It will be denoted by*

$$\textbf{resultant}(P-\alpha, Q-\beta, X),$$

thus **resultant**$(P-\alpha, Q-\beta, X) \in \mathbb{C}[Y, \alpha, \beta, \{a_{ij}\}_{i+j\le n}, \{b_{ij}\}_{i+j\le n}]$.

Definition 8. *Let n be a positive integer. The Jacobian variety of degree n is the variety*

$$V_C(J_F(X, Y, \{a_{ij}\}_{i+j\le n}, \{b_{ij}\}_{i+j\le n}) - 1)$$

where J_F is considered to be a polynomial of (X, Y).

Thus $V_C(J_F - 1)$ is given by $\binom{2(n-1)+2}{2} = \binom{2n}{2}$ equations which are bilinear forms in $(\{a_{ij}\}_{i+j\le n}, \{b_{ij}\}_{i+j\le n})$ (so in only $2\binom{n+2}{2} - 2$ indeterminates). All the equations are homogeneous except for the equation

$$a_{10}b_{01} - a_{01}b_{10} = 1.$$

Definition 9. *Let n be a positive integer. Consider a generic polynomial map of degree n, $F = (P, Q)$. If we consider* **resultant**$(P-\alpha, Q-\beta, Y)$ *as a polynomial in X then each of it's coefficients is a polynomial in $\mathbb{C}[\alpha, \beta, \{a_{ij}\}_{i+j\le n}, \{b_{ij}\}_{i+j\le n}]$. We consider these coefficients as polynomials in (α, β) with coefficients which are in $\mathbb{C}[\{a_{ij}\}_{i+j\le n}, \{b_{ij}\}_{i+j\le n}]$. The highest coefficient of* **resultant**$(P-\alpha, Q-\beta, Y)$ *is the first coefficient (highest degree of X) which is not the zero polynomial in (α, β) when it's coefficients are considered as regular functions on the Jacobian variety $V_C(J_F - 1)$. We shall denote the highest coefficient of* **resultant**$(P-\alpha, Q-\beta, Y)$ *by $R_n(\alpha, \beta, \{a_{ij}\}_{i+j\le n}, \{b_{ij}\}_{i+j\le n})$. Similarly, we denote the highest coefficient of* **resultant**$(P-\alpha, Q-\beta, X)$ *by $S_n(\alpha, \beta, \{a_{ij}\}_{i+j\le n}, \{b_{ij}\}_{i+j\le n})$.*

We can now state and prove our theorem.

Theorem 5. *The Jacobian conjecture is equivalent to*

$$\forall n, \ R_n, S_n \in \mathbb{C}^*.$$

Moreover, the Jacobian conjecture is valid for all maps of degree at most N iff

$$R_N, S_N \in \mathbb{C}^*.$$

Proof. We shall prove the second equivalence which clearly implies the first. Let N be a positive integer.

The Jacobian conjecture is valid for all maps of degree at most $N \iff$ (by Hadamard's condition)

$\forall(\{a_{ij}\}_{i+j\le n}, \{b_{ij}\}_{i+j\le n}) \in V_C(J_F - 1)$, F which is generated by $(\{a_{ij}\}_{i+j\le n}, \{b_{ij}\}_{i+j\le n})$ has no asymptotic values $\iff$ (by Theorem 2)

R_N and S_N do not vanish $\iff$ (by the Fundamental Theorem of Algebra)

$R_N, S_N \in \mathbb{C}^*$. □

Remark 4. By symmetricity it is clear that it suffices to prove only $R_N \in \mathbb{C}^*$.

Example 2.

$$F(X,Y) = (X + (Y + X^2)^3, Y + X^2).$$

This map is tame since it factors as follows

$$(X,Y) \to (X, Y + X^2) \to F(X,Y).$$

Here $\partial(P,Q)/\partial(X,Y) \equiv 1$ and also $\mathbf{resultant}(P-\alpha, Q-\beta, Y) = -X + (\alpha - \beta^3)$. Hence $R_N = -1$ and it never vanishes which is consistent with Theorem 2.

Example 3.

$$F(X,Y) = (Y^2 + XY, Y^3 + X^2Y^2 + X^2Y + X).$$

This map has coordinates that belong to

$$\mathbb{C}[Y, XY, X^2Y + X] = I(U^{-1}, VU^2 - U),$$

see [21, 22]. So that it must have asymptotic values. Here

$$\partial(P,Q)/\partial(X,Y) = (3 - 4X)Y^3 - 4XY^2 - (2 + X^2)Y - X$$

and also

$$\begin{aligned}\mathbf{resultant}(P-\alpha, Q-\beta, Y) &= (1+\alpha)X^5 + (-2 + \alpha^2 - 2\alpha - \beta)X^4\\ &+ (2\alpha + 2\beta + \alpha^2)X^3 + (1 - 2\alpha\beta - 3\alpha - 2\alpha^2)X^2\\ &+ (-2\beta + 3\beta\alpha)X + (\beta^2 - \alpha^3).\end{aligned}$$

So that $R_N = 1 + \alpha$ and hence by Theorem 3 (a,b) is an X-infinite asymptotic value of $F(X,Y)$ iff $1 + a = 0$. We note that the dual map (see [20–22]) to $F(X,Y)$ is:

$$G(U,V) = ((VU^2 - U)^2 + VU - 1, (VU^2 - U)^3 + (VU - 1)^2 + V).$$

Hence $\lim_{U\to 0} F(U^{-1}, VU^2 - U) = G(0,V) = (-1, 1 + V)$ so that the set

$$\{(-1,T) |\ T \in C\}$$

is contained in the variety of asymptotic values of $F(X,Y)$ which agrees with $a = -1$ above.

Example 4. We shall now prove the Jacobian conjecture for maps of degree at most 2 using Theorem 5.

$$\begin{aligned}P(X,Y) &= aX^2 + bXY + cY^2 + dX + eY\\ &= cY^2 + (bX + e)Y + (aX^2 + dX),\\ Q(X,Y) &= AX^2 + BXY + CY^2 + DX + EY\\ &= CY^2 + (BX + E)Y + (AX^2 + DX).\end{aligned}$$

We have

$$\begin{aligned}\partial(P,Q)/\partial(X,Y) = 2(aB - Ab)X^2 + 4(aC - Ac)XY \\ + 2(bC - Bc)Y^2 + [2(aE - Ae) + (dB - Db)]X \\ + [(bE - Be) + 2(dC - Dc)]Y + (dE - De).\end{aligned}$$

So the Jacobian variety of degree 2 is given by

$$\begin{aligned}V_C((aB - Ab, aC - Ac, bC - Bc, 2(aE - Ae) + (dB - Db), \\ (bE - Be) + 2(dC - Dc), dE - De - 1)).\end{aligned}$$

We now compute **resultant**$(P - \alpha, Q - \beta, Y)$ and reduce the coefficients of this polynomial (as a polynomial in X) modulo the above Jacobian variety. The result is

$$\textbf{resultant}(P - \alpha, Q - \beta, Y) = (cE - eC)X + Const.$$

Thus in our notations we have

$$R_2 = cE - eC.$$

We need to prove that $R_2 \in \mathbb{C}^*$. However on the Jacobian variety of degree 2 we have

$$aB - Ab = aC - Ac = bC - Bc = 0,$$

so if also $cE - eC = 0$ then this would have implied $dE - De = 0$. This cannot be for on the Jacobian variety we have $dE - De = 1$. Hence $R_2 = cE - eC \in \mathbb{C}^*$ which proves the Jacobian conjecture for all maps of degree 2 at most.

6 The Jacobian Conjecture in Dimension 2 is Decidable

We are now going to generalize the last example in the sense that we shall present an algorithm that gets as an input a positive integer n and decides whether the Jacobian conjecture in dimension 2 holds true for all the polynomial maps of degree at most n. More precisely we have the following:

Theorem 6. *There exists an algorithm such that for any given positive integer n it decides (after finitely many steps) if the following statement is true or not. If it is not true, then the algorithm outputs a counterexample (in fact, it generates all the counterexamples). The statement is:*

Let $P(X,Y), Q(X,Y) \in \mathbb{C}[X,Y]$ and let $\deg P, \deg Q \leq n$. If $\partial(P,Q)/\partial(X,Y) \equiv 1$ then the polynomial map $F = (P,Q)$ is invertible.

Proof. Given n we shall describe the algorithm step by step.

Step 1. Generate the Jacobian variety of degree n. To do this generate the $\binom{2n}{2}$ equations that define the variety. These are equations in $2\binom{n+2}{2} - 2$ indeterminates, $(\{a_{ij}\}_{i+j\leq n}, \{b_{ij}\}_{i+j\leq n})$, which are determined by bilinear forms in those variables. All the equations are homogeneous except for the equation $a_{10}b_{01} - a_{01}b_{10} = 1$.

Computations:

$$P \leftarrow \sum_{1 \leq i+j \leq n} a_{ij} X^i Y^j,$$

$$Q \leftarrow \sum_{1 \leq i+j \leq n} b_{ij} X^i Y^j,$$

$$J \leftarrow \partial(P,Q)/\partial(X,Y)$$

$$= \sum_{0 \leq i+j \leq 2(n-1)} \left[\sum_{\substack{i_1+i_2-1=i \\ j_1+j_2-1=j}} i_1 j_2 (a_{i_1 j_1} b_{i_2 j_2} - a_{i_2 j_2} b_{i_1 j_1}) \right] X^i Y^j.$$

So the Jacobian variety of degree n is given by the formula

$$V_n \leftarrow V_{\mathbb{C}} \left(\left\{ \sum_{\substack{i_1+i_2-1=i \\ j_1+j_2-1=j}} i_1 j_2 (a_{i_1 j_1} b_{i_2 j_2} - a_{i_2 j_2} b_{i_1 j_1}) | 1 \leq i+j \leq 2(n-1) \right\}, \right.$$
$$\left. a_{10}b_{01} - a_{01}b_{10} - 1 \right).$$

Step 2. Calculate **resultant**$(P - \alpha, Q - \beta, Y)$.

Computations:

$$P - \alpha \leftarrow \sum_{j=0}^{n} \left(\sum_{i=0}^{n-j} a_{ij} X^i \right) Y^j - \alpha,$$

$$Q - \beta \leftarrow \sum_{j=0}^{n} \left(\sum_{i=0}^{n-j} b_{ij} X^i \right) Y^j - \beta.$$

To compute the resultant we do not expand the determinant of a $2(n+1) \times 2(n+1)$ matrix. Instead we use the following:

Fact 1: If $\deg f(Y) = n$ and $g(Y) = c \equiv Const.$ then

$$\textbf{resultant}(f(Y), g(Y), Y) = \textbf{resultant}(f(Y), c, Y) = \mathbb{C}^n.$$

Fact 2: If $\deg f(Y) = n$ and $\deg g(Y) = m$ then

$$\textbf{resultant}(f(Y), g(Y), Y) = (-1)^{nm} \textbf{resultant}(g(Y), f(Y), Y).$$

Fact 3: If $f = A_n Y^n + \cdots + A_0$ and $g = B_m Y^m + \cdots + B_0$ and $m \leq n$ we let

$$h = f - (A_n / B_m) Y^{n-m} g,$$

and then we have

$$\mathbf{resultant}(f(Y), g(Y), Y) = B_m^{n-\deg h}\mathbf{resultant}(h(Y), g(Y), Y).$$

We note that $\deg h \leq n-1$.

Fact 4: Follows from Fact 3:

If we use the division algorithm to write $f = qg + r$ where $\deg r < \deg g$ then

$$\mathbf{resultant}(f(Y), g(Y), Y) = B_m^{n-\deg r}\mathbf{resultant}(r(Y), g(Y), Y).$$

We can now iterate using Facts 4 and 2 in order to reduce degrees. For example if using the division algorithm we write $f = q_1 g + r_1,\ \ g = q_2 r_1 + r_2$ then

$$\begin{aligned}\mathbf{resultant}(f, g, Y) &= B_m^{\deg f - \deg r_1}\mathbf{resultant}(r_1, g, Y)\\ &= (-1)^{\deg g \deg r_1} B_m^{\deg f - \deg r_1}\mathbf{resultant}(g, r_1, Y)\\ &= (-1)^{\deg g \deg r_1} B_m^{\deg f - \deg r_1} C_l^{\deg g - \deg r_2}\\ &\quad \times \mathbf{resultant}(r_2, r_1, Y)\\ &= (-1)^{\deg g \deg r_1 + \deg r_1 \deg r_2} B_m^{\deg f - \deg r_1} C_l^{\deg g - \deg r_2}\\ &\quad \times \mathbf{resultant}(r_1, r_2, Y),\end{aligned}$$

where B_m, C_l are the leading coefficients of g, r_1. To put that in a form of a pseudocode:
Input: f, g
Output: $\mathbf{resultant}(f, g, Y)$

$$h \leftarrow f$$
$$s \leftarrow g$$
$$res \leftarrow 1$$

WHILE $\deg s > 0$ DO

$$r \leftarrow remainder(h, s)$$
$$Res \leftarrow (-1)^{\deg r \deg s} lead(s)^{\deg h - \deg r} Res$$
$$h \leftarrow s$$
$$s \leftarrow r$$

IF $h = 0$ OR $s = 0$ THEN $Res \leftarrow 0$ ELSE
IF $\deg h > 0$ THEN $Res \leftarrow s^{\deg h} Res$
END

Step 3. Reduce the coefficients of $\mathbf{resultant}(P - \alpha, Q - \beta, Y)$ modulo the Jacobian variety of degree n, V_n. Start with the highest coefficient and proceed till you get to the first coefficient R_n which is different from 0.

Computations:

By Step 1 we have

$$V_n \leftarrow V_{\mathbb{C}}\Bigg(\Bigg\{ \sum_{\substack{i_1+i_2-1=i\\ j_1+j_2-1=j}} i_1 j_2 (a_{i_1 j_1} b_{i_2 j_2} - a_{i_2 j_2} b_{i_1 j_1}) | 1 \le i+j \le 2(n-1)\Bigg\},$$

$$a_{10}b_{01} - a_{01}b_{10} - 1\Bigg).$$

Find a Gröbner basis G_n for the ideal generated by the defining equations of V_n. Consider the coefficient C_N of X^N in **resultant**$(P-\alpha, Q-\beta, Y)$. C_N is a polynomial in

$$(\alpha, \beta, \{a_{ij}\}_{1\le i+j\le n}, \{b_{ij}\}_{1\le i+j\le n}).$$

Express

$$C_N = \sum A_{ij}(\{a_{kl}\}, \{b_{kl}\})\alpha^i \beta^j,$$

where $A_{ij} \in \mathbb{C}[\{a_{ij}\}_{i+j\le n}, \{b_{ij}\}_{i+j\le n}]$. For each (i,j) divide the polynomial A_{ij} by the Gröbner basis $G_n = \{g_1, \dots, g_t\}$ of degree n to get

$$A_{ij} = q_1 g_1 + \cdots + q_t g_t + r_{ij}.$$

We have

$$A_{ij} \equiv 0 \pmod{V_n} \iff r_{ij} \equiv 0.$$

This algorithm, obviously, terminates after finitely many steps. If $R_n \in \mathbb{C}^*$ then the Jacobian conjecture is established up to (including) degree n. Otherwise it is clear how to generate all the counterexamples for degree n. □

Remark 5. For Step 3, references to the Buchberger algorithm for computing Gröbner basis are [3, 4, 8].

The complexity of the Buchberger type algorithms is huge. There are some results on uniform upper bounds on the degrees of the Gröbner basis elements in terms of the original generators [4, 13, 16].

The bounds on the degrees are large. That is a necessity in some sense, for in [13] there are examples of ideals generated by polynomials of degree n at most, whose Gröbner basis involve polynomials of degree proportional to 2^{2^n}.

Remark 6. The algorithm above was implemented by the author (while visiting the University of Michigan) on a Sun platform. He used the standard **Gröbner package in Maple**. The input to the program was a positive integer d and the output was a **true/false** flag. If **true** then the Jacobian conjecture was found to be valid for all the complex polynomial mappings in dimension two and of degree d or less. A **false** flag meant the other alternative and a counterexample of degree d or less would have been given. The package was able to automatically prove the conjecture for mappings of degree 15 or less. The running time was about 40 h. It was instrumental to observe how fast the running time of the program went up with the degree of the polynomials. For low degrees (up to degree 5) it finished within less than 5 min. Increasing the degree further had a very significant effect on the elapsed running time. It quickly climbed to about one hour in degree 7 to about three hours up to degree 10. It than went to an overnight running time in degree 13 and degrees 14, 15 quickly reached the few days running time. The author did not check the memory size that was involved and it is quite possible that the elapsed running time had a significant portion due to memory paging. It hints that very quickly one expects the combination of the max degree of the reduced Gröbner basis and its size (in terms of how many polynomials it contains) to get very large.

Remark 7. After the Linz conference Massimiliano (Max) Sala went into programming the algorithm more professionally. It is now an ongoing project of Max and his students. The project started while Max was in Ireland and it continues now in 2009 while he is in Trento, Italy. In the very first experiments Max used the degrevlex. For $n = 2$ the Jacobian ideal was a radical. It had 13 elements, 9 of degree 2 and 4 of degree 3. The 9 degree 2 polynomials formed a minimal basis and so the 4 degree 3 polynomials were inessential. The largest coefficient was 4 which was a good sign (usually the coefficients of the members of the basis might get very large). The 9 polynomials seemed to be sparse. Four of them had very nice structure: $b_{11}^2 - 4b_{20}b_{02}$, $a_{11}^2 - 4a_{20}a_{02}$, $a_{11}b_{11} - 4a_{20}b_{02}$ and $a_{02}b_{20} - a_{20}b_{02}$. However, already in degree $n = 3$ the Gröbner basis had about 3000 polynomials and much less structure was visible.

7 A Straight Forward Inductive Approach Fails

The results of the previous section motivate an inductive approach to deal with the Jacobian conjecture. The purpose of this section is to describe the simplest possible such a trial and to show that it does not work. This may indicate a negative evidence for the validity of the conjecture in dimension two.

Let $n > 1$ be an integer. We are going to make the following **induction hypothesis:**

Let $P(X,Y), Q(X,Y) \in \mathbb{C}[X,Y]$ satisfy

(1) $\deg_Y P = \deg_Y Q = \deg P = \deg Q = n - 1$,
(2) $\partial(P,Q)/\partial(X,Y) \in \mathbb{C}^*$.

Then

$$R_{n-1} \in \mathbb{C}^*.$$

We now proceed to the proof of the following:

Let $P(X,Y), Q(X,Y) \in \mathbb{C}[X,Y]$ satisfy

(1) $\deg_Y P = \deg_Y Q = \deg P = \deg Q = n$,
(2) $\partial(P,Q)/\partial(X,Y) \in \mathbb{C}^*$.

Then

$$P(X,Y) = cY^n + \cdots, \quad Q(X,Y) = CY^n + \cdots$$

for some $c, C \in \mathbb{C}^*$. Let

$$P_1(X,Y) = CP(X,Y) - cQ(X,Y), \quad Q_1(X,Y) = cQ(X,Y).$$

Then

$$\begin{aligned} P_1(X,Y) &= C[cY^n + (bX+e)Y^{n-1} + \cdots] - c[CY^n + (BX,E)Y^{n-1} + \cdots] \\ &= [(Cb - cB)X + (Ce - cE)]Y^{n-1} + \cdots . \end{aligned}$$

Now

$$\begin{aligned} \partial(P,Q)/\partial(X,Y) &= \begin{vmatrix} bY^{n-1} + \cdots + ncY^{n-1} + \cdots \\ BY^{n-1} + \cdots + nCY^{n-1} + \cdots \end{vmatrix} \\ &= n(bC - Bc)Y^{2(n-1)} + \cdots + \in \mathbb{C}^*. \end{aligned}$$

So that $Cb - cB = 0$ and we obtain

$$P_1(X,Y) = (Ce - cE)Y^{n-1} + \cdots .$$

We note that

$$\begin{aligned} \partial(P_1,Q_1)/\partial(X,Y) &= \partial(CP - cQ, cQ)/\partial(X,Y) = \partial(CP, cQ)/\partial(X,Y) \\ &= cC\partial(P,Q)/\partial(X,Y) \in \mathbb{C}^*. \end{aligned}$$

Also

$$\begin{aligned} &\mathbf{resultant}(P_1 - \alpha, Q_1 - \beta, Y) = \mathbf{resultant}(CP - cQ - \alpha, cQ - \beta, Y) \\ &\quad = \{(-1)^n/cC\}\mathbf{resultant}(CP - (\alpha+\beta), cQ - \beta, Y). \end{aligned}$$

This shows that

$$R_n^{P_1,Q_1}(\alpha,\beta) = \{(-1)^n/cC\}R_n^{CP,cQ}(\alpha+\beta,\beta).$$

Here R_n^F denotes the highest coefficient of the Y-resultant that corresponds to the map F. We have $\deg_Y P_1 \le n-1$ so we could have used the induction hypothesis if $\deg_Y Q_1 \le n-1$. However $\deg_Y Q_1 = \deg_Y Q = n$ unfortunately.

Looking at

$$\begin{aligned} P_1(X,Y) &= (Ce - cE)Y^{n-1} + \cdots , \\ Q_1(X,Y) &= cCY^n + c(BX+E)Y^{n-1} + \cdots , \end{aligned}$$

and assuming for a moment that $Ce - cE \in \mathbb{C}^*$ it could have made sense to divide Q_1 by P_1 and consider:

$$P_2(X,Y) = cCP_1(X,Y),$$
$$Q_2(X,Y) = (Ce - cE)Q_1(X,Y) - cCP_1(X,Y)Y,$$

for then we would get $\deg_Y Q_2 \le n-1$ and the resultants of (P_1, Q_1) and of (P_2, Q_2) would have differ by a constant non-zero factor. However, this transformation destroys the Jacobian condition for

$$\partial(P_2,Q_2)/\partial(X,Y) = cC(Ce - cE)\partial(P_1,Q_1)/\partial(X,Y) - (cC)^2 P_1 P_{1X}.$$

Going back to

$$P_1(X,Y) = (Ce - cE)Y^{n-1} + \cdots,$$
$$Q_1(X,Y) = cCY^n + c(BX + E)Y^{n-1} + \cdots,$$

we might look for a more general transformation on $Q_1(X,Y)$ than just the division transform by P_1. Namely, we might want to find a polynomial

$$S(X,Y) = -cCY^n + \cdots$$

such that $\partial(P_1,S)/\partial(X,Y) \equiv 0$ for then if we consider

$$P_2(X,Y) = P_1(X,Y),$$
$$Q_2(X,Y) = Q_1(X,Y) - S(X,Y),$$

then also $\deg_Y Q_2 \le n-1$ but this time the Jacobian condition is preserved for

$$\begin{aligned}\partial(P_2,Q_2)/\partial(X,Y) &= \partial(P_1,Q_1+S)/\partial(X,Y)\\ &= \partial(P_1,Q_1)/\partial(X,Y) + \partial(P_1,S)/\partial(X,Y)\\ &= \partial(P_1,Q_1)/\partial(X,Y) \in \mathbb{C}^*.\end{aligned}$$

There is also a hope to find a relation between $R_{n-1}^{P_2,Q_2}$ and $R_{n-1}^{P_1,Q_1}$.

However, there is no hope in finding such a $S(X,Y)$ except in the case $n = 2$ in which the Jacobian conjecture is true. The reason lies in the fact that if $P(X,Y) \in \mathbb{C}[X,Y]$ and if we define the following derivation on $\mathbb{C}[X,Y]$

$$D_P : \mathbb{C}[X,Y] \to \mathbb{C}[X,Y],$$
$$D_P(g) = \partial(P,g)/\partial(X,Y),$$

then, as we shall see in the next four sections we have

$$\ker(D_P, \mathbb{C}[X,Y]) = \mathbb{C}[P].$$

Thus if $P(X,Y) = Y^{n-1} + \cdots$ and $S(X,Y) = Y^n + \cdots$ satisfy $D_P(S) = 0$ then $S(X,Y) = G(P(X,Y))$ for some $G(T) \in \mathbb{C}[T]$ so $Y^n + \cdots = G(Y^{n-1} + \cdots)$ which implies that

$$n = (n-1)\deg G$$

so that $n = \deg G = 2$ is the only solution. It is indeed a solution for we take $G(T) = T^2$ and $S = P^2$.

We will discuss this property of the Jacobian derivation D_P as well as other properties on the next four sections.

8 Elementary Properties of Resultants of Jacobian Pairs

Since we have a resultant formulation for the Jacobian conjecture it makes sense to establish some properties of resultants of Jacobian pairs. This section is dedicated to the more elementary such properties.

Proposition 2. *Let $P(X,Y), Q(X,Y) \in \mathbb{C}[X,Y]$ be a Jacobian pair. Let α and β be indeterminates. Then the highest coefficient of* **resultant**$(P(X,Y)-\alpha, Q(X,Y)-\beta, Y)$ *cannot be the coefficient of X^0.*

Proof. Assume that the proposition is false. Then

$$\mathbf{resultant}(P(X,Y)-\alpha, Q(X,Y)-\beta, Y) = R_0(\alpha,\beta) \in \mathbb{C}[\alpha,\beta].$$

We consider 2 cases:
Case 1: $R_0(\alpha,\beta) \in \mathbb{C}^*$.
Then for any pair (α,β) the equations

$$P(X,Y)-\alpha = 0,$$
$$Q(X,Y)-\beta = 0,$$

have no solution which is a contradiction.
Case 2: Let $(\alpha_0,\beta_0) \in \mathbb{C}^2$ satisfy $R_0(\alpha_0,\beta_0)=0$.
Then for any X there exists a $Y(X)$ such that

$$P(X,Y(X)) = \alpha_0,$$
$$Q(X,Y(X)) = \beta_0.$$

Differentiations of these equations with respect to X give

$$P_X(X,Y(X)) + Y'(X)P_Y(X,Y(X)) = 0,$$
$$Q_X(X,Y(X)) + Y'(X)Q_Y(X,Y(X)) = 0.$$

We consider these as two homogeneous equations in $(1, Y'(X))$. Then we get

$$\begin{vmatrix} P_X(X,Y(X)) P_Y(X,Y(X)) \\ Q_X(X,Y(X)) Q_Y(X,Y(X)) \end{vmatrix} = 0$$

which contradicts the Jacobian condition. □

Remark 8. We shall call the argument that was used in the proof of case 2 above, "the level curve principle". It is merely the obvious fact that a Jacobian pair cannot share a common level curve.

Remark 9. In case 2 above we could have argued differently:
If $R_0(\alpha,\beta) \equiv 0$ then $P(X,Y)-\alpha$, $Q(X,Y)-\beta$ had to have a common factor of a positive degree in Y. That is clearly impossible for α,β are independent (also because $P(X,Y)-\alpha, Q(X,Y)-\beta$ is a Jacobian pair so these polynomials must be coprime).

If $\deg R_0(\alpha, \beta) > 0$ then the equations

$$P(X,Y) - \alpha = 0,$$
$$Q(X,Y) - \beta = 0,$$

had a solution only along the algebraic curve $R_0(\alpha, \beta) = 0$. So the image of the map $F(X,Y) = (P(X,Y), Q(X,Y))$ collapses to the algebraic curve $R_0(\alpha, \beta) = 0$. That cannot be for F is étale and hence in particular an open map.

Proposition 3. *Let $P(X,Y), Q(X,Y) \in \mathbb{C}[X,Y]$ be a Jacobian pair. Let α and β be indeterminates. Then there cannot exist a pair (α_0, β_0) such that*

$$\forall X \ \mathbf{resultant}(P(X,Y) - \alpha_0, Q(X,Y) - \beta_0, Y) \equiv 0.$$

Proof. Assume that there exists such a pair $(\alpha_0, \beta_0) \in \mathbb{C}^2$. Then we use, as before, the level curve principle to arrive at a contradiction, namely:

for any X there exists a $Y(X)$ such that

$$P(X, Y(X)) = \alpha_0,$$
$$Q(X, Y(X)) = \beta_0,$$

and we arrive at a contradiction. □

Remark 10. This proposition generalizes case 2 in the proof of Proposition 2.

Corollary 1. *Let $P(X,Y), Q(X,Y) \in \mathbb{C}[X,Y]$ be a Jacobian pair. Let α and β be indeterminates. Let*

$$\mathbf{resultant}(P(X,Y) - \alpha, Q(X,Y) - \beta, Y) = R_N(\alpha,\beta)X^N + \cdots + R_0(\alpha,\beta),$$

where $R_N(\alpha, \beta)$ is the highest non-vanishing (identically) coefficient of $\mathbf{resultant}(P(X,Y) - \alpha, Q(X,Y) - \beta, Y)$. *Then $R_j(\alpha, \beta)$, $0 \le j \le N$, cannot share a zero. In particular the highest common divisor of $(R_N(\alpha, \beta), \ldots, R_0(\alpha, \beta))$ is 1.*

Corollary 2. *Let $P(X,Y), Q(X,Y) \in \mathbb{C}[X,Y]$ be a Jacobian pair. Let α and β be indeterminates. Let*

$$\mathbf{resultant}(P(X,Y) - \alpha, Q(X,Y) - \beta, Y) = R_1(\alpha,\beta)X + R_0(\alpha,\beta),$$

where $R_1(\alpha, \beta)$ is the highest non-vanishing (identically) coefficient of $\mathbf{resultant}(P(X,Y) - \alpha, Q(X,Y) - \beta, Y)$. *Then the image of the map $F(X,Y) = (P(X,Y), Q(X,Y))$ is*

$$\mathbb{C}^2 - \{(\alpha,\beta) | R_1(\alpha,\beta) = 0\}.$$

Proof. If $R_1(\alpha_0, \beta_0) = 0$ then by the previous corollary $R_0(\alpha_0, \beta_0) \neq 0$ so that the resultant is not 0 and the equations

$$P(X,Y) = \alpha_0,$$
$$Q(X,Y) = \beta_0,$$

have no solution.

If, on the other hand, $R_1(\alpha_0, \beta_0) \neq 0$ then $X_0 = -R_0(\alpha_0, \beta_0)/R_1(\alpha_0, \beta_0)$ is the zero of the resultant and so there exists a Y_0 such that

$$P(X_0, Y_0) = \alpha_0,$$
$$Q(X_0, Y_0) = \beta_0.$$

So $(\alpha_0, \beta_0) \in F(\mathbb{C}^2)$. □

Corollary 3. *Let $P(X,Y), Q(X,Y) \in \mathbb{C}[X,Y]$ be a Jacobian pair. Let α and β be indeterminates. If*

$$\textbf{resultant}(P(X,Y) - \alpha, Q(X,Y) - \beta, Y) = R_1 X + R_0(\alpha, \beta),$$

where $R_1 \in \mathbb{C}^*$ then $F(X,Y) = (P(X,Y), Q(X,Y))$ is onto $\mathbb{C}^2$.

In fact we can do better than in the last two corollaries. Let us prove the following covering theorem.

Theorem 7. *Let $P(X,Y), Q(X,Y) \in \mathbb{C}[X,Y]$ and denote $F(X,Y) = (P(X,Y), Q(X,Y))$. Then there are finitely many polynomials $R_N(\alpha, \beta), \ldots, R_1(\alpha, \beta) \in \mathbb{C}[\alpha, \beta]$ such that*

$$\mathbb{C}^2 - F(\mathbb{C}^2) = \{(a,b) \in \mathbb{C}^2 | R_N(a,b) = \cdots = R_1(a,b) = 0\}$$
$$= V_{\mathbb{C}}(R_N, \ldots, R_1).$$

Moreover, $\mathbb{C}^2 - F(\mathbb{C}^2)$ is infinite iff $\partial(P,Q)/\partial(X,Y) \equiv 0$, provided that the highest coefficients of X, Y are non-zero constants.

Remark 11. This theorem is the analog of Picard's Little Theorem for analytic functions.

Proof. Let α, β be two new indeterminates and let us form

$$R(X) = \textbf{resultant}(P(X,Y) - \alpha, Q(X,Y) - \beta, Y) \in \mathbb{C}[X, \alpha, \beta].$$

Suppose that we have the following representation of $R(X)$ as a polynomial in X

$$R(X) = R_N(\alpha, \beta) X^N + \cdots + R_1(\alpha, \beta) + R_0(\alpha, \beta),$$

where $R_N(\alpha, \beta), \ldots, R_0(\alpha, \beta) \in \mathbb{C}[\alpha, \beta]$. Then we have $(a,b) \in \mathbb{C}^2 - F(\mathbb{C}^2) \iff$ there is no solution in (X,Y) to $P(X,Y) - a = Q(X,Y) - b = 0 \iff$ for any X_0, $R(X_0) \neq 0$ where $(\alpha, \beta) \leftarrow (a,b) \iff R(X) \in \mathbb{C}^*$ where $(\alpha, \beta) \leftarrow (a,b)$

$\iff R(X) = R_N(a,b)X^N + \cdots + R_0(a,b) = R_0(a,b) \neq 0 \iff R_N(a,b) = \cdots, R_1(a,b) = 0$

Moreover, by the Bezout Theorem $\mathbb{C}^2 - F(\mathbb{C}^2)$ is infinite $\iff L(\alpha,\beta) = (R_N(\alpha,\beta),\ldots,R_1(\alpha,\beta)) \in \mathbb{C}[\alpha,\beta] - \mathbb{C}^* \iff L(P(X_0,Y_0),Q(X_0,Y_0)) \neq 0$ $\forall (X_0,Y_0) \iff L(P(X,Y),Q(X,Y)) = c \in \mathbb{C}^* \iff \partial(P,Q)/\partial(X,Y) \equiv 0.$ □

Corollary 4. *If*

$$\deg_X \mathbf{resultant}(P(X,Y)-\alpha, Q(X,Y)-\beta, Y) = 1,$$

then either $\partial(P,Q)/\partial(X,Y) \equiv 0$ *or* $F(X,Y) = (P(X,Y),Q(X,Y))$ *is onto (for* $R_1(\alpha,\beta) = 0$ *is either empty or infinite over* $\mathbb{C}$*).*

Remark 12. We already know that if the Jacobian conjecture is true then the Jacobian condition should imply that $R_n(\alpha,\beta) \in \mathbb{C}^*$ where $R_n(\alpha,\beta)$ is the highest coefficient of **resultant**$(P(X,Y)-\alpha, Q(X,Y)-\beta, Y)$. It is natural to ask if the following more concrete statement holds true:
If $R_N(\alpha_0,\beta_0) = 0$ **then there exists** (X_0,Y_0) **such that**

$$P(X_0,Y_0) = \alpha_0,$$
$$Q(X_0,Y_0) = \beta_0,$$

for which $\partial(P,Q)/\partial(X,Y)(X_0,Y_0) = 0$?

The answer is **negative.**

Example 5. We have considered this example before:

$$P(X,Y) = Y^2 + XY, \quad Q(X,Y) = Y^3 + X^2Y^2 + X^2Y + X.$$

Then we computed

$$\mathbf{resultant}(P(X,Y)-\alpha, Q(X,Y)-\beta, Y) = (1+\alpha)X^5 + \cdots.$$

Thus $R_5(-1,\beta_0) = 0$ for any $\beta_0 \in C$ and we need to check if

$$\partial(P,Q)/\partial(X,Y)(X_0,Y_0) = 0$$

for a point (X_0,Y_0) for which

$$P(X_0,Y_0) = Y_0^2 + X_0Y_0 = -1,$$
$$Q(X_0,Y_0) = Y_0^3 + X_0^2Y_0^2 + X_0^2Y_0 + X_0 = \beta_0,$$

by the first equation we have $Y_0^2 = -1 - X_0Y_0$. So using the second equation we get

$$Y_0^3 = \beta_0$$

so $Y_0 = \beta_0^{1/3}$ are the three possible solutions and $X_0 = -(1+Y_0^2)/Y_0$ or $X_0 = -\beta_0^{-1/3} - \beta_0^{1/3}$ are the three corresponding X_0's.

We now compute

$$\begin{aligned}&\partial(P,Q)/\partial(X,Y)(-\beta_0^{-1/3}-\beta_0^{1/3},\beta_0^{1/3})\\&=(3-4X)Y^3-4XY^2-(2+X^2)Y-X|_{(X_0,Y_0)}\\&=4\beta_0^{4/3}+6\beta_0+4\beta_0^{2/3}+\beta_0^{1/3}.\end{aligned}$$

Since $\beta_0\neq 0$ we see that in general this is not zero. It vanishes exactly for $T_0=\beta_0^{1/3}$ which satisfies:

$$4T_0^3+6T_0^2+4T_0+1=0.$$

Thus, maybe, only the following is true:

Conjecture: $\partial(P,Q)/\partial(X,Y)$ has zero on $F^{-1}(R_n(\alpha,\beta)=0)$.

This is an equivalent formulation for the Jacobian conjecture because of the Hadamard condition on a local diffeomorphism to be a global one, and because for a polynomial automorphism the leading coefficient $R_n(\alpha,\beta)$ belongs to $\mathbb{C}^{\times}$. However, this formulation is sharper than the standard formulation in that it gives a location for a zero of the determinant of the Jacobian in the case of a polynomial mapping which is not an automorphism.

9 Grading an Algebra with a Derivation – Introduction

This section serves as an introductory section for the next three sections.

If k is a field of characteristic 0 and A is a k-algebra with 1 and D a derivation of A, then there is a natural gradation of A with respect to D: the elements of class n in the gradation are those that are annihilated by n iterations of D.

If we assume, further, the existence of an element $q\in A$ for which $D(q)=1$ then in "reasonable" algebras the multiplication of an element in class n by q is an element of class $n+1$ and conversely (Theorem 8).

As a consequence we obtain a structure theorem for the classes of this D-gradation in terms of $\ker(D,A)$ and of q (Theorem 9).

This leads immediately to the well known fact that $Nil(D)=\ker(D,A)[q]$ which was noted by several people [23, 26] and [17] (Corollary 1).

It also gives a necessary and sufficient condition for D to be a locally nilpotent derivation of A (Corollary 6).

One application of the above is for the polynomial ring over k in n variables, $A=k[x_1,\ldots,x_n]$. Given n elements $f_1,\ldots,f_n\in A$ that satisfy the Jacobian condition, namely, that $\partial(f_1,\ldots,f_n)/\partial(x_1,\ldots,x_n)=1$ one can take $D=\partial(f_1,\ldots,f_{n-1},.)/\partial(x_1,\ldots,x_n)$, $q=f_n$ and apply the above. This is closely related to the results on the derivations $d_1,\ldots,d_n$ that were introduced in [17], however, the methods of proofs are different.

10 Grading an Algebra According to a Derivation

Let k be a field of characteristic 0, A a k-algebra with 1 and D a derivation of A.

Definition 10. *For $n \in \mathbb{Z}^+ \bigcup\{0\}$ we denote $A_n(D) = \{f \in A \mid D^n(f) = 0\}$.*

Remark 13. By the definition we have $A_0(D) = \{0\}$, $A_1(D) = \ker(D, A)$. Also $A_n(D) \subseteq A_{n+1}(D)$ for any $n \in \mathbb{Z}^+ \bigcup\{0\}$.

Definition 11. $A_\infty(D) = A - \bigcup_{n=0}^{\infty} A_n(D)$.

Let $q \in A$ satisfy $D(q) = 1$. In this section we shall see that, roughly, the action of multiplication by q on the classes $A_n(D)$ is to advance the class index by 1.

It will be convenient to state two lemmas first.

Lemma 1. *For any $f \in A$ and for any $n \in \mathbb{Z}^+ \bigcup\{0\}$ we have*

$$(n+1)D^n(f) = D^{n+1}(fq) - D^{n+1}(f)q.$$

Proof. By $D(fq) = D(f)q + fD(q) = D(f)q + f$ we get $f = D(fq) - D(f)q$ which takes care of the case $n = 0$. We proceed by induction on n. We assume that

$$nD^{n-1}(f) = D^n(fq) - D^n(f)q, \quad n \geq 1.$$

Then

$$D(nD^{n-1}(f)) = D(D^n(fq) - D^n(f)q),$$
$$nD^n(f) = D^{n+1}(fq) - D^{n+1}(f)q - D^n(f).$$

So $(n+1)D^n(f) = D^{n+1}(fq) - D^{n+1}(f)q$. □

Lemma 2. *If $f \in A$ satisfies $D^{n+1}(fq) = 0$ then for every $k \in \mathbb{Z}^+$, $\mathbb{Q}^k$ is a right divisor of $D^n(f)$.*

Proof. We will see by induction on k that $(n+k)D^{n+k-1}(f) = -D^{n+k}(f)q$. By $D^{n+1}(fq) = 0$ and by Lemma 1 we have the case $k = 1$. We now assume that

$$(n+k)D^{n+k-1}(f) = -D^{n+k}(f)q.$$

Then

$$D((n+k)D^{n+k-1}(f)) = D(-D^{n+k}(f)q),$$
$$(n+k)D^{n+k}(f) = -D^{n+k+1}(f)q - D^{n+k}(f).$$

So $(n+k+1)D^{n+k}(f) = -D^{n+k+1}(f)q$. This identity shows that

$$(n+1)D^n(f) = (-1)^k D^{n+k}(f)\mathbb{Q}^k/(n+2)(n+3)\cdots(n+k). \quad \square$$

We are now ready to prove the relation between multiplication by q and the D-gradation of A.

Theorem 8.

(i) *If $f \in A$ satisfies $D^n(f) = 0$ and $D^{n-1}(f) \neq 0$ then $D^{n+1}(fq) = 0$ and $D^n(fq) \neq 0$.*
(ii) *If for any $g \in A - \{0\}$ there exists an $n(g)$ such that $\mathbb{Q}^{n(g)}$ is not a right divisor of g, then if $f \in A$ satisfies $D^{n+1}(fq) = 0$ and $D^n(fq) \neq 0$ then $D^n(f) = 0$ and $D^{n-1}(f) \neq 0$.*

Proof.

(i) By $D^n(f) = 0$ we get $D^{n+1}(f) = 0$ and so Lemma 1 implies that $D^{n+1}(fq) = 0$. Also by Lemma 1 $nD^{n-1}(f) = D^n(fq)$ and so $D^n(fq) \neq 0$.
(ii) By Lemma 2 $D^n(f) = 0$. By Lemma 1 $nD^{n-1}(f) = D^n(fq)$ and so $D^{n-1}(f) \neq 0$. □

11 The Structure of the D-classes

We now give a structure theorem for the classes $A_n(D)$ in terms of $\ker(D, A)$ and of q (it has been noted before, see e.g. [23, 26] and [17]). Our proof is completely elementary.

Definition 12. *Let B be a k-algebra with 1. For $n \in \mathbb{Z}^+$ we will denote*

$$B^{(n)}[q] = \{F(q) \mid F(x) \in B[x], \deg F < n\}.$$

Theorem 9.

$$A_n(D) = (A_1(D))^{(n)}[q] = (\ker(D, A))^{(n)}[q].$$

Proof. The assertion holds trivially for $n = 1$. We proceed by induction on n. We assume that $A_m(D) = (A_1(D))^{(m)}[q]$ for $1 \leq m \leq n$. By Theorem 8(i) we have the inclusion

$$(A_1(D))^{(n+1)}[q] = A_1(D) + A_n(D)q \subseteq A_{n+1}(D).$$

For the inverse inclusion we take an $f \in A_{n+1}(D)$. We compute

$$\begin{aligned} D\Big(f - \sum_{m=1}^{n}(-1)^{m+1}D^m(f)\mathbb{Q}^m/m!\Big) \\ &= D(f) - \sum_{m=1}^{n}(-1)^{m+1}D(D^m(f)\mathbb{Q}^m)/m! \\ &= D(f) - \sum_{m=1}^{n}(-1)^{m+1}(mD^m(f)\mathbb{Q}^{m-1} + D^{m+1}(f)\mathbb{Q}^m)/m! = 0. \end{aligned}$$

Hence $f - \sum_{m=1}^{n}(-1)^{m+1}D^m(f)\mathbb{Q}^m/m! \in A_1(D)$. Also by the induction hypothesis $D^m(f) \in (A_1(D))^{(n-m+1)}[q]$ and so $f \in (A_1(D))^{(n+1)}[q]$. □

Corollary 5.

$$\ker(D, A)[q] = A_1(D)[q] = \left\{f \in A \mid \exists n \in \mathbb{Z}^+ \bigcup\{0\}, D^n(f) = 0\right\}.$$

Corollary 6. *D is a locally nilpotent derivation of A iff*

$$A_1(D)[q] = \ker(D, A)[q] = A.$$

Corollary 7.

$$A_\infty(D) = A - A_1(D)[q] = A - \ker(D, a)[q].$$

12 Application to Automorphisms of Polynomial Rings

We apply the results of the previous section to derive some properties of automorphisms of polynomial rings (in characteristic 0). These are strongly related to Theorem 3.3 in [17].

Let k be, as before, a field of characteristic 0. We will take $A = k[x_1, \ldots, x_n]$ the polynomial algebra in n variables over k. Any n-tuple $(f_1, \ldots, f_n) \in A^n$ determines a k-endomorphism $\varphi = \varphi(f_1, \ldots, f_n)$ of A by $\varphi(x_j) = f_j$, $1 \leq j \leq n$. Also the operator

$$D_{(f_1,\ldots,f_{n-1})} : A \longrightarrow A,$$
$$D_{(f_1,\ldots,f_{n-1})}(g) = \partial(f_1, \ldots, f_{n-1}, g)/\partial(x_1, \ldots, x_n),$$

is a k-derivation of A which we call the $(f_1, \ldots, f_{n-1})$-Jacobian derivation. Clearly we have

$$A_1(D_{(f_1,\ldots,f_{n-1})}) = \ker(D_{(f_1,\ldots,f_{n-1})}, A) \supseteq k[f_1, \ldots, f_{n-1}],$$

for $n = 2$ the inclusion is known to be an equality.

Theorem 10. *Let k be a field of characteristic 0 and let $f_1, \ldots, f_n \in k[x_1, \ldots, x_n]$ satisfy $D_{(f_1,\ldots,f_{n-1})}(f_n) = 1$.*

(i) *If $\varphi(f_1, \ldots, f_n)$ is invertible then $\ker(D_{(f_1,\ldots,f_{n-1})}, A) = k[f_1, \ldots, f_{n-1}]$ and $D_{(f_1,\ldots,f_{n-1})}$ is a locally nilpotent derivation of $k[x_1, \ldots, x_n]$.*
(ii) *If $n = 2$ then $\varphi(f_1, f_2)$ is invertible iff $D_{(f_1)}$ is a locally nilpotent derivation of $k[x_1, x_2]$.*

Proof. (i) $\varphi(f_1, \ldots, f_n)$ is invertible iff $k[f_1, \ldots, f_n] = k[x_1, \ldots, x_n]$. On the other hand, as noted above

$$A_1(D_{(f_1,\ldots,f_{n-1})}) \supseteq k[f_1, \ldots, f_{n-1}].$$

Hence $A_1(D_{(f_1,\ldots,f_{n-1})})[f_n] \supseteq k[f_1, \ldots, f_n]$. Thus we have $A_1(D_{(f_1,\ldots,f_{n-1})}) = k[f_1, \ldots, f_{n-1}]$ and so by Corollary 6 $D_{(f_1,\ldots,f_{n-1})}$ is locally nilpotent.

(ii) For $n = 2$ we have $A_1(D_{(f_1)}) = k[f_1]$ and so we have the desired equivalence. □

13 Invertible Morphisms, Their Resultants and Inversion Formulas

Some of the results of this section are well known results. The purpose of this section is to bring up all these results in a unified manner with the aid of the notion of the resultant of a map. Also we will discover some new results, as well.

Theorem 11. *Let $P(X,Y), Q(X,Y) \in \mathbb{C}[X,Y]$ and suppose that $F(X,Y) = (P(X,Y), Q(X,Y))$ is an invertible morphism. Then there exist two positive integers N and M such that*

$$\mathbf{resultant}(P(X,Y)-\alpha, Q(X,Y)-\beta, Y) = R_N(X + r_0(\alpha,\beta))^N,$$
$$\mathbf{resultant}(P(X,Y)-\alpha, Q(X,Y)-\beta, X) = S_M(X + s_0(\alpha,\beta))^M,$$

where $R_N, S_M \in \mathbb{C}^*$ and where $r_0(\alpha,\beta), s_0(\alpha,\beta) \in \mathbb{C}[\alpha,\beta]$.

Proof. Let

$$\begin{aligned}&\mathbf{resultant}(P(X,Y)-\alpha, Q(X,Y)-\beta, Y)\\ &\quad = R_N X^N + R_{N-1}(\alpha,\beta)X^{N-1} + \cdots + R_0(\alpha,\beta).\end{aligned}$$

Since $F(X,Y)$ is a morphism it follows that $P(X,Y), Q(X,Y)$ is a Jacobian pair so by the results on the previous section, $N \geq 1$. Also since $F(X,Y)$ cannot have asymptotic values it follows by the resultant formulation of the Jacobian conjecture that $R_N \in \mathbb{C}^*$. Clearly, $R_{N-1}(\alpha,\beta), \ldots, R_0(\alpha,\beta) \in \mathbb{C}[\alpha,\beta]$. Given any (α_0,β_0) and any X_0 which is a zero of

$$R_N X^N + \cdots + R_0(\alpha_0,\beta_0), \tag{3}$$

there exists a Y_0 such that $F(X_0,Y_0) = (\alpha_0,\beta_0)$. Since $F(X,Y)$ is injective it follows that all the zeros of (3) must coincide. This proves that

$$\begin{aligned}&R_N X^N + \cdots + R_0(\alpha_0,\beta_0)\\ &\quad = R_N(X^N + R_{N-1}(\alpha,\beta)/R_N X^{N-1} + \cdots + R_0(\alpha,\beta)/R_N)\\ &\quad = R_N(X + (R_0(\alpha,\beta)/R_N)^{1/N})^N.\end{aligned} \tag{4}$$

In particular we obtain the following $N+1$ relations

$$R_j(\alpha,\beta)/R_N = \binom{N}{j}(R_0(\alpha,\beta)/R_N)^{(N-j)/N}, \quad 0 \leq j \leq N. \tag{5}$$

In particular for $j = N-1$ we get

$$(R_0(\alpha,\beta)/R_N)^{1/N} = R_{N-1}(\alpha,\beta)/(NR_N) \in \mathbb{C}[\alpha,\beta]. \tag{6}$$

A similar argument proves the result for

$$\mathbf{resultant}(P(X,Y)-\alpha, Q(X,Y)-\beta, X). \quad \square$$

More Analysis:

Let us assume that we have the following standard representations

$$\begin{aligned} P(X,Y) &= a_n(X)Y^n + \cdots + a_0(X), \quad \deg_Y P(X,Y) = n, \\ Q(X,Y) &= b_m(X)Y^m + \cdots + b_0(X), \quad \deg_Y Q(X,Y) = m. \end{aligned} \tag{7}$$

Then by the Sylvester's formula we get

$$\mathbf{resultant}(P(X,Y)-\alpha, Q(X,Y)-\beta, Y)$$
$$= \begin{vmatrix} a_n(X) & \dots & 0 & b_m(X) & \dots & 0 \\ \vdots & \dots & a_n(X) & \vdots & \dots & b_m(X) \\ a_0(X)-\alpha & \dots & \vdots & b_0(X)-\beta & \dots & \vdots \\ \vdots & \dots & \vdots & \vdots & \dots & \vdots \\ 0 & \dots & a_0(X)-\alpha & 0 & \dots & b_0(X)-\beta \end{vmatrix}. \tag{8}$$

So in particular we have

$$R_0(\alpha,\beta) = \begin{vmatrix} a_n(0) & \dots & 0 & b_m(0) & \dots & 0 \\ \vdots & \dots & a_n(0) & \vdots & \dots & b_m(0) \\ a_0(0)-\alpha & \dots & \vdots & b_0(0)-\beta & \dots & \vdots \\ \vdots & \dots & \vdots & \vdots & \dots & \vdots \\ 0 & \dots & a_0(0)-\alpha & 0 & \dots & b_0(0)-\beta \end{vmatrix}. \tag{9}$$

Hence $\deg R_0(\alpha,\beta) \le \max(n,m)$ or equivalently

$$\deg R_0(\alpha,\beta) \le \max(\deg_Y P(X,Y), \deg_Y Q(X,Y)) = \deg_Y F(X,Y). \tag{10}$$

By (4) we obtain the following conclusion:

If $P(X,Y)=\alpha$, $Q(X,Y)=\beta$ then $X(\alpha,\beta) = -(R_0(\alpha,\beta)/R_N)^{1/N}$. By (6) $X(\alpha,\beta) \in \mathbb{C}[\alpha,\beta]$ and by (10) $\deg X(\alpha,\beta) \le \deg_Y F(X,Y)$. Similarly if

$$\begin{aligned} &\mathbf{resultant}(P(X,Y)-\alpha, Q(X,Y)-\beta, X) \\ &= S_M Y^M + S_{M-1}(\alpha,\beta)Y^{M-1} + \cdots + S_0(\alpha,\beta), \end{aligned}$$

then we have:

$$S_M Y^M + \cdots + S_0(\alpha,\beta) = S_M(Y + (S_0(\alpha,\beta)/S_M)^{1/M})^M, \tag{11}$$

$$S_j(\alpha,\beta)/S_M = \binom{M}{j} (S_0(\alpha,\beta)/S_M)^{(M-j)/M}, \quad 0 \le j \le M, \tag{12}$$

$$(S_0(\alpha,\beta)/S_M)^{1/M} = S_{M-1}(\alpha,\beta)/(MS_M) \in \mathbb{C}[\alpha,\beta], \tag{13}$$

$$\deg S_0(\alpha,\beta) \le \max(\deg_X P(X,Y),\ \deg_X Q(X,Y)) = \deg_X F(X,Y). \tag{14}$$

So by (11) we obtain the following conclusion:

If $P(X,Y)=\alpha\ Q(X,Y)=\beta$ then $Y(\alpha,\beta)=-(S_0(\alpha,\beta)/S_M)^{1/M}$. By (13) $Y(\alpha,\beta)\in\mathbb{C}[\alpha,\beta]$ and by (14) $\deg Y(\alpha,\beta)\le\deg_X F(X,Y)$.

Remark 14. It is possible to show that $N=M=1$.

We now can conclude more results that emerge from the proof of the last theorem.

Theorem 12. *Let $P(X,Y),Q(X,Y)\in\mathbb{C}[X,Y]$ and suppose that $F(X,Y)=(P(X,Y),Q(X,Y))$ is an invertible morphism. Let us assume that we have the following standard representations:*

$$P(X,Y)=a_n(X)Y^n+\cdots+a_0(X)=A_N(Y)X^N+\cdots+A_0(Y),$$

where $\deg_Y P(X,Y)=n$ and $\deg_X P(X,Y)=N$.

$$Q(X,Y)=b_m(X)Y^m+\cdots+b_0(X)=B_M(Y)X^M+\cdots+B_0(Y),$$

where $\deg_Y Q(X,Y)=m$ and $\deg_X Q(X,Y)=M$. Let $F^{-1}(\alpha,\beta)=(X(\alpha,\beta),Y(\alpha,\beta))$. Then there are $R_1,S_1\in\mathbb{C}^*$ so that we have the following formulas for the inverse map

$$-R_1X(\alpha,\beta)=\begin{vmatrix} a_n(0) & \dots & 0 & b_m(0) & \dots & 0\\ \vdots & \dots & a_n(0) & \vdots & \dots & b_m(0)\\ a_0(0)-\alpha & \dots & \vdots & b_0(0)-\beta & \dots & \vdots\\ \vdots & \dots & \vdots & \vdots & \dots & \vdots\\ 0 & \dots & a_0(0)-\alpha & 0 & \dots & b_0(0)-\beta \end{vmatrix},$$

and

$$-S_1Y(\alpha,\beta)=\begin{vmatrix} A_N(0) & \dots & 0 & B_M(0) & \dots & 0\\ \vdots & \dots & A_N(0) & \vdots & \dots & B_M(0)\\ A_0(0)-\alpha & \dots & \vdots & B_0(0)-\beta & \dots & \vdots\\ \vdots & \dots & \vdots & \vdots & \dots & \vdots\\ 0 & \dots & A_0(0)-\alpha & 0 & \dots & B_0(0)-\beta \end{vmatrix}.$$

Proof. This follows by the discussion after (10) and by (9). □

Example 6. **Degree 1.**

$$P(X,Y) = aX + bY, \quad Q(X,Y) = AX + BY,$$

$$R(\alpha,\beta) = \begin{vmatrix} b & B \\ -\alpha & -\beta \end{vmatrix} = B\alpha - b\beta, \qquad S(\alpha,\beta) = \begin{vmatrix} A & a \\ -\beta & -\alpha \end{vmatrix} = a\beta - A\alpha.$$

On the other hand calculation of the inverse gives: $aX+bY = \alpha$, $AX+BY = \beta \implies X = (aB-bA)^{-1}R(\alpha,\beta)$, $Y = (aB-bA)^{-1}S(\alpha,\beta)$

Example 7. **Degree 2.**

$$P(X,Y) = 4Y^2 + (1+4X)Y + X^2 + X = X^2 + (1+4Y)X + 4Y^2 + Y,$$
$$Q(X,Y) = -4Y^2 + (1-4X)Y - X^2 = -X^2 - 4XY - 4Y^2 + Y,$$

$$R(\alpha,\beta) = \begin{vmatrix} 4 & 0 & -4 & 0 \\ 1 & 4 & 1 & -4 \\ -\alpha & 1 & -\beta & 1 \\ 0 & -\alpha & 0 & -\beta \end{vmatrix} = 16\alpha^2 + 32\alpha\beta + 16\beta^2 - 8\alpha + 8\beta.$$

So $R(\alpha,\beta) = 16(\alpha+\beta)^2 - 8(\alpha-\beta)$. Similarly we have $S(\alpha,\beta) = (\alpha+\beta)^2 + \beta$. On the other hand calculation of the inverse gives:

$$\begin{aligned} &\alpha = X + Y + (X+2Y)^2, \\ &\beta = Y - (X+2Y)^2, \\ \implies \quad &X = -8R(\alpha,\beta), \; Y = S(\alpha,\beta). \end{aligned}$$

Theorem 13. *Let $P(X,Y), Q(X,Y) \in \mathbb{C}[X,Y]$ and suppose that $F(X,Y) = (P(X,Y), Q(X,Y))$ is an invertible morphism. Let us denote $F^{-1}(\alpha,\beta) = (X(\alpha,\beta), Y(\alpha,\beta))$. Then we have the following formulas for the inverse map*

$$X(\alpha,\beta) = -R(X,\alpha,\beta)/(d/dX\{R(X,\alpha,\beta)\})(0),$$
$$Y(\alpha,\beta) = -S(Y,\alpha,\beta)/(d/dY\{S(Y,\alpha,\beta)\})(0),$$

where

$$R(X,\alpha,\beta) = \mathbf{resultant}(P(X,Y)-\alpha, Q(X,Y)-\beta, Y),$$
$$S(Y,\alpha,\beta) = \mathbf{resultant}(P(X,Y)-\alpha, Q(X,Y)-\beta, X).$$

Proof. This follows by the previous theorem and by (9). □

Theorem 14. *Let $P(X,Y), Q(X,Y) \in \mathbb{C}[X,Y]$ and suppose that $F(X,Y) = (P(X,Y), Q(X,Y))$ is an invertible morphism. Let us denote $F^{-1}(\alpha,\beta) = (X(\alpha,\beta), Y(\alpha,\beta))$. Then*

$$\deg X(\alpha,\beta) \le \max(\deg_Y P(X,Y), \deg_Y Q(X,Y)) = \deg_Y F(X,Y),$$
$$\deg Y(\alpha,\beta) \le \max(\deg_X P(X,Y), \deg_X Q(X,Y)) = \deg_X F(X,Y),$$
$$\deg P(X,Y) \le \max(\deg_\beta X(\alpha,\beta), \deg_\beta Y(\alpha,\beta)) = \deg_\beta F^{-1}(\alpha,\beta),$$
$$\deg Q(X,Y) \le \max(\deg_\alpha X(\alpha,\beta), \deg_\alpha Y(\alpha,\beta)) = \deg_\alpha F^{-1}(\alpha,\beta),$$
$$\deg F(X,Y) = \deg F^{-1}(\alpha,\beta).$$

Proof. The first two inequalities were proved in (10) and (14) and the discussion that followed. The second two inequalities follow from the first two by changing the roles of F and F^{-1}. The fifth equality is a conclusion of the previous 4 inequalities. □

Remark 15. The fifth equality appears in [2] on p. 292. It is proved there for any dimension (with the appropriate modification). It was communicated to the authors of [2] by Ofer Gaber who attributed it to an unrecalled colloquium lecturer at Harvard. The authors mention that John Tyrrell (Kings College, Univ. of London) has indicated that this equality was well known to classical geometers.

Theorem 15. *Let $P(X,Y), Q(X,Y) \in \mathbb{C}[X,Y]$ and suppose that $F(X,Y) = (P(X,Y), Q(X,Y))$ is an invertible morphism. Let us denote $F^{-1}(\alpha,\beta) = (X(\alpha,\beta), Y(\alpha,\beta))$. Let the coefficients of $P(X,Y)$ be $\{a_i\}$, i.e., $P(X,Y) \in Z(\{a_i\})[X,Y]$ and let the coefficients of $Q(X,Y)$ be $\{b_i\}$, i.e., $Q(X,Y) \in Z(\{b_i\})[X,Y]$. Then*

$$X(\alpha,\beta), Y(\alpha,\beta) \in Q(\{a_i\},\{b_i\})[\alpha,\beta].$$

Proof. This follows from Theorem 12 and the fact that

$$R_1 \in Z(\{a_i\}),\ S_1 \in Z(\{b_i\}).\quad \square$$

Corollary 8. *Let $P(X,Y), Q(X,Y) \in \mathbb{Q}[X,Y]$ and suppose that $F(X,Y) = (P(X,Y), Q(X,Y))$ is an invertible morphism, then $F(\mathbb{Q}^2) = \mathbb{Q}^2$.*

14 The Rigidity of Morphisms

A remarkable property of invertible morphisms is the fact that the inverse map depends on very few of the coefficients of the map. If the map is invertible and if we are given these very few coefficients then we can reconstruct the map and hence there is only one morphism with such data.

We call this property **rigidity** and we now describe it's details. We shall need the following standard

Definition 13. *Let $P(X,Y) \in \mathbb{C}[X,Y]$. The two face polynomials of $P(X,Y)$ are the following polynomials of a single indeterminate $P(X,0)$ and $P(0,Y)$.*

Let $P(X,Y), Q(X,Y) \in \mathbb{C}[X,Y]$ and let $F(X,Y) = (P(X,Y), Q(X,Y))$ be the corresponding polynomial map. The four face polynomials of $F(X,Y)$ are $P(0,Y), Q(0,Y), P(X,0), Q(X,0)$.

We shall also need the following notation

Definition 14. *Let $f(T), g(T) \in \mathbb{C}[T]$. The induced polynomial of f and g is defined as follows*

$$R_{f,g}(\alpha, \beta) = \mathbf{resultant}(f(T) - \alpha, g(T) - \beta),$$

where α, β are two new indeterminates.

Remark 16. Let $f(T), g(T) \in \mathbb{C}[T]$ and let α, β be two indeterminates.

(1) Clearly $R_{f,g}(\alpha, \beta) \in \mathbb{C}[\alpha, \beta]$.
(2) Let us assume that we have the following standard representations

$$f(T) = a_n T^n + \cdots + a_0, \ \deg f(T) = n,$$
$$g(T) = b_m T^m + \cdots + b_0, \ \deg g(T) = m,$$

then by Sylvester's formula we have

$$R_{f,g}(\alpha, \beta) = \begin{vmatrix} a_n & \dots & 0 & b_m & \dots & 0 \\ \vdots & \dots & a_n & \vdots & \dots & b_m \\ a_0 - \alpha & \dots & \vdots & b_0 - \beta & \dots & \vdots \\ \vdots & \dots & \vdots & \vdots & \dots & \vdots \\ 0 & \dots & a_0 - \alpha & 0 & \dots & b_0 - \beta \end{vmatrix}.$$

(3) Using the representation of $R_{f,g}(\alpha, \beta)$ in (2) above we obtain the following:

$$\deg_\alpha R_{f,g}(\alpha, \beta) = m = \deg g(T),$$

and the α-leading term is

$$+- b_m^n \alpha^m = +- \text{(leading coeff of } g)^{\deg f(T)} \alpha^{\deg g(T)},$$
$$\deg_\beta R_{f,g}(\alpha, \beta) = n = \deg f(T),$$

and the β-leading term is

$$a_n^m (-\beta)^n = (-1)^{\deg f(T)} \text{(leading coeff of } f)^{\deg g(T)} \beta^{\deg f(T)},$$
$$\deg R_{f,g}(\alpha, \beta) = \max(n, m) = \max(\deg f(T), \deg g(T)).$$

We can now restate the previous results in terms of the rigidity property.

Theorem 16 (The rigidity of morphisms). *Let $P(X,Y), Q(X,Y) \in \mathbb{C}[X,Y]$ and let $F(X,Y) = (P(X,Y), Q(X,Y))$ be the corresponding polynomial map. Then we have the following:*

(a) *If $F(X,Y)$ is an invertible morphism then the polynomial map*

$$G(\alpha,\beta)=(R_{P(0,Y),Q(0,Y)}(\alpha,\beta),R_{P(X,0),Q(X,0)}(\alpha,\beta)),$$

which is induced by the four face polynomials of $F(X,Y)$ is also an invertible morphism.

(b) *If $F(X,Y)$ is an invertible morphism then $F^{-1}(\alpha,\beta)$ differs from $G(\alpha,\beta)$ by a dilation, i.e., there are $R_1,S_1\in\mathbb{C}^*$ so that*

$$F^{-1}(\alpha,\beta)=(-R_{P(0,Y),Q(0,Y)}(\alpha,\beta)/R_1,-R_{P(X,0),Q(X,0)}(\alpha,\beta)/S_1).$$

(c) *If $F(X,Y)$ is an invertible morphism and $G(\alpha,\beta)=(X(\alpha,\beta),Y(\alpha,\beta))$ then $F(X,Y)$ is an inner dilation of the polynomial map which is induced by the four face polynomials of $G(\alpha,\beta)$, i.e., there exist $R_1,S_1\in\mathbb{C}^*$ so that*

$$F(X,Y)=(R_{X(0,\beta),Y(0,\beta)}(R_1X,S_1Y),R_{X(\alpha,0),Y(\alpha,0)}(R_1X,S_1Y)).$$

Proof. All the three statements follow in a straight forward manner from Theorem 12. □

Remark 17. The core of the rigidity property lies in part (c) of the last theorem: for if we are given the four face polynomials of a morphism, namely, $P_1(X)=P(X,0),P_2(Y)=P(0,Y),Q_1(X)=Q(X,0),Q_2(Y)=Q(0,Y)$ then the process

$$\begin{aligned}(P_2(Y),Q_2(Y))\to X(\alpha,\beta)=R_{P_2,Q_2}(\alpha,\beta)&\to(X(0,\beta),Y(0,\beta))\\&\to P(R_1X,S_1Y),\end{aligned}$$

and

$$\begin{aligned}(P_1(Y),Q_1(Y))\to Y(\alpha,\beta)=R_{P_1,Q_1}(\alpha,\beta)&\to(X(\alpha,0),Y(\alpha,0))\\&\to Q(R_1X,S_1Y),\end{aligned}$$

reconstructs morphism from it's four face polynomials up to dilations. Note that the number of coefficients of $F(X,Y)$ is

$$\binom{\deg P(X,Y)+2}{2}+\binom{\deg Q(X,Y)+2}{2},$$

while the number of coefficients of the four face polynomials is only

$$\deg P(X,0)+\deg P(0,Y)+\deg Q(X,0)+\deg Q(0,Y)+4,$$

hence only a linear number (in the degrees) of the coefficients out of the quadratic number of their total is used for the reconstruction of the morphism.

It would have been nice if the rigidity property were a characterization of invertibility. Unfortunately, it is not.

Example 8. We shall take a single polynomial $P(X,Y) \in \mathbb{C}[X,Y]$ and generate the map $F(X,Y) = (P(X,Y), P(X,Y))$ which is clearly not invertible. However, we shall see that for certain choices of $P(X,Y)$ the reconstruction process of Theorem 16(c) when applied to $F(X,Y)$ reconstructs $F(X,Y)$.

Take $P(X,Y) = (X-Y)^2 = Y^2 - 2XY + X^2 = X^2 - 2YX + Y^2$. Then

$$X(\alpha,\beta) = R_{P(0,Y),P(0,Y)}(\alpha,\beta) = (\alpha-\beta)^2 = P(\alpha,\beta).$$

Similarly $Y(\alpha,\beta) = P(\alpha,\beta)$. We note that in this case

$$\mathbf{resultant}((X-Y)^2-\alpha, (X-Y)^2-\beta, Y) = (\alpha-\beta)^2,$$

and

$$(\beta-\alpha)^2 = (\beta-\alpha)((X-Y)^2-\alpha) - (\beta-\alpha)((X-Y)^2-\beta).$$

One might suspect that when the process of Theorem 16(c) is performed enough times then it might stabilize either on an invertible morphism or on the other extreme, a map whose coordinates are algebraically dependent. Again, unfortunately this is not the case.

Example 9. Let

$$P(X,Y) = X^2 + Y^2 + X = Y^2 + (X^2 + X) = X^2 + X + Y^2,$$
$$Q(X,Y) = XY + Y = (1+X)Y.$$

Then

$$R_{P(0,Y),Q(0,Y)}(\alpha,\beta) = \begin{vmatrix} 1 & 1 & 0 \\ 0 & -\beta & 1 \\ -\alpha & 0 & -\beta \end{vmatrix} = \beta^2 - \alpha,$$

and

$$R_{P(X,0),Q(X,0)}(\alpha,\beta) = \begin{vmatrix} 1 & 0 & 0 \\ 1 & -\beta & 0 \\ -\alpha & 0 & -\beta \end{vmatrix} = \beta^2.$$

So the next stage is to work with

$$P_1(X,Y) = Y^2 - X, \quad Q_1(X,Y) = Y^2,$$
$$R_{P_1(0,Y),Q_1(0,Y)}(\alpha,\beta) = (\alpha-\beta)^2,$$
$$R_{P_1(X,0),Q_1(X,0)}(\alpha,\beta) = \beta^2.$$

So the next stage is to work with

$$P_2(X,Y) = (Y - X)^2 = Y^2 - 2XY + X^2, \quad Q_2(X,Y) = Y^2,$$
$$R_{P_2(0,Y),Q_2(0,Y)}(\alpha,\beta) = (\alpha - \beta)^2,$$
$$R_{P_2(X,0),Q_2(X,0)}(\alpha,\beta) = \beta^2.$$

So $P_3(X,Y) = P_2(X,Y)$ and $Q_3(X,Y) = Q_2(X,Y)$. The process had stabilized, however:

$$\partial(P_3, Q_3)/\partial(X,Y) = 4Y(X - Y) \not\equiv 0.$$

Remark 18. The examples suggest that iterations of the process of Theorem 16(c) stabilize. If so, what are the possible periods?

Theorem 12 gives an inversion formula in terms of the face polynomials. Theorem 13 gives an inversion formula in terms of the logarithmic derivatives of the two resultants at the origin. We now add one more inversion formula in terms of the polynomial coefficients in the resultant representation, namely:

Theorem 17. *Let $P(X,Y), Q(X,Y)) \in \mathbb{C}[X,Y]$ and suppose that $F(X,Y) = (P(X,Y), Q(X,Y))$ is an invertible morphism normalized by $F(0,0) = (0,0)$. Let*

$$\begin{aligned}&\mathbf{resultant}(P(X,Y) - \alpha, Q(X,Y) - \beta, Y)\\ &\quad = A(X,Y,\alpha,\beta)(P(X,Y) - \alpha) + B(X,Y,\alpha,\beta)(Q(X,Y) - \beta),\end{aligned}$$

and

$$\begin{aligned}&\mathbf{resultant}(P(X,Y) - \alpha, Q(X,Y) - \beta, X)\\ &\quad = A_1(X,Y,\alpha,\beta)(P(X,Y) - \alpha) + B_1(X,Y,\alpha,\beta)(Q(X,Y) - \beta).\end{aligned}$$

Then

$$\begin{aligned}F^{-1}(X,Y) = (&(XA(0,0,X,Y) + YB(0,0,X,Y))/R_1,\\ &(XA_1(0,0,X,Y) + YB_1(0,0,X,Y))/S_1),\end{aligned}$$

for some $R_1, S_1 \in \mathbb{C}^*$.

Proof. The polynomial $A(X,Y,\alpha,\beta)(P(X,Y)-\alpha)+B(X,Y,\alpha,\beta)(Q(X,Y)-\beta)$ is independent of Y and so equals to $A(X,0,\alpha,\beta)(P(X,0) - \alpha) + B(X,0,\alpha,\beta)(Q(X,0) - \beta)$. To obtain the X-free term $R_0(\alpha,\beta)$ of $\mathbf{resultant}(P(X,Y) - \alpha, Q(X,Y) - \beta, Y)$ we substitute in this polynomial $X = 0$ and obtain

$$\begin{aligned}R_0(\alpha,\beta) &= A(0,0,\alpha,\beta)(P(0,0) - \alpha) + B(0,0,\alpha,\beta)(Q(0,0) - \beta)\\ &= -\alpha A(0,0,\alpha,\beta) - \beta B(0,0,\alpha,\beta).\end{aligned}$$

We do the same for the second resultant and obtain

$$\alpha A_1(0,0,\alpha,\beta) - \beta B_1(0,0,\alpha,\beta),$$

and now the result follows by Theorem 12. □

15 The Fibre Theorem

The following result is well known.

Theorem (The Fibre Theorem). *If k is an algebraically closed field, $F : k^n \to k^n$ a polynomial map such that* $\det J(F)$ *never vanishes, then the cardinality of the fibre $F^{-1}(\{b\})$ over any $b \in k^n$ is less than or equal to the geometric degree of F, $\dim_{k(F)} k(X)$ which is finite.*

See [12]. This was generalized by Arno van den Essen to fields k of characteristic 0 [6].

For the field $k = \mathbb{R}$ and in dimension $n = 2$ the finiteness of the fibres was proved in 1988 [14]. See also [1, 2].

This problem was handled in a geometrical fashion for maps that are not even polynomial. See [18, 19].

The purpose of this small section is to give one more proof for the finiteness result of the fibres of polynomial étale maps in dimension two and over an algebraically closed field, using the ideas that were developed in the previous sections.

Theorem 18 (The Fibre Theorem). *If $F : \mathbb{C}^2 \to \mathbb{C}^2$ is a polynomial map such that* $\det J(F)$ *never vanishes, then for each $b \in \mathbb{C}^2$ we have*

$$F^{-1}(\{b\}) \leq (\deg_X F)(\deg_Y F).$$

Proof. Let $F(X,Y) = (P(X,Y), Q(X,Y))$ where $P(X,Y), Q(X,Y) \in \mathbb{C}[X,Y]$. Let α, β be two new indeterminates. We use our standard notation

$$\mathbf{resultant}(P(X,Y) - \alpha, Q(X,Y) - \beta, Y) = R_N(\alpha,\beta)X^N + \cdots + R_0(\alpha,\beta),$$

where we know (by our previous results) that $N \geq 1$ and that $R_j(\alpha,\beta) \in \mathbb{C}[\alpha,\beta]$ for $0 \leq j \leq N$. There exist $A(X,Y,\alpha,\beta), B(X,Y,\alpha,\beta) \in \mathbb{C}[X,Y,\alpha,\beta]$ such that the following holds true

$$\begin{aligned} &R_N(\alpha,\beta)X^N + \cdots + R_0(\alpha,\beta) \\ &\quad = A(X,Y,\alpha,\beta)(P(X,Y) - \alpha) + B(X,Y,\alpha,\beta)(Q(X,Y) - \beta). \end{aligned}$$

For any $b = (\alpha_0, \beta_0) \in \mathbb{C}^2$ the X-fibre of $F(X,Y) = b$ equals the X-zero set of $R_N(\alpha_0,\beta_0)X^N + \cdots + R_0(\alpha_0,\beta_0)$. This set contains at most N points. Let us recall once more that by the Sylvester's formula for $\mathbf{resultant}(P(X,Y) - \alpha, Q(X,Y) - \beta, Y)$ we have

$$N \leq \max(\deg_Y P(X,Y), \deg_Y Q(X,Y)) = \deg_Y F(X,Y).$$

Similarly the set of Y-fibre of $F(X,Y) = b$ contains at most $\deg_X F(X,Y)$ points. Hence we obtain

$$F^{-1}(\{b\}) \leq (\deg_X F)(\deg_Y F). \quad \square$$

16 One more Inversion Formula and an Equivalent Formulation to the Jacobian Conjecture

We start by recalling well known facts about resultants:

Let $f(X), g(X) \in \mathbb{C}[X]$. Let us assume that we have the following standard representations:

$$f(X) = a_m X^m + \cdots + a_0 = a_m \prod_{i=1}^{m} (X - \alpha_i),$$

and

$$g(X) = b_n X^n + \cdots + b_0 = b_n \prod_{j=1}^{n} (X - \beta_j).$$

Then the following holds true:

$$\begin{aligned}\textbf{resultant}(f, g, X) &= a_m^n \prod_{i=1}^{m} g(\alpha_i) \\ &= (-1)^{mn} b_n^m \prod_{j=1}^{n} f(\beta_j) = a_m^n b_n^m \prod_{i=1}^{m} \prod_{j=1}^{n} (\alpha_i - \beta_j).\end{aligned}$$

Suppose now that $P(X,Y), Q(X,Y) \in \mathbb{C}[X,Y]$ is a Jacobian pair, i.e.,

$$\partial(P,Q)/\partial(X,Y) = P_X Q_Y - P_Y Q_X \equiv 1.$$

Let us use the following assumptions and notations:

$$P(X,Y) - \alpha = a_n Y^n + \cdots + a_0 - \alpha = a_n \prod_{i=1}^{n} (Y - \alpha_i),$$

$$Q(X,Y) - \beta = b_n Y^n + \cdots + b_0 - \beta = b_n \prod_{j=1}^{n} (Y - \beta_j),$$

where $a_i, b_j \in \mathbb{C}[X]$, $0 \leq i, j \leq n$, $a_n, b_n \in \mathbb{C}^*$, $\alpha_i = \alpha_i(X, \alpha)$ are algebraic over $\mathbb{C}[X, \alpha]$, $\beta_j = \beta_j(X, \beta)$ are algebraic over $\mathbb{C}[X, \beta]$.

$$\begin{aligned}\partial/\partial X\{P(X,Y) - \alpha\} &= -a_n \sum_{i=1}^{n} (Y - \alpha_1) \cdots (Y - \alpha_{i-1}) \\ &\quad \times \alpha_i'(Y - \alpha_{i+1}) \ldots (Y - \alpha_n),\end{aligned}$$

$$\partial/\partial Y\{P(X,Y) - \alpha\} = a_n \sum_{i=1}^{n} (Y - \alpha_1) \cdots (Y - \alpha_{i-1})(Y - \alpha_{i+1}) \cdots (Y - \alpha_n),$$

$$\partial/\partial X\{Q(X,Y)-\beta\} = -b_n \sum_{j=1}^{n}(Y-\beta_1)\cdots(Y-\beta_{j-1}) \times \beta_j'(Y-\beta_{j+1})\cdots(Y-\beta_n),$$

$$\partial/\partial Y\{Q(X,Y)-\beta\} = b_n \sum_{j=1}^{n}(Y-\beta_1)\cdots(Y-\beta_{j-1})(Y-\beta_{j+1})\cdots(Y-\beta_n).$$

Hence

$$P_X(X,\alpha_i) = -a_n(\alpha_i-\alpha_1)\cdots(\alpha_i-\alpha_{i-1})\alpha_i'(\alpha_i-\alpha_{i+1})\cdots(\alpha_i-\alpha_n),$$

and

$$P_Y(X,\alpha_i) = a_n(\alpha_i-\alpha_1)\cdots(\alpha_i-\alpha_{i-1})(\alpha_i-\alpha_{i+1})\cdots(\alpha_i-\alpha_n).$$

Proposition 4.

$$\begin{aligned} 1 &= -P_Y(X,\alpha_i)d/dX\{Q(X,\alpha_i)\} \\ &= Q_Y(X,\beta_j)d/dX\{P(X,\beta_j)\}, \quad 1 \le i,j \le n. \end{aligned}$$

Proof.

$$\begin{aligned} 1 &= P_X(X,\alpha_i)Q_Y(X,\alpha_i) - P_Y(X,\alpha_i)Q_X(X,\alpha_i) \\ &= [-a_n(\alpha_i-\alpha_1)\cdots(\alpha_i-\alpha_{i-1})\alpha_i'(\alpha_i-\alpha_{i+1})\cdots(\alpha_i-\alpha_n)]Q_Y(X,\alpha_i) \\ &\quad - [a_n(\alpha_i-\alpha_1)\cdots(\alpha_i-\alpha_{i-1})(\alpha_i-\alpha_{i+1})\cdots(\alpha_i-\alpha_n)]Q_X(X,\alpha_i) \\ &= -[a_n(\alpha_i-\alpha_1)\cdots(\alpha_i-\alpha_{i-1})(\alpha_i-\alpha_{i+1})\cdots(\alpha_i-\alpha_n)]\{Q_X(X,\alpha_i) \\ &\quad + \alpha_i'Q_Y(X,\alpha_i)\} \\ &= -P_Y(X,\alpha_i)d/dX\{Q(X,\alpha_i)\}. \end{aligned}$$

A similar computation gives

$$1 = Q_Y(X,\beta_j)d/dX\{P(X,\beta_j)\}. \quad \square$$

Theorem 19. *Let* $R(X) = \textbf{resultant}(P(X,Y)-\alpha, Q(X,Y)-\beta, Y)$. *Then*

$$\begin{aligned} -\mathbb{R}'(X)/R(X) &= \sum_{i=1}^{n}\{1/P_Y(X,\alpha_i)[Q(X,\alpha_i)-\beta]\} \\ &= -\sum_{i=1}^{n}\{1/Q_Y(X,\beta_j)[P(X,\beta_j)-\alpha]\}. \end{aligned}$$

Proof. By the previous proposition we have

$$d/dX\{Q(X,\alpha_i)-\beta\}/\{Q(X,\alpha_i)-\beta\} = -1/P_Y(X,\alpha_i)[Q(X,\alpha_i)-\beta], \quad 1 \le i \le n.$$

Hence

$$\sum_{i=1}^{n} d/dX\{Q(X,\alpha_i)-\beta\}/\{Q(X,\alpha_i)-\beta\}$$
$$= -\sum_{i=1}^{n} 1/P_Y(X,\alpha_i)[Q(X,\alpha_i)-\beta],$$

or

$$d/dX\left\{a_n^n\prod_{i=1}^{n}(Q(X,\alpha_i)-\beta)\right\}\Big/a_n^n\prod_{i=1}^{n}(Q(X,\alpha_i)-\beta)$$
$$= -\sum_{i=1}^{n} 1/P_Y(X,\alpha_i)[Q(X,\alpha_i)-\beta].$$

So

$$-\mathbb{R}'(X)/R(X) = \sum_{i=1}^{n}\{1/P_Y(X,\alpha_i)[Q(X,\alpha_i)-\beta]\},$$

where in the last step we used the identity

$$\textbf{resultant}(P(X,Y)-\alpha, Q(X,Y)-\beta, Y) = a_n^n\prod_{i=1}^{n}(Q(X,\alpha_i)-\beta).$$

A similar computation gives the following:

$$d/dX\{P(X,\beta_j)-\alpha\}/\{P(X,\beta_j)-\alpha\} = 1/Q_Y(X,\beta_j)[P(X,\beta_j)-\alpha],$$
$$1 \le j \le n,$$

$$\sum_{j=1}^{n} d/dX\{P(X,\beta_j)-\alpha\}/\{P(X,\beta_j)-\alpha\}$$
$$= \sum_{j=1}^{n} 1/Q_Y(X,\beta_j)[P(X,\beta_j)-\alpha],$$

$$d/dX\left\{(-1)^{n^2}b_n^n\prod_{j=1}^{n}(P(X,\beta_j)-\alpha)\right\}\Big/(-1)^{n^2}b_n^n\prod_{j=1}^{n}(P(X,\beta_j)-\alpha)$$
$$= \sum_{j=1}^{n} 1/Q_Y(X,\beta_j)[P(X,\beta_j)-\alpha],$$

$$\mathbb{R}'(X)/R(X) = \sum_{j=1}^{n} 1/Q_Y(X,\beta_j)[P(X,\beta_j)-\alpha]. \quad \square$$

Theorem 20. *Let $P(X,Y), Q(X,Y) \in \mathbb{C}[X,Y]$ and suppose that $F(X,Y) = (P(X,Y), Q(X,Y))$ is an invertible morphism. Let us denote $F^{-1}(\alpha,\beta) = (X(\alpha,\beta), Y(\alpha,\beta))$. Then we have the following formulas for the inverse map*

$$X(\alpha,\beta) = \left\{ \sum_{i=1}^{n} J/P_Y(0,\alpha_i(0,\alpha))[Q(0,\alpha_i(0,\alpha)) - \beta] \right\}^{-1}$$
$$= \left\{ -\sum_{j=1}^{n} J/Q_Y(0,\beta_j(0,\beta))[P(0,\beta_j(0,\beta)) - \alpha] \right\}^{-1},$$

where $J = \partial(P,Q)/\partial(X,Y) \in \mathbb{C}^*$. *An analogous pair of expressions hold for* $Y(\alpha,\beta)$.

Proof. By Theorem 13 we have

$$X(\alpha,\beta) = -R(X)/d/dX\{R(X)\}(0),$$

where

$$R(X) = \textbf{resultant}(P(X,Y) - \alpha, Q(X,Y) - \beta, Y),$$

and now the result follows by Theorem 19 above with J replacing 1 in the numerators. □

Theorem 21. *The Jacobian conjecture is true iff the following holds true:*
If $P(X,Y), Q(X,Y) \in \mathbb{C}[X,Y]$ *satisfy following two conditions*

(1) $\partial(P,Q)/\partial(X,Y) \equiv 1$
and
(2) $P(X,Y) = a_n \prod_{i=1}^{n}(Y - \alpha_i(X))$, $Q(X,Y) = b_n \prod_{j=1}^{n}(Y - \beta_j(X))$, *where* $a_n, b_n \in \mathbb{C}^*$
then

$$\sum_{i=1}^{n} \left\{ \left[\prod_{k=1}^{n} Q(X,\alpha_k(X)) \right] \Big/ P_Y(X,\alpha_i(X))Q(X,\alpha_i(X)) \right\} \in \mathbb{C}^*.$$

Moreover, this condition is equivalent to each of the following 3 conditions

$$\sum_{i=1}^{n} \left\{ \left[\prod_{k=1}^{n} P(X,\beta_k(X)) \right] \Big/ P_Y(X,\alpha_i(X))Q(X,\alpha_i(X)) \right\} \in \mathbb{C}^*,$$
$$\sum_{j=1}^{n} \left\{ \left[\prod_{k=1}^{n} Q(X,\alpha_k(X)) \right] \Big/ Q_Y(X,\beta_j(X))P(X,\beta_j(X)) \right\} \in \mathbb{C}^*,$$

and

$$\sum_{j=1}^{n} \left\{ \left[\prod_{k=1}^{n} P(X,\beta_k(X)) \right] \Big/ Q_Y(X,\beta_j(X))P(X,\beta_j(X)) \right\} \in \mathbb{C}^*.$$

Proof. The Jacobian conjecture is true $\iff$
If $P, Q \in \mathbb{C}[X,Y]$ satisfy $\partial(P,Q)/\partial(X,Y) \equiv 1$ then $F = (P,Q)$ is an invertible morphism $\iff$

If $P, Q \in \mathbb{C}[X,Y]$ satisfy $\partial(P,Q)/\partial(X,Y) \equiv 1$ then

$$R(X) = \textbf{resultant}(P-\alpha, Q-\beta, Y) = R_1 X + R_0(\alpha,\beta),$$

where $R_1 \in \mathbb{C}^*$ (by Theorem 11 and Remark 12) $\Longleftrightarrow$
If $P, Q \in \mathbb{C}[X,Y]$ satisfy $\partial(P,Q)/\partial(X,Y) \equiv 1$ then $R'(X) \in \mathbb{C}^* \Longleftrightarrow$
If $P, Q \in \mathbb{C}[X,Y]$ satisfy $\partial(P,Q)/\partial(X,Y) \equiv 1$ then

$$R(X) \times \sum_{i=1}^{n} 1/P_Y(X,\alpha_i)[Q(X,\alpha_i) - \beta] \in \mathbb{C}^*.$$

Equivalently (by Theorem 19 above)

$$R(X) \times \sum_{j=1}^{n} 1/Q_Y(X,\beta_j)[P(X,\beta_j) - \alpha] \in \mathbb{C}^*.$$

However, for any (α,β), and any $P, Q \in \mathbb{C}[X,Y]$ we have:
(P,Q) is a Jacobian pair $\Longleftrightarrow$ $(P-\alpha, Q-\beta)$ is a Jacobian pair
Hence we may take above $\alpha = \beta = 0$.

Finally, $R(X)$ equals $\prod_{k=1}^{n} P(X,\beta_k)$ and also $\prod_{k=1}^{n} Q(X,\alpha_k)$ up to a $\mathbb{C}^*$-factor. This completes the proof. □

Theorem 22. *The Jacobian conjecture is true iff the following holds true:*
If $P(X,Y), Q(X,Y) \in \mathbb{C}[X,Y]$ satisfy the following two conditions

(1) $\partial(P,Q)/\partial(X,Y) \equiv 1$
and
(2) $P(X,Y) = a_n \prod_{i=1}^{n}(Y - \alpha_i(X))$, $Q(X,Y) = b_n \prod_{j=1}^{n}(Y - \beta_j(X))$, *where* $a_n, b_n \in \mathbb{C}^*$
then

$$\lim_{X\to\infty} \sum_{i=1}^{n} X/P_Y(X,\alpha_i)Q(X,\alpha_i) = -1.$$

Moreover, this condition is equivalent to the following condition

$$\lim_{X\to\infty} \sum_{j=1}^{n} X/Q_Y(X,\beta_j)P(X,\beta_j) = -1.$$

Proof. By Theorem 19 above we have

$$\begin{aligned} -R'(X)/R(X) &= \sum_{i=1}^{n} 1/P_Y(X,\alpha_i)Q(X,\alpha_i) \\ &= \sum_{j=1}^{n} 1/Q_Y(X,\beta_j)P(X,\beta_j), \end{aligned}$$

where we picked $\alpha = \beta = 0$ (we use for that the same argument that was used in the proof of the previous theorem). If

$$R(X) = R_l X^l + \cdots + R_0, \quad R_l \neq 0,$$

then

$$\lim_{X \to \infty} XR'(X)/R(X) = l.$$

Hence $R(X) = R_1 X + R_0,\ R_1 \neq 0$ iff $\lim_{X\to\infty} XR'(X)/R(X) = 1$ and the proof is completed as the proof of the previous theorem. □

Theorem 23. *Let* $P(X,Y) = a_n \prod_{i=1}^{n}(Y - \alpha_i(X)), Q(X,Y) = b_n \prod_{j=1}^{n}(Y - \beta_j(X)) \in \mathbb{C}[X,Y]$. *Let* $R(X) = \textbf{resultant}(P(X,Y), Q(X,Y), Y)$. *Then the following are true:*

(a) If $R(X_0) = 0$ *then there are* $1 \leq i_0, j_0 \leq n$ *so that*

$$\alpha_{i_0}(X_0) = \beta_{j_0}(X_0).$$

(b) If $P(X,Y)$ *is regular and if* $P_Y(X_0, \alpha_{i_0}(X_0)) = 0$ *then*

$$\lim_{X \to X_0} |\alpha'_{i_0}(X)| = \infty.$$

(c) If (P,Q) *is a Jacobian pair and if* $P_Y(X_0, \alpha_{i_0}(X_0)) = 0$ *but* $R(X_0) \neq 0$ *then there exists a* $j_0 \neq i_0$ *so that*

$$P_Y(X_0, \alpha_{j_0}(X_0)) = 0.$$

Proof.

(a) We have $R(X) = A(X,Y)P(X,Y) + B(X,Y)Q(X,Y)$ and as is well known, if $R(X_0) = 0$ then there exists a Y_0 so that $P(X_0, Y_0) = Q(X_0, Y_0) = 0$. Thus $\alpha_{i_0}(X_0) = \beta_{j_0}(X_0)$ for some $1 \leq i_0, j_0 \leq n$.

(b) For an open Zariski set $\alpha'_{i_0}(X)$ exists. By $P(X, \alpha_{i_0}(X)) \equiv 0$ we obtain on this open set

$$P_X(X, \alpha_{i_0}(X)) + \alpha'_{i_0}(X) P_Y(X, \alpha_{i_0}(X)) \equiv 0.$$

Letting $X \to X_0$ we have:

$$\lim_{X \to X_0} P_Y(X, \alpha_{i_0}(X)) = P_Y(X_0, \alpha_{i_0}(X_0)) = 0.$$

So by regularity

$$\lim_{X \to X_0} P_X(X, \alpha_{i_0}(X)) = P_X(X_0, \alpha_{i_0}(X_0)) \neq 0.$$

But then by

$$\lim_{X \to X_0} \{P_X(X, \alpha_{i_0}(X)) + \alpha'_{i_0}(X) P_Y(X, \alpha_{i_0}(X))\} = 0,$$

we deduce that

$$\lim_{X \to X_0} |\alpha'_{i_0}(X)| = \infty,$$

as desired.

(c) In this case, by Theorem 19 above we have

$$R'(X)/R(X) = -\sum_{i=1}^{n} 1/P_Y(X,\alpha_i(X))Q(X,\alpha_i(X)).$$

Since $R(X_0) \neq 0$ it follows that

$$\lim_{X \to X_0} \sum_{i=1}^{n} 1/P_Y(X,\alpha_i(X))Q(X,\alpha_i(X)) = -R'(X_0)/R(X_0) \in \mathbb{C}.$$

It also follows by $R(X_0) \neq 0$ that for each $i = 1, \ldots, n$ we have $Q(X_0, \alpha_i(X_0)) \neq 0$. Since $P_Y(X_0, \alpha_{i_0}(X_0)) = 0$ it follows from the above that there must be some cancellation which must be of the form $P_Y(X_0, \alpha_{j_0}(X_0)) = 0$ for some $j_0 \neq i_0$. This concludes the proof of part (c). □

17 Parametrization of the Jacobian Variety

In this section we address in a very elementary manner the problem of parametrizing the Jacobian variety. If such a parametrization is available it will usually carry with it some useful information about the variety. For example it will tell us it's dimension.

In general we have $2(\binom{n+2}{2} - 1)$ coefficients for two polynomials of degree n (at most) and the Jacobian variety is determined by $\binom{2n}{2}$ equations on these coefficients. All the equations are quadratic-expressible linearly in terms of the Grassmann coordinates of the line that passes through the two points that are determined by the coefficients of the two polynomials. All the equations except for one are homogeneous. Let us denote:

$$P(X,Y) = \sum_{1 \leq i+j \leq n} a_{ij} X^i Y^j,$$

$$Q(X,Y) = \sum_{1 \leq i+j \leq n} b_{ij} X^i Y^j.$$

Then

$$\partial P/\partial X = \sum_{1 \leq i+j \leq n} i a_{ij} X^{i-1} Y^j, \quad \partial P/\partial Y = \sum_{1 \leq i+j \leq n} j a_{ij} X^i Y^{j-1},$$

$$\partial Q/\partial X = \sum_{1 \leq i+j \leq n} i b_{ij} X^{i-1} Y^j, \quad \partial Q/\partial Y = \sum_{1 \leq i+j \leq n} j b_{ij} X^i Y^{j-1}.$$

So

$$\begin{aligned}\partial(P,Q)/\partial(X,Y) &= \Big(\sum_{1\le i+j\le n} ia_{ij}X^{i-1}Y^j\Big)\Big(\sum_{1\le i+j\le n} jb_{ij}X^iY^{j-1}\Big)\\ &\quad - \Big(\sum_{1\le i+j\le n} ja_{ij}X^iY^{j-1}\Big)\Big(\sum_{1\le i+j\le n} ib_{ij}X^{i-1}Y^j\Big)\\ &= \sum_{\substack{1\le i_1+j_1\le n\\ 1\le i_2+j_2\le n}} i_1j_2(a_{i_1j_1}b_{i_2j_2} - a_{i_2j_2}b_{i_1j_1})X^{i_1+i_2-1}Y^{j_1+j_2-1}.\end{aligned}$$

We note that if we put $i = i_1+i_2-1, j = j_1+j_2-1$ then $i_2 = i-i_1+1, j_2 = j-j_1+1$. Also $1 \le i_1+j_1 \le n$ and $1 \le i_2+j_2 \le n$. The last double inequality is

$$1 \le (i-i_1+1)+(j-j_1+1) \le n,$$

or

$$i+j+2-n \le i_1+j_1 \le i+j+1.$$

Combining that with $1 \le i_1+j_1 \le n$ we obtain

$$\max(1, i+j+2-n) \le i_1+j_1 \le \min(n, i+j+1).$$

This proves the following.

Theorem 24. *The Jacobian variety of degree n is given by the following $\binom{2n}{2}$ equations:*

$$\sum_{\max(1,i+j+2-n)\le i_1+j_1\le\min(n,i+j+1)} i_1(j-j_1+1)$$
$$\times\,[a_{i_1j_1}b_{(i-i_1+1)(j-j_1+1)} - a_{(i-i_1+1)(j-j_1+1)}b_{i_1j_1}] = 0$$
$$a_{10}b_{01} - a_{01}b_{10} = J,$$

where $1 \le i+j \le 2(n-1)$.

Definition 15. *Let us arrange the coefficients $\{b_{ij}\}_{0\le i+j\le n}$ in a certain order, say, the lexicographical order. Then we may view the $\binom{2n}{2}$ equations in Theorem 24 as a linear system which is non-homogeneous, of $p = \binom{2n}{2}$ equations in the $q = \binom{n+2}{2} - 1$ unknowns $\{b_{ij}\}_{0\le i+j\le n}$. The matrix of the coefficients of the system has for it's entries integral multiples of certain $\{a_{ij}\}$ such that the sum of the coefficients in each row is 0. We shall denote this matrix that depends only on $P(X,Y)$ by M_P.*

We can now give a linear algebraic characterization of polynomials that have Jacobian mates.

Theorem 25. *Let $P(X,Y) \in \mathbb{C}[X,Y]$ and denote $n = \deg P(X,Y)$ and $e_{\binom{2n}{2}} = \begin{pmatrix} 0\\ \vdots\\ 0\\ 1\end{pmatrix}$ the unit vector of dimensions $\binom{2n}{2} \times 1$. $P(X,Y)$ has a Jacobian mate iff*

$$rank\{M_P\} = rank\{M_P e_{\binom{2n}{2}}\}.$$

Here $M_P e_{\binom{2n}{2}}$ is the matrix M_P augmented by the column $e_{\binom{2n}{2}}$.

Proof. The following fact is a well-known, see Theorem X on p. 85 of [10]:

A necessary and sufficient condition that the non-homogeneous equations

$$\sum_{j=1}^{q} \alpha_{ij} X_j = C_i \quad (i = 1, \ldots, p),$$

have a solution is that the matrices

$$\begin{pmatrix} a_{11} & \ldots & a_{1q} \\ \vdots & \vdots & \vdots \\ a_{p1} & \ldots & a_{pq} \end{pmatrix} \quad \text{and} \quad \begin{pmatrix} a_{11} & \ldots & a_{1q} & C_1 \\ \vdots & \vdots & \vdots & \vdots \\ a_{p1} & \ldots & a_{pq} & C_p \end{pmatrix},$$

have the same rank.

Clearly $P(X, Y)$ has a Jacobian mate iff the system of equations of Theorem 24 has a solution $\{b_{ij}\}$, $1 \le i + j \le n$. However, this is a linear system in the $\{b_{ij}\}$ and it's coefficient matrix is M_P while it's column of free elements is $Je_{\binom{2n}{2}}$. This proves the assertion. □

As noted above and as is clear from Theorem 24 the Jacobian variety is determined by $\binom{2n}{2} - 1$ homogeneous equations plus one more non-homogeneous equation, namely, $a_{10}b_{01} - a_{01}b_{10} = J$. With the lexicographical order this last equation corresponds to the last row of M_P. This row is:

$$(0, \ldots, 0, -a_{01}, a_{10}).$$

Thus M_P has the following structure: it consists of an $(\binom{2n}{2} - 1)$ by an $(\binom{n+2}{2} - 1)$ matrix H_P that corresponds to the $\binom{2n}{2} - 1$ homogeneous equations, sitting on the top of the row $(0, \ldots, 0, -a_{01}, a_{10})$.

Definition 16. *The $(\binom{2n}{2} - 1) \times (\binom{n+2}{2} - 1)$ matrix H_P is called the homogeneous matrix of the polynomial $P(X, Y)$.*

We can now give a second linear algebraic characterization of polynomials that have Jacobian mates.

Theorem 26. *Let $P(X, Y) \in \mathbb{C}[X, Y]$. $P(X, Y)$ has a Jacobian mate iff*

$$rank\{M_P\} = 1 + rank\{H_P\}.$$

Proof. Let $n = \deg P(X, Y)$ and $r = rank\{M_P\}$. By Theorem 25 $P(X, Y)$ has a Jacobian mate iff

$$rank\{M_P\} = rank\{M_P e_{\binom{2n}{2}}\}.$$

Since $r = rank\{M_P\}$ it follows that M_P contains an $r \times r$ submatrix which is regular and that any $(r+1) \times (r+1)$ submatrix of M_P is singular.

We claim that the condition

$$rank\{M_P\} = rank\{M_P e_{\binom{2n}{2}}\},$$

is equivalent to the following condition:

Any $r \times r$ regular submatrix of M_P must contain entries in the last row of M_P, namely, in $(0, \ldots, 0, -a_{01}, a_{10})$.

For if there was an $r \times r$ regular submatrix of M_P containing no entries in the last row then this submatrix had also to be a submatrix of H_P and so clearly $M_P e_{\binom{2n}{2}}$ had to contain a regular $(r+1) \times (r+1)$ submatrix (one of whose entries is the 1 in $(\binom{2n}{2}, \binom{n+1}{2})$). So that

$$rank\{M_P e_{\binom{2n}{2}}\} \geq r + 1,$$

which is not possible.

Hence H_P contains no submatrix of order $r \times r$ which is regular but contains regular submatrices of order $(r-1) \times (r-1)$. Hence

$$rank\{H_P\} = r - 1 = rank\{M_P\} - 1. \quad \square$$

Acknowledgments

The author is pleased to acknowledge Prof. Bruno Buchberger and the coordinators of the Special Semester on Gröbner Bases (February 1 – July 31, 2006), organized by RICAM, Austrian Academy of Sciences, and RISC, Johannes Kepler University, Linz Austria for partial support of this project. The author would also like to thank the anonymous referees for helpful criticism as well as the editor of these proceedings, Misha Klin for the wonderful editorial job he did on this paper.

References

1. K. Adjamagbo, H. Derksen, and A. Van den Essen, *On Polynomial Maps in Positive Characteristic and the Jacobian Conjecture*, Report 9208, Univ. of Nijmegen, 1992
2. H. Bass, E. Connell, and D. Wright, The Jacobian conjecture: reduction of degree and formal expansion of the inverse, *Bull. Amer. Math. Soc.* (2), **7** (1982), 287–330.
3. B. Buchberger, Gröbner Bases: an algorithmic method in polynomial ideal theory, in N. K. Bose (ed.) *Multidimensional Systems Theory*, pp. 184–232, Reidel, Dordrecht, 1985.

4. T. W. Dube, *The Structure of Polynomial Ideals and Gröbner Bases*, Preprint, 1990
5. A. van den Essen, A note on Meisters and Olech's proof of the global asymptotic stability Jacobian conjecture, *Pac. J. Math.* (2), **151** (1991), 351–357.
6. A. van den Essen, *Polynomial Automorphisms and the Jacobian Conjecture*, Progress in Mathematics, Vol. 190, Birkhäuser, Basel, 2000.
7. C. Gutierrez, and N. Van Chau, Properness and the Jacobian conjecture in $\mathbb{R}^2$, *Vietnam J. Math.* (4), **31** (2003), 421–427.
8. R. Gebauer and H. M. Moller, On an installation of Buchberger's algorithm, in L. Robbiano (ed.) *Computational Aspects of Commutative Algebra*, pp. 141–152, Academic Press, New York, 1988.
9. J. Hadamard, Sur les transformations pontuelles, *Bull. de la Soc. Math. de France*, **34** (1906), 71–81.
10. W. V. D. Hodge, and D. Pedoe, *Methods of Algebraic Geometry*, Vol. 1, Cambridge University Press, Cambridge, 1947.
11. O. H. Keller, Ganze Cremona-Transformationen, *Monatsh. Math. Phys.*, **47** (1939), 299–306.
12. K. Kurdyka and K. Rusek, Surjectivity of certain injective semialgebraic transformations of $\mathbb{R}^n$, *Math. Z.*, **200** (1988), 141–148.
13. E. Mayr and A. Meyer, The complexity of the word problem for commutative semigroups and polynomial ideals, *Adv. Math.*, **46** (1982), 305–329.
14. G. Meisters and C. Olech, Solution of the global asymptotic stability Jacobian conjecture for the polynomial case, in *Analyse Mathematique et applications*, pp. 373–381, Gautier-Villars, Paris, 1998.
15. T. Moh, On the global Jacobian Conjecture and the configuration of roots, *J. Reine Angew. Math.*, **340** (1983), 140–212.
16. H. M. Moller and F. Mora, Upper and lower bounds for the degree of Gröbner bases, in *EUROSAM 1984*, Lecture Notes in Computer Science, Vol. 174, pp. 172–183, Springer, Berlin, 1984.
17. P. Nousiainen and M. E. Sweedler, Automorphisms of polynomial and power series rings, *J. Pure Appl. Algebra*, **29** (1983), 93–97.
18. R. Peretz, *The Topology of Maximal Domains for Local Homeomorphism Mappings on $\mathbb{R}^2$ and an Application to the Jacobian Conjecture*, Technion Preprint Series NO-MT, 833, 1988.
19. R. Peretz, Maximal domains for entire functions, *J. Anal. Math.*, **61** (1993), 1–28.
20. R. Peretz, The variety of the asymptotic values of a real polynomi al ètale map, *J. Pure Appl. Algebra*, **01/106** (1996), 103–112.
21. R. Peretz, On counterexamples to Keller's problem, *Illinois J. Math.* (02), **40** (1996), 293–303.
22. R. Peretz, The geometry of the asymptotics of polynomial maps, *Israel J. Math.*, **105** (1998), 1–59.
23. S. Pinchuk, A counterexample to the strong real Jacobian conjecture, *Math. Z.*, **217** (1994), 1–4.

24. M. Razar, Polynomial maps with constant Jacobian, *Israel J. Math.*, **32** (1979), 97–106.
25. N. Van Chau, Two remarks on non-zero constant Jacobian polynomial maps of $\mathbb{C}^2$, *Ann. Pol. Math.* (1), **82** (2003), 39–44.
26. D. Wright, On the Jacobian conjecture, *Illinois J. Math.*, **25** (1981), 423–440.

Part B

Research Papers

Some Meeting Points of Gröbner Bases and Combinatorics

Bálint Felszeghy[1,2] and Lajos Rónyai[1,2]

[1] Computer and Automation Institute, Hungarian Academy of Science, Budapest, Hungary. fbalint@math.bme.hu
[2] Institute of Mathematics, Budapest University of Technology and Economics, Budapest, Hungary. lajos@ilab.sztaki.hu

Summary. Let $\mathbb{F}$ be a field, $V \subseteq \mathbb{F}^n$ be a set of points, and denote by $I(V)$ the vanishing ideal of V in the polynomial ring $\mathbb{F}[x_1, \ldots, x_n]$. Several interesting algebraic and combinatorial problems can be formulated in terms of some finite V, and then Gröbner bases and standard monomials of $I(V)$ yield a powerful tool for solving them.

We present the Lex Game method, which allows one to efficiently compute the lexicographic standard monomials of $I(V)$ for any finite set $V \subseteq \mathbb{F}^n$. We apply this method to determine the Gröbner basis of $I(V)$ for some V of combinatorial and algebraic interest, and present four applications of this type. We give a new easy proof of a theorem of Garsia on a generalization of the fundamental theorem of symmetric polynomials. We also reprove Wilson's theorem concerning the modulo p rank of some inclusion matrices. By examining the Gröbner basis of the vanishing ideal of characteristic vectors of some specific set systems, we obtain results in extremal combinatorics. Finally, we point out a connection among the standard monomials of $I(V)$ and $I(V^c)$, where $V \subseteq \{0,1\}^n$ and $V^c = \{0,1\}^n \setminus V$. This has immediate consequences in combinatorial complexity theory.

The main results have appeared elsewhere in several papers. We collected them into a unified account to demonstrate the usefulness of Gröbner basis methods in combinatorial settings.

Key words: Gröbner basis, Standard monomial, Lexicographic order, Vanishing ideal, Hilbert function, Inclusion matrix, Rank formula

1 Introduction

Throughout the paper n will be a positive integer, and $[n]$ stands for the set $\{1, 2, \ldots, n\}$. The set of all subsets of $[n]$ is denoted by $2^{[n]}$. Subsets of $2^{[n]}$ are called *set families* or *set systems*. Let $\binom{[n]}{m}$ denote the family of all *m-subsets* of $[n]$ (subsets which have cardinality m), and $\binom{[n]}{\leq m}$ is the family of those subsets that have at most m elements. By $\mathbb{N}$ we mean the nonnegative

M. Klin et al. (eds.), Algorithmic Algebraic Combinatorics and Gröbner Bases, DOI 10.1007/978-3-642-01960-9_6, © Springer-Verlag Berlin Heidelberg 2009

integers, $\mathbb{Z}$ is the set of integers, $\mathbb{Q}$ is the field of rational numbers, and $\mathbb{F}_p$ is the field of p elements, where p is a prime.

Let $\mathbb{F}$ be a field. As usual, we denote by $\mathbb{F}[x_1, \ldots, x_n] = \mathbb{F}[\mathbf{x}]$ the ring of polynomials in variables $x_1, \ldots, x_n$ over $\mathbb{F}$. To shorten our notation, we write $f(\mathbf{x})$ for $f(x_1, \ldots, x_n)$. Vectors of length n are denoted by boldface letters, for example $\mathbf{y} = (y_1, \ldots, y_n) \in \mathbb{F}^n$. If $\mathbf{w} \in \mathbb{N}^n$, we write $\mathbf{x}^{\mathbf{w}}$ for $x_1^{w_1} \ldots x_n^{w_n} \in \mathbb{F}[\mathbf{x}]$. For a subset $M \subseteq [n]$, the monomial x_M is $\prod_{i \in M} x_i$ (and $x_\emptyset = 1$). We say that a polynomial is *multilinear* if it is a linear combination of some x_M ($M \subseteq [n]$).

Suppose that $V \subseteq \mathbb{F}^n$. Then the *vanishing ideal* $I(V)$ *of* V consists of polynomials in $\mathbb{F}[\mathbf{x}]$, which as functions vanish on V. In our applications, we consider finite sets V, and use the Gröbner bases, or standard monomials of $I(V)$ (see the next subsection for the definitions) to prove claims on V.

Let $\mathbf{v}_F \in \{0,1\}^n$ denote the *characteristic vector of a set* $F \subseteq [n]$, that is the ith coordinate of $\mathbf{v}_F$ is 1 iff $i \in F$. For a system of sets $\mathcal{F} \subseteq 2^{[n]}$, let us put $V_{\mathcal{F}}$ for the set of the characteristic vectors of elements of $\mathcal{F}$. By $I(\mathcal{F})$ we understand the vanishing ideal $I(V_{\mathcal{F}})$, as it will make no confusion.

In Sect. 2 we collected the definitions and basic facts we need about Gröbner bases and Hilbert functions.

We develop a combinatorial description of the lexicographic standard monomials of $I(V)$ in the subsequent section via a two player game. Lea and Stan play the Lex Game with some fixed parameters $V \subseteq \mathbb{F}^n$ and $\mathbf{w} \in \mathbb{N}^n$. We show that $\mathbf{x}^{\mathbf{w}}$ is a lexicographic standard monomial of $I(V)$ if and only if Stan has a winning strategy in the game. This description proves to be more than just a toy. It yields a fast algorithm to determine the standard monomials of $I(V)$ for an arbitrary finite V. On the other hand, it is also applicable in the 'symbolic' computation of the standard monomials for some particular sets V. We shall see several examples of such calculations in Sect. 4, which is devoted to combinatorial and algebraic applications.

We give a new easy proof of a theorem of Garsia on a generalization of the fundamental theorem of symmetric polynomials. We also reprove Wilson's theorem concerning the modulo p rank of some inclusion matrices. In the direction of extremal combinatorics, we obtain results on the maximal cardinality of some set systems. To be a bit more specific, we will consider modulo q L-avoiding L-intersecting families, and families that do not shatter large sets. The last application is to point out a connection among the standard monomials and Hilbert functions of $I(V)$ and $I(V^c)$, where $V \subseteq \{0,1\}^n$ and $V^c = \{0,1\}^n \setminus V$. An immediate consequence of this in combinatorial complexity theory is shown.

Much of the results described here have already appeared elsewhere, most notably in [12, 19, 16, 11, 15, 22]. In some cases the way of exposition, which is based primarily on the Lex Game, is new and considerably simpler than the original one. We collected the material to point out interesting combinatorial applications of Gröbner basis methods.

2 Preliminaries

2.1 Gröbner Bases and Standard Monomials

We recall now some basic facts concerning Gröbner bases in polynomial rings over fields. More detailed exposition can be found in the classic papers by prof. Bruno Buchberger [3–5], and in the textbook [9].

A total order $\prec$ on the monomials composed from variables $x_1, x_2, \ldots, x_n$ is a *term order*, if 1 is the minimal element of $\prec$, and $\prec$ is compatible with multiplication with monomials. Two important term orders are the *lexicographic* (*lex* for short) and the *degree compatible lexicographic* (*deglex*) orders. We have $\mathbf{x}^{\mathbf{w}} \prec_{\text{lex}} \mathbf{x}^{\mathbf{u}}$ if and only if $w_i < u_i$ holds for the smallest index i such that $w_i \neq u_i$. As for deglex, we have that a monomial of smaller degree is smaller in deglex, and among monomials of the same degree lex decides the order. Also in general, $\prec$ is *degree compatible*, if $\deg(\mathbf{x}^{\mathbf{w}}) < \deg(\mathbf{x}^{\mathbf{u}})$ implies $\mathbf{x}^{\mathbf{w}} \prec \mathbf{x}^{\mathbf{u}}$.

The *leading monomial* (or *leading term*) $\mathrm{lm}(f)$ of a nonzero polynomial $f \in \mathbb{F}[\mathbf{x}]$ is the largest monomial (with respect to $\prec$) which appears with nonzero coefficient in f, when written as the usual linear combination of monomials. It is easy to verify that the leading monomial of a product $f \cdot g$ of nonzero polynomials is $\mathrm{lm}(f) \cdot \mathrm{lm}(g)$. We denote the set of all leading monomials of polynomials of a given ideal $I \trianglelefteq \mathbb{F}[\mathbf{x}]$ by $\mathrm{Lm}(I) = \{\mathrm{lm}(f) : f \in I\}$, and we simply call them the *leading monomials of* I.

A monomial is called a *standard monomial* of I, if it is not a leading monomial of any $f \in I$. Let $\mathrm{Sm}(I)$ denote the set of standard monomials of I.

Obviously, $\mathrm{Sm}(I)$ is a *downset* with respect to division, that is, a divisor of a standard monomial is again in $\mathrm{Sm}(I)$.

A finite subset $G \subseteq I$ is a *Gröbner basis* of I, if for every $f \in I$ there exists a $g \in G$ such that $\mathrm{lm}(g)$ divides $\mathrm{lm}(f)$.

Using that $\prec$ is a well founded order, it follows that G is actually a basis of I, that is, G generates I as an ideal of $\mathbb{F}[\mathbf{x}]$. It is a fundamental fact that every nonzero ideal I of $\mathbb{F}[\mathbf{x}]$ has a Gröbner basis.

A Gröbner basis $G \subseteq I$ is *reduced* if for all $g \in G$, the *leading coefficient* of g (i.e. the coefficient of $\mathrm{lm}(g)$) is 1, and $g \neq h \in G$ implies that no nonzero monomial in g is divisible by $\mathrm{lm}(h)$. This is clearly equivalent to saying that every $g \in G$ has leading coefficient 1, $\{\mathrm{lm}(g) : g \in G\}$ is the set of minimal elements of $\mathrm{Lm}(I)$ (with respect to division), and the polynomial $g - \mathrm{lm}(g)$ is a linear combination of standard monomials. For any fixed term order and any nonzero ideal of $\mathbb{F}[\mathbf{x}]$ there exists a unique reduced Gröbner basis.

Suppose that $f \in \mathbb{F}[\mathbf{x}]$ contains a monomial $\mathbf{x}^{\mathbf{w}} \cdot \mathrm{lm}(g)$, where g is some other polynomial with leading coefficient c. Then we can *reduce f with g* (and get $\hat{f}$), that is, we can replace $\mathbf{x}^{\mathbf{w}} \cdot \mathrm{lm}(g)$ in f with $\mathbf{x}^{\mathbf{w}} \cdot (\mathrm{lm}(g) - \frac{1}{c}g)$. Clearly if $g \in I$, then f and $\hat{f}$ represent the same coset in $\mathbb{F}[\mathbf{x}]/I$. Also note that $\mathrm{lm}(\mathbf{x}^{\mathbf{w}} \cdot (\mathrm{lm}(g) - \frac{1}{c}g)) \prec \mathbf{x}^{\mathbf{w}} \cdot \mathrm{lm}(g)$. As $\prec$ is a well founded order, this guarantees that if we reduce f repeatedly with a set of polynomials G, then we end up with a *reduced* $\hat{f}$ in finitely many steps, that is a polynomial such that none of its monomials is divisible by any $\mathrm{lm}(g)$ $(g \in G)$.

Assume now that G is a Gröbner basis of some ideal I. In this case, it can be shown that the reduction of any polynomial with G is unique. We see from the definitions that the reduction $\hat{f}$ of a polynomial f is a linear combination of standard monomials of I. From these, it follows directly that for a nonzero ideal I the set $\mathrm{Sm}(I)$ is a linear basis of the $\mathbb{F}$-vectorspace $\mathbb{F}[\mathbf{x}]/I$. If $I(V)$ is a vanishing ideal of a finite set V of points in $\mathbb{F}^n$, then $\mathbb{F}[\mathbf{x}]/I(V)$ can be interpreted as the space of functions $V \to \mathbb{F}$. An immediate consequence is that the number of standard monomials of $I(V)$ is $|V|$. In particular for every family of sets we have $|\mathcal{F}| = |\mathrm{Sm}(I(\mathcal{F}))|$.

Another property of the standard monomials of $I(\mathcal{F})$ will be needed several times: for an arbitrary set family $\mathcal{F}$, one has $x_i^2 - x_i \in I(\mathcal{F})$, therefore all the elements of $\mathrm{Sm}(I(\mathcal{F}))$ are multilinear monomials.

2.2 The Hilbert Function

We write $\mathbb{F}[\mathbf{x}]_{\le m}$ for the vector space of polynomials over $\mathbb{F}$ with degree at most m. Similarly, if $I \trianglelefteq \mathbb{F}[\mathbf{x}]$ is an ideal then $I_{\le m} = I \cap \mathbb{F}[\mathbf{x}]_{\le m}$ stands for the linear subspace of polynomials from I with degree at most m. The *Hilbert function* of the $\mathbb{F}$-algebra $\mathbb{F}[\mathbf{x}]/I$ is $H_I : \mathbb{N} \to \mathbb{N}$, where

$$H_I(m) = \dim_{\mathbb{F}}\left(\mathbb{F}\left[\mathbf{x}\right]_{\le m} / I_{\le m}\right).$$

Let $\prec$ be any degree compatible term ordering (deglex for instance). One can easily see that the set of standard monomials with respect to $\prec$ of degree at most m forms a linear basis of $\mathbb{F}[\mathbf{x}]_{\le m}/I_{\le m}$. Hence we can obtain $H_I(m)$ by determining the set $\mathrm{Sm}(I)$ with respect to any degree compatible term ordering.

When $\mathcal{F}$ is a system of sets, we call $H_{I(\mathcal{F})}(m)$ the Hilbert function of $\mathcal{F}$ and denote it by $H_{\mathcal{F}}(m)$, as it makes no confusion. In the combinatorial literature $H_{\mathcal{F}}(m)$ is usually given in terms of inclusion matrices.

For two families $\mathcal{F}, \mathcal{G} \subseteq 2^{[n]}$ the *inclusion matrix* $I(\mathcal{F}, \mathcal{G})$ is a matrix of size $|\mathcal{F}| \times |\mathcal{G}|$, whose rows and columns are indexed by the elements of $\mathcal{F}$ and $\mathcal{G}$, respectively. The entry at position (F, G) is 1 if $G \subseteq F$ and 0 otherwise $(F \in \mathcal{F}, G \in \mathcal{G})$.

It is a simple matter to verify that the Hilbert function of $\mathcal{F}$ is given by

$$H_{\mathcal{F}}(m) = \dim_{\mathbb{F}}\left(\mathbb{F}\left[\mathbf{x}\right]_{\le m} / I(\mathcal{F})_{\le m}\right) = \mathrm{rank}_{\mathbb{F}}\, I\left(\mathcal{F}, \binom{[n]}{\le m}\right). \tag{1}$$

We will benefit from a similar statement in Sect. 4.2, which claims that

$$\dim_{\mathbb{F}}\left(\mathcal{P}_{\mathcal{F},m}\right) = \mathrm{rank}_{\mathbb{F}}\, I\left(\mathcal{F}, \binom{[n]}{m}\right), \tag{2}$$

where $\mathcal{P}_{\mathcal{F},m}$ is the linear space of functions from $V_{\mathcal{F}}$ to $\mathbb{F}$ which can be represented as homogeneous multilinear polynomials of degree m. (With a slight abuse of notation we could have written $\mathcal{P}_{\mathcal{F},m} = \mathbb{F}[\mathbf{x}]_{=m}/I(\mathcal{F})_{=m}$.)

Incidence matrices and their ranks are important in the study of finite geometries as well. Standard monomials and Hilbert functions are also useful in that setting. The reader is referred to Moorhouse [20] in the present volume for an account on applications of this type.

3 Computation of the Lex Standard Monomials

In this section we sketch a purely combinatorial description of the lexicographic standard monomials of vanishing ideals of finite sets of points. This is the main tool which can be applied to compute lex standard monomials of sets of combinatorial interest. The original source is [12], and the interested reader can find an extension to general zero dimensional ideals in [13].

Throughout the section, we use the lexicographic ordering, so – even if it is not stated explicitly – $\mathrm{Sm}(I)$ and $\mathrm{Lm}(I)$ is defined with respect to lex.

As before, let $\mathbb{F}$ be a field, $V \subseteq \mathbb{F}^n$ a finite set and $\mathbf{w} = (w_1, \ldots, w_n) \in \mathbb{N}^n$ an n dimensional vector of natural numbers. With these data fixed, we define the Lex Game $\mathrm{Lex}(V; \mathbf{w})$, which is played by two persons, Lea and Stan.

Both Lea and Stan know V and $\mathbf{w}$.

1 Lea chooses w_n elements of $\mathbb{F}$.
Stan picks a value $y_n \in \mathbb{F}$, different from Lea's choices.
2 Lea now chooses w_{n-1} elements of $\mathbb{F}$.
Stan picks a $y_{n-1} \in \mathbb{F}$, different from Lea's (last w_{n-1}) choices.
... (The game goes on in this same fashion.)
n Lea chooses w_1 elements of $\mathbb{F}$.
Stan finally picks a $y_1 \in \mathbb{F}$, different from Lea's (last w_1) choices.

The winner is Stan if he could pick $\mathbf{y} = (y_1, \ldots, y_n)$ such that $\mathbf{y} \in V$, otherwise Lea wins the game. (Also, if in any step there is no proper choice y_i for Stan, then Lea wins.)

Example 1. Let $n = 5$, and $\alpha, \beta \in \mathbb{F}$ be different elements. Let V be the set of all α-β sequences in $\mathbb{F}^5$ in which the number of the α coordinates is 1, 2 or 3. We claim that Lea can win with the question vector $\mathbf{w} = (11100)$, but with $\mathbf{w} = (01110)$ Stan has a chance to win.

Indeed, let $\mathbf{w} = (11100)$. To have $\mathbf{y} \in V$, Stan is forced to select values from $\{\alpha, \beta\}$. If Stan gives only β for the last 2 coordinates, then Lea will choose α in the first three, therefore $\mathbf{y}$ cannot contain any α coordinates. However if Stan gives at least one α for the last 2 coordinates, then Lea, by keeping on choosing β, can prevent $\mathbf{y}$ to have at least two β coordinates.

In the case $\mathbf{w} = (01110)$ Stan's winning strategy is to pick $y_5 = \beta$, and choose from $\{\alpha, \beta\}$ (for the 4th, 3rd and 2nd coordinates). One can easily check that y_1 then can always be taken such that $\mathbf{y} \in V$.

It is quite clear that, being a finite deterministic game, in $\mathrm{Lex}(V;\mathbf{w})$ either Lea or Stan has a winning strategy. We will simply say that Lea or Stan wins $\mathrm{Lex}(V;\mathbf{w})$ accordingly. The main theorem of this section is the following.

Theorem 1. *Let $V \subseteq \mathbb{F}^n$ be a finite set and $\mathbf{w} \in \mathbb{N}^n$. Stan wins $\mathrm{Lex}(V;\mathbf{w})$ if and only if $\mathbf{x}^{\mathbf{w}} \in \mathrm{Sm}(I(V))$.*

An immediate consequence is that Lea wins the game iff $\mathbf{x}^{\mathbf{w}}$ is a leading monomial for $I(V)$.

There is a fast algorithm[3] which lists those $\mathbf{w} \in \mathbb{N}^n$, for which Stan wins $\mathrm{Lex}(V;\mathbf{w})$ for a given V. In view of Theorem 1, it actually computes the lex standard monomials of $I(V)$. In this paper we intend to use the Theorem to obtain explicit combinatorial description of $\mathrm{Sm}(I(V))$ for some interesting sets V.

Also, note that the game does not use anything more from the properties of the base field than its cardinality. That is, we can conclude that the set of lex standard monomials of a vanishing ideal is rather a combinatorial object, than an algebraic one.

In line with the recursive nature of the game, we will use induction on n to prove the theorem. The following notation will be useful.

For $\mathbf{y} = (y_1, \ldots, y_n) \in \mathbb{F}^n$ we set $\overline{\mathbf{y}} = (y_1, \ldots, y_{n-1})$, if $n \geq 2$. We shall also use $\overline{\mathbf{y}}$ for denoting a vector of length $n-1$, even if it is not a prefix of a vector of length n. Similarly we shall write sometimes $\overline{\mathbf{w}}$, or even $\overline{\mathbf{x}}^{\overline{\mathbf{w}}}$ instead of $x_1^{w_1} \ldots x_{n-1}^{w_{n-1}}$.

Let $y \in \mathbb{F}$, suppose that $n \geq 2$, and set

$$V_y = \left\{\overline{\mathbf{y}} \in \mathbb{F}^{n-1} : (\overline{\mathbf{y}}, y) \in V\right\}.$$

It is clear that if Stan picks $y_n = y$ in the first step, then they continue as if they have just started a $\mathrm{Lex}(V_y;\overline{\mathbf{w}})$ game.

Proof of Theorem 1. We prove the statement by induction on n.

The case $n = 1$ is easy. Let $w \geq 0$ be an integer. Then $x^w \in \mathrm{Sm}(I(V))$ if and only if $w < |\mathrm{Sm}(I(V))| = |V|$ by the fact that $\mathrm{Sm}(I(V))$ is a downset with respect to division. But this means precisely that there has to be a $y \in V$ which is not among Lea's guesses, thus Stan wins the game by picking that y.

Suppose that $n \geq 2$, and that the theorem is true for $n-1$. Set

$$Z = \left\{y \in \mathbb{F} : \overline{\mathbf{x}}^{\overline{\mathbf{w}}} \in \mathrm{Sm}(I(V_y))\right\}.$$

The inductive hypothesis yields that Stan wins $\mathrm{Lex}(I(V_y);\overline{\mathbf{w}})$ if and only if $y \in Z$. From what we said about the connection between the games $\mathrm{Lex}(V;\mathbf{w})$ and $\mathrm{Lex}(V_y;\overline{\mathbf{w}})$ it follows that Stan wins $\mathrm{Lex}(V;\mathbf{w})$ if and only if $w_n < |Z|$. Therefore it is enough to show that

[3] It uses constant times $|V|nk$ comparisons of field elements in the worst case, where k is the maximum number of different elements which appear in a fixed coordinate of points of V; see [12].

$$\mathbf{x}^{\mathbf{w}} \in \mathrm{Sm}\left(I(V)\right) \iff w_n < \left|\left\{y \in \mathbb{F} : \overline{\mathbf{x}}^{\overline{\mathbf{w}}} \in \mathrm{Sm}\left(I(V_y)\right)\right\}\right|.$$

Suppose first that $\mathbf{x}^{\mathbf{w}} \in \mathrm{Lm}(I(V))$, and let $f(\mathbf{x}) \in I(V)$ be a witness of this fact, that is $\mathrm{lm}(f) = \mathbf{x}^{\mathbf{w}}$. By collecting together the terms of the form $\overline{\mathbf{x}}^{\overline{\mathbf{w}}} x_n^i$ $(i \in \mathbb{N})$ we get a decomposition $f(\mathbf{x}) = \overline{\mathbf{x}}^{\overline{\mathbf{w}}} g(x_n) + h(\mathbf{x})$, where all monomials of $h(\mathbf{x})$ are lexicographically smaller than $\overline{\mathbf{x}}^{\overline{\mathbf{w}}}$, and $\deg(g) = w_n$.

If $y \in \mathbb{F}$ is not a root of $g(x_n)$, then $\hat{f}(\overline{\mathbf{x}}) = \overline{\mathbf{x}}^{\overline{\mathbf{w}}} g(y) + h(\overline{\mathbf{x}}, y)$ is a polynomial which vanishes on V_y, and has the property that $\mathrm{lm}(\hat{f}) = \overline{\mathbf{x}}^{\overline{\mathbf{w}}}$. Thus, if y is not a root of g, then $\overline{\mathbf{x}}^{\overline{\mathbf{w}}} \in \mathrm{Lm}(I(V_y))$. In other words there are at most $\deg(g) = w_n$ elements $y \in \mathbb{F}$ such that $\overline{\mathbf{x}}^{\overline{\mathbf{w}}} \in \mathrm{Sm}(I(V_y))$.

For the other direction, assume that $\mathbf{x}^{\mathbf{w}} \in \mathrm{Sm}(I(V))$. First note that by the finiteness of V, we have $V_y = \emptyset$ (and then $\mathrm{Sm}(I(V_y)) = \emptyset$) with finitely many exceptions $y \in \mathbb{F}$, hence $|Z| < \infty$. Now, it suffices to show that $\overline{\mathbf{x}}^{\overline{\mathbf{w}}} x_n^{|Z|} \in \mathrm{Lm}(I(V))$, since in this case $\overline{\mathbf{x}}^{\overline{\mathbf{w}}} x_n^{|Z|}$ cannot be a divisor of $\mathbf{x}^{\mathbf{w}}$, that is $w_n < |Z|$.

Set $F = \{y \in \mathbb{F} : V_y \neq \emptyset\}$ and $y \in F \setminus Z$. On one hand, $\overline{\mathbf{x}}^{\overline{\mathbf{w}}} \in \mathrm{Lm}(I(V_y))$ implies the existence of a polynomial $f_y(\overline{\mathbf{x}})$ such that all monomials of $f(\overline{\mathbf{x}})$ are less than $\overline{\mathbf{x}}^{\overline{\mathbf{w}}}$, and $\overline{\mathbf{x}}^{\overline{\mathbf{w}}} + f_y(\overline{\mathbf{x}}) \in I(V_y)$. On the other hand, let $\chi_y(x_n)$ be a polynomial such that for $y' \in F \setminus Z$

$$\chi_y(y') = \begin{cases} 1, & y' = y, \\ 0, & \text{otherwise.} \end{cases} \tag{3}$$

Since F is finite, such a polynomial does exist.

And finally let

$$s(\mathbf{x}) = \left(\overline{\mathbf{x}}^{\overline{\mathbf{w}}} + \sum_{y \in F \setminus Z} \chi_y(x_n) f_y(\overline{\mathbf{x}})\right) \cdot \prod_{y \in Z} (x_n - y).$$

By the properties of the lex order $\mathrm{lm}(\overline{\mathbf{x}}^{\overline{\mathbf{w}}} + \sum_{y \in F} \chi_y(x_n) f_y(\overline{\mathbf{x}})) = \overline{\mathbf{x}}^{\overline{\mathbf{w}}}$, therefore we have that the leading monomial of $s(\mathbf{x})$ is $\overline{\mathbf{x}}^{\overline{\mathbf{w}}} x_n^{|Z|}$. It remains to verify $s(\mathbf{x}) \in I(V)$.

Let $\mathbf{y} = (\overline{\mathbf{y}}, y) \in V$ be arbitrary. Clearly $V_y \neq \emptyset$, that is $y \in F$. We may suppose that $y \notin Z$ for otherwise the second term of $s(\mathbf{x})$ vanishes on $\mathbf{y}$. Property (3) of the polynomials $\chi_{y'}(x_n)$ gives (for some $\alpha \in \mathbb{F}$)

$$s(\overline{\mathbf{x}}, y) = \left(\overline{\mathbf{x}}^{\overline{\mathbf{w}}} + \sum_{y' \in F \setminus Z} \chi_{y'}(y) f_{y'}(\overline{\mathbf{x}})\right) \cdot \alpha = \left(\overline{\mathbf{x}}^{\overline{\mathbf{w}}} + f_y(\overline{\mathbf{x}})\right) \cdot \alpha,$$

which vanishes on $\overline{\mathbf{y}} \in V_y$ by the definition of f_y, thus $s(\mathbf{x})$ is zero on $\mathbf{y}$. This completes the proof. □

For those, who do not like playing whilst doing math, we emphasize below the main point of Theorem 1, a fact first noted by Cerlienco and Mureddu [8].

Corollary 1. *If $V \subseteq \mathbb{F}^n$ is finite, $n \geq 2$, and $\mathbf{w} \in \mathbb{N}^n$ then*

$$\mathbf{x}^{\mathbf{w}} \in \mathrm{Sm}_{\mathrm{lex}}\left(I(V)\right) \iff w_n < \left|\left\{y \in \mathbb{F} : \overline{\mathbf{x}}^{\overline{\mathbf{w}}} \in \mathrm{Sm}_{\mathrm{lex}}\left(I(V_y)\right)\right\}\right|.$$

Theorem 1 has the immediate consequence that the standard monomials are largely independent of the base field $\mathbb{F}$ and of the precise embedding of V into $\mathbb{F}^n$. As here we consider more than one field, let us temporarily put $I_{\mathbb{F}}(V)$ for the polynomial ideal $I(V)$ in $\mathbb{F}[\mathbf{x}]$.

Corollary 2. *Assume that $V \subseteq V_1 \times \cdots \times V_n$ for some finite sets $V_i \subseteq \mathbb{F}$. Let $\hat{\mathbb{F}}$ be any field and suppose that $\varphi_i \colon V_i \to \hat{\mathbb{F}}$ are injective maps for $i \in [n]$. Let $\hat{V}$ be the image of V, that is*

$$\hat{V} = \{(\varphi_1(y_1), \ldots, \varphi_n(y_n)) : \mathbf{y} \in V\}.$$

Then $\mathrm{Sm}(I_{\mathbb{F}}(V)) = \mathrm{Sm}(I_{\hat{\mathbb{F}}}(\hat{V}))$. In particular, if $V \subseteq \{0,1\}^n$ then the set $\mathrm{Sm}(I_{\mathbb{F}}(V))$ is independent of the base field $\mathbb{F}$.

Proof. The $\mathrm{Lex}(V;\mathbf{w})$ game is essentially the same as the $\mathrm{Lex}(\hat{V};\mathbf{w})$ game since we have changed only the names of the elements (bijectively). The second part follows from the first, because $0 \neq 1$ in $\mathbb{F}$ for any field $\mathbb{F}$. □

The second part of the corollary concerning sets $V \subseteq \{0,1\}^n$ has been proven in [2] by a different method. We now show that the reduced lexicographic Gröbner basis of $I_{\mathbb{F}}(V)$ for a set $V \subseteq \{0,1\}^n$ is essentially the same over any field. We remark that this can be generalized to finite sets with more than two integer coordinate values.

If $f \in \mathbb{Z}[\mathbf{x}]$, then for all fields $\mathbb{F}$ of characteristic 0 we clearly have $f \in \mathbb{F}[\mathbf{x}]$, but also if the characteristic of $\mathbb{F}$ is $p > 0$, we can still consider f as an element of $\mathbb{F}[\mathbf{x}]$ by reducing its integer coefficients modulo p.

Corollary 3. *If $V \subseteq \{0,1\}^n$, then the reduced lex Gröbner basis G of $I_{\mathbb{Q}}(V)$ has integer coefficients. For an arbitrary field $\mathbb{F}$, the set in $\mathbb{F}[\mathbf{x}]$ corresponding to G is the reduced lex Gröbner basis of the ideal $I_{\mathbb{F}}(V)$.*

Proof. Let $\mathbf{x}^{\mathbf{w}} + g(\mathbf{x})$ be an element of the reduced lex Gröbner basis of $I_{\mathbb{Q}}(V)$, where every monomial of $g \in \mathbb{Q}[\mathbf{x}]$ is smaller than $\mathbf{x}^{\mathbf{w}}$, and is contained in $\mathrm{Sm}(I_{\mathbb{Q}}(V))$. Suppose by contradiction that $g \notin \mathbb{Z}[\mathbf{x}]$.

Let $z \in \mathbb{Z}$ such that $zg(\mathbf{x})$ has relatively prime integer coefficients. If a prime p divides z, then reduce $zg \in \mathbb{Z}[\mathbf{x}]$ modulo p to get a polynomial over $\mathbb{F}_p$. It is a nonzero polynomial which (modulo p) vanishes on V, as $z\mathbf{x}^{\mathbf{w}} + zg(\mathbf{x})$ vanishes on V and $p \mid z$. Thus the leading monomial of $zg(\mathbf{x})$ is in $\mathrm{Lm}(I_{\mathbb{F}_p}(V)) = \mathrm{Lm}(I_{\mathbb{Q}}(V))$, by Corollary 2. That is a contradiction.

For the second statement, let $\mathbb{F}$ be an arbitrary field and let us think of G as a subset of $\mathbb{F}[\mathbf{x}]$. Obviously $G \subseteq I_{\mathbb{F}}(V)$ is still true and the leading monomials of G remain the same. By $\mathrm{Lm}(I_{\mathbb{F}}(V)) = \mathrm{Lm}(I_{\mathbb{Q}}(V))$, we have that G is a Gröbner basis of $I_{\mathbb{F}}(V)$. As the elements of G, except for their leading monomials, are linear combinations of standard monomials, G is also reduced. □

Before going on to present mathematical (mostly combinatorial) applications of the Lex Game, we briefly comment on the algorithmic problem of actually computing standard monomials, or more generally a basis of $\mathrm{Sm}(I_{\mathbb{F}}(V))$ over $\mathbb{F}$. The problem has had a long history starting with the outstanding paper by Buchberger and Möller [7]. Their algorithm, as well as the subsequent methods of Marinari, Möller and Mora [21] and Abbott, Bigatti, Kreuzer and Robbiano [1] give also a Gröbner basis of $I_{\mathbb{F}}(V)$. For the arithmetic complexity of these methods we have the bound $O(n^2m^3)$ when V is a subset of $\mathbb{F}^n$ and $|V| = m$ (see Sect. 3 in [10] for a related discussion). The Lex Game provides only the standard monomials, but in return it appears to lead to a much faster algorithm (see [12] for the details). In general we have the bound $O(nm^2)$. In some important special cases, such as the case of small finite ground fields which appear naturally in coding applications, one can even have a linear bound $O(nm)$ on the time demand of the algorithm.

4 Applications

4.1 Generalization of the Fundamental Theorem of Symmetric Polynomials

Following [19], we present an easy proof of a theorem by Garsia [17], which is a generalization of the fundamental theorem of symmetric polynomials.

The *ith elementary symmetric polynomial* is

$$\sigma_i(\mathbf{x}) = \sum_{\substack{\mathbf{w}\in\{0,1\}^n \\ \deg(\mathbf{x}^{\mathbf{w}})=i}} \mathbf{x}^{\mathbf{w}},$$

provided that $0 \le i \le n$. Later we will also use the *complete symmetric polynomial of degree* $i \ge 0$, which is

$$h_i(\mathbf{x}) = \sum_{\substack{\mathbf{w}\in\mathbb{N}^n \\ \deg(\mathbf{x}^{\mathbf{w}})=i}} \mathbf{x}^{\mathbf{w}}.$$

The fundamental theorem of symmetric polynomials claims that if $f(\mathbf{x})$ is a symmetric polynomial, then it can be written uniquely as a finite sum

$$f(\mathbf{x}) = \sum_{\mathbf{u}\in\mathbb{N}^n} \alpha_{\mathbf{u}} \boldsymbol{\sigma}(\mathbf{x})^{\mathbf{u}},$$

where $\alpha_{\mathbf{u}} \in \mathbb{F}$, and $\boldsymbol{\sigma}(\mathbf{x})^{\mathbf{u}}$ stands for $\prod_{i=1}^n \sigma_i(\mathbf{x})^{u_i}$.

We intend to prove the following generalization, which was obtained by A. Garsia [17].

Theorem 2. *Any polynomial $f(\mathbf{x}) \in \mathbb{F}[\mathbf{x}]$ can be written uniquely as a finite sum*

$$f(\mathbf{x}) = \sum_{\substack{\mathbf{w}\in\mathbb{N}^n \\ \mathbf{w}\le\mathbf{v}}} \sum_{\mathbf{u}\in\mathbb{N}^n} \alpha_{\mathbf{w},\mathbf{u}} \mathbf{x}^{\mathbf{w}} \boldsymbol{\sigma}(\mathbf{x})^{\mathbf{u}},$$

where $\mathbf{v} = (0, 1, \ldots, n-1)$, $\mathbf{w} \le \mathbf{v}$ is understood coordinatewise, and $\alpha_{\mathbf{w},\mathbf{u}} \in \mathbb{F}$.

We need some preparations before the proof. Let $z_1, \ldots, z_n$ be different elements of a field and set

$$V = \{(z_{\pi(1)}, \ldots, z_{\pi(n)}) : \pi \in S_n\}$$

the set of all permutations of the sequence $z_1, \ldots, z_n$.

We first show that the lexicographic standard monomials of $I(V)$ are exactly the divisors of $x_2 x_3^2 \ldots x_n^{n-1}$. In other words, the minimal lex leading monomials are of the form x_i^i for $i \in [n]$.

Proposition 1. *For the set of points V defined above, we have that $\mathbf{x}^{\mathbf{w}}$ is a lexicographic standard monomial of $I(V)$ if and only if $\mathbf{w} \le (0, 1, \ldots, n-1)$.*

Proof. One can get the lexicographic standard monomials of V using the Lex Game (Theorem 1). Suppose that $\mathbf{w} \le (0, 1, \ldots, n-1)$. Then Stan's strategy will be to pick in the $(n-i+1)$th step (for y_i) any element from the set $\{z_1, \ldots, z_n\} \setminus \{y_n, \ldots, y_{i+1}\}$. This set has exactly i elements, so $w_i < i$ guarantees that Lea cannot choose all of them, that is there will always be a proper choice for Stan.

On the other hand, if for example $w_i \ge i$, then in the $(n-i+1)$th step Lea can choose all the elements of $\{z_1, \ldots, z_n\} \setminus \{y_n, \ldots, y_{i+1}\}$, thus y_i will either be the same as a previously selected y_j (and then $\mathbf{y} \notin V$) or an element different from all z_j (again $\mathbf{y} \notin V$). □

We use the following easy fact without proof (see for example [9]) which holds for all $i \in [n]$:

$$\sum_{j=0}^{i} (-1)^j h_{i-j}(x_i, \ldots, x_n) \sigma_j(\mathbf{x}) = 0. \tag{4}$$

Let $i \in [n]$ and set

$$f_i(\mathbf{x}) = \sum_{j=0}^{i} (-1)^j h_{i-j}(x_i, \ldots, x_n) \sigma_j(\mathbf{z}).$$

Proposition 2. *The set of polynomials $\{f_i : i \in [n]\}$ is the reduced Gröbner basis of V for all term orders, such that the order of the variables is $x_1 \succ x_2 \succ \cdots \succ x_n$.*

Proof. Clearly, if $x_1 \succ x_2 \succ \cdots \succ x_n$ holds for a term order, then $\mathrm{lm}(f_i) = x_i^i$. It is also obvious by Proposition 1 that every monomial of $f_i(\mathbf{x}) - x_i^i$ is a lex standard monomial. Equation 4 implies that f_i vanishes on V. As the minimal lex leading monomials (again by Proposition 1) are $\{x_i^i : i \in [n]\}$, we have that $\{f_i : i \in [n]\}$ is indeed a reduced lex Gröbner basis. But the leading monomials of the f_i for all term orders $\prec$ considered in the statement are the same, thus $\mathrm{Sm}_{\mathrm{lex}}(I(V)) \supseteq \mathrm{Sm}_{\prec}(I(V))$. Due to the equality of the cardinalities of the two sides, we have that the standard monomials are the same for all term orders considered. We conclude that $\{f_i : i \in [n]\}$ is a reduced Gröbner basis also with respect to $\prec$. □

Proof of Theorem 2. We had a good reason for not choosing base field for V until now. Let $\mathbb{F}(\mathbf{z})$ be the function field over $\mathbb{F}$ in n variables $z_1, \dots, z_n$ and let $V \subseteq \mathbb{F}(\mathbf{z})$ be the set of all permutations of these variables, as before.

Let $f(\mathbf{x}) \in \mathbb{F}[\mathbf{x}] \subseteq \mathbb{F}(\mathbf{z})[\mathbf{x}]$ be any polynomial, and reduce $f(\mathbf{x})$ by the Gröbner basis $\{f_i(\mathbf{x}) \in \mathbb{F}(\mathbf{z})[\mathbf{x}] : i \in [n]\}$ of V. The result is an $\mathbb{F}(\mathbf{z})$-linear combination of monomials $\mathbf{x}^{\mathbf{w}} \in \mathrm{Sm}(I(V))$. Furthermore, since actually $f_i \in \mathbb{F}[\mathbf{z}][\mathbf{x}]$, and f_i is symmetric in the variables $z_1, \dots, z_n$, the coefficients of the $\mathbf{x}^{\mathbf{w}} \in \mathrm{Sm}(I(V))$ in this $\mathbb{F}(\mathbf{z})$-linear combination are symmetric polynomials from $\mathbb{F}[\mathbf{z}]$. Thus as functions on V, we have an equality

$$f(\mathbf{x}) = \sum_{\mathbf{x}^{\mathbf{w}} \in \mathrm{Sm}(I(V))} \mathbf{x}^{\mathbf{w}} g_{\mathbf{w}}(\mathbf{z}),$$

where $g_{\mathbf{w}}(\mathbf{z}) \in \mathbb{F}[\mathbf{z}]$ is a symmetric polynomial. Therefore putting $\mathbf{z}$ in the place of $\mathbf{x}$ (since $\mathbf{z} \in V$) we get the equation

$$f(\mathbf{z}) = \sum_{\mathbf{z}^{\mathbf{w}} \in \mathrm{Sm}(I(V))} \mathbf{z}^{\mathbf{w}} g_{\mathbf{w}}(\mathbf{z})$$

of elements of $\mathbb{F}(\mathbf{z})$. An application of the fundamental theorem of symmetric polynomials, together with $\mathrm{Sm}(I(V)) = \{\mathbf{x}^{\mathbf{w}} : \mathbf{w} \le (0, 1, \dots, n-1)\}$ yields the existence of the required form for f.

Uniqueness now follows: suppose that

$$f(\mathbf{x}) = \sum_{\mathbf{x}^{\mathbf{w}} \in \mathrm{Sm}(I(V))} \sum_{\mathbf{u} \in \mathbb{N}^n} \alpha_{\mathbf{w},\mathbf{u}} \mathbf{x}^{\mathbf{w}} \boldsymbol{\sigma}(\mathbf{x})^{\mathbf{u}}.$$

Then as functions on V we have

$$f(\mathbf{x}) = \sum_{\mathbf{x}^{\mathbf{w}} \in \mathrm{Sm}(I(V))} \sum_{\mathbf{u} \in \mathbb{N}^n} \alpha_{\mathbf{w},\mathbf{u}} \mathbf{x}^{\mathbf{w}} \boldsymbol{\sigma}(\mathbf{z})^{\mathbf{u}} = \sum_{\mathbf{x}^{\mathbf{w}} \in \mathrm{Sm}(I(V))} \mathbf{x}^{\mathbf{w}} \tilde{g}_{\mathbf{w}}(\mathbf{z}),$$

for some polynomials $\tilde{g}_{\mathbf{w}}(\mathbf{z}) \in \mathbb{F}[\mathbf{z}]$. We expressed $f(\mathbf{x})$ as an $\mathbb{F}(\mathbf{z})$-linear combination of standard monomials. But this is unique, hence $\tilde{g}_{\mathbf{w}}(\mathbf{z}) = g_{\mathbf{w}}(\mathbf{z})$, and so (using the uniqueness part of the fundamental theorem of symmetric polynomials) the claim follows. □

It is instructive to compare our approach here to the one followed by Buchberger and Elias in [6]. They used Gröbner bases to detect and guess identities among polynomials, which involved Fermat polynomials and elementary symmetric polynomials. Subsequently they went on, generalized these to obtain conjectures and then proved these conjectures by traditional inductive means. Here we employ Gröbner bases as a proof technique to establish the generalized identity constituting Theorem 2.

4.2 Wilson's Rank Formula

Consider the inclusion matrix $A = I\left(\binom{[n]}{d}, \binom{[n]}{m}\right)$, where $m \leq d \leq n-m$.

A famous theorem of Richard M. Wilson [23, Theorem 2] describes a diagonal form of A over $\mathbb{Z}$. A is shown to be row-column equivalent over $\mathbb{Z}$ to a diagonal matrix with diagonal entries $\binom{d-i}{m-i}$ with multiplicity $\binom{n}{i} - \binom{n}{i-1}$ for $0 \leq i \leq m$. As a corollary, the following rank formula holds:

Theorem 3. *Let p be a prime. Then*

$$\operatorname{rank}_{\mathbb{F}_p}(A) = \sum_{\substack{0 \leq i \leq m \\ p \nmid \binom{d-i}{m-i}}} \binom{n}{i} - \binom{n}{i-1}.$$

We shall outline a simple proof which uses polynomial functions, and some simple notions related to Gröbner bases. We note first that the rank of A is exactly the dimension of the linear space $\mathcal{P}_{d,m}$ over $\mathbb{F}_p$ of the functions from $V_{\binom{[n]}{d}}$ to $\mathbb{F}_p$ which are spanned by the monomials x_M with $|M| = m$.

Let P_m denote the subspace of homogeneous multilinear polynomials in $\mathbb{F}_p[\mathbf{x}]$ of degree m. Suppose that $m \leq n/2$, and for a set $M \subseteq [n]$, $|M| \leq m$ we define the multilinear polynomial

$$y_M = \sum_{\substack{M' \supseteq M \\ |M'| = m}} x_{M'} \in P_m.$$

To simplify our notation, we write I for the vanishing ideal $I\left(\binom{[n]}{m}\right)$ of $\binom{[n]}{m}$.

Lemma 1. *The collection of polynomials y_M, where $x_M \in \mathrm{Sm}(I)$, is a linear basis of P_m over $\mathbb{F}_p$.*

Proof. Since $\{x_M + I : x_M \in \mathrm{Sm}(I)\}$ is a linear basis of $\mathbb{F}_p[\mathbf{x}]/I$, and $x_M + I = y_M + I$ (they represent the same function on $V_{\binom{[n]}{m}}$), we obtain that $\{y_M + I : x_M \in \mathrm{Sm}(I)\}$ is a basis of $\mathbb{F}_p[\mathbf{x}]/I$. As $y_M \in P_m$, it is also clear that $P_m + I = \mathbb{F}_p[\mathbf{x}]$. From the fact that $P_m \cap I = \{0\}$, we have a natural isomorphism $P_m \to \mathbb{F}_p[\mathbf{x}]/I$ which sends y_M to $y_M + I$. We conclude that $\{y_M : x_M \in \mathrm{Sm}(I)\}$ is indeed a basis of P_m. □

We can state Wilson's rank formula in this setting as follows.

Theorem 4. *Let p be a prime, suppose that $m \leq d \leq n-m$ and put $I = I(\binom{[n]}{m})$. A basis of the space $\mathcal{P}_{d,m}$ of $\mathbb{F}_p$-valued functions on $V_{\binom{[n]}{d}}$, which are $\mathbb{F}_p$-linear combinations of monomials x_M, $|M| = m$ is*

$$B = \left\{ y_M : x_M \in \mathrm{Sm}\,(I)\,, p \nmid \binom{d-|M|}{m-|M|} \right\}.$$

In particular,

$$\dim_{\mathbb{F}_p} \mathcal{P}_{d,m} = |B| = \sum_{\substack{0 \leq i \leq m \\ p \nmid \binom{d-i}{m-i}}} \binom{n}{i} - \binom{n}{i-1}.$$

Proof. Let $\mathbf{v}_F$ be the characteristic vector of a d-subset of $[n]$. It is immediate that

$$y_M(\mathbf{v}_F) = \binom{d-|M|}{m-|M|} \cdot x_M(\mathbf{v}_F). \tag{5}$$

We obtain that, as a function on $V_{\binom{[n]}{d}}$, y_M is a scalar multiple of x_M. This, together with the linear independence of the x_M gives that B is an independent set. Also, B spans $\mathcal{P}_{d,m}$, because P_m spans $\mathcal{P}_{d,m}$ by definition, and the y_M span P_m by Lemma 1. To conclude, it remains to verify that for $0 \leq i \leq m$ there are exactly $\binom{n}{i} - \binom{n}{i-1}$ monomials of degree i in $\mathrm{Sm}(I)$. This will be proven in Lemma 2. □

Lemma 2. *For an arbitrary term order and any integers $0 \leq i \leq m \leq \frac{n}{2}$, there are exactly $\binom{n}{i} - \binom{n}{i-1}$ monomials of degree i in $\mathrm{Sm}(I(\binom{[n]}{m}))$.*

Proof. We will restrict ourselves to the lex order. Note that this is enough for completing the proof of Theorem 4. The full proof could be carried out by the same ideas we use in Proposition 2 or outline after Theorem 5.

We say that a vector $\mathbf{w} \in \{0,1\}^n$ is a *ballot sequence* if in every prefix of $\mathbf{w}$ there are at least as many 0, as 1 coordinates. We shall prove that $\mathbf{x}^{\mathbf{w}}$ is a lex standard monomial for $I = I(\binom{[n]}{m})$ iff $\deg(\mathbf{x}^{\mathbf{w}}) \leq m$ and $\mathbf{w}$ is a ballot sequence. By Theorem 1, we can use the Lex Game $\mathrm{Lex}(V_{\binom{[n]}{m}}; \mathbf{w})$ to show this.

If the number of 1 coordinates in $\mathbf{w}$ is more than m, then Lea will choose 0 at each of her guesses. This way, Stan has to put $y_i = 1$ for more than m times, therefore $\mathbf{y} \notin V_{\binom{[n]}{m}}$ at the end, and Lea wins. That is, if $\deg(\mathbf{x}^{\mathbf{w}}) > m$, then $\mathbf{x}^{\mathbf{w}} \in \mathrm{Lm}(I)$.

Suppose now, that $\deg(\mathbf{x}^{\mathbf{w}}) \leq m$ and $\mathbf{w}$ is not a ballot sequence. Let $i \in [n]$ be such that $(w_1, \ldots, w_i)$ is the shortest prefix of $\mathbf{w}$ that violates the ballot condition. It is easy to see that i is odd, and there are exactly $\frac{i+1}{2}$ coordinates equal to 1. Assume that when in the game Stan picked y_{i+1} then there are $m-k$ ones among $y_n, \ldots, y_{i+1}$. Stan would win only if he could pick the remaining $y_i, \ldots, y_1$, such that k of them was 1, $i-k$ of them was 0. But if

$k \le \frac{i-1}{2}$, then Lea always chooses 0, thus there will be at least $\frac{i+1}{2} > k$ ones among $y_i, \dots, y_1$. And when $k > \frac{i-1}{2}$, then $i - k \le \frac{i-1}{2}$, so if Lea keeps on choosing 1, then Stan has to claim at least $\frac{i+1}{2} > i - k$ zero coordinates, and hence he loses the game.

Next we show how Stan can win if $\mathbf{w}$ is a ballot sequence with at most m ones. Set $J = \{j \in [n] : w_j = 1\}$. For all $j \in J$ let us pick an $\ell(j) \in [n]$, such that $w_{\ell(j)} = 0$, $\ell(j) < j$, and $\ell : J \to [n]$ is injective. (This can be done if $\mathbf{w}$ is a ballot sequence.) Let us put $L = \{\ell(j) : j \in J\}$, and $K = [n] \setminus (J \cup L)$. Stan's strategy to choose y_i is the following. If $i \in J$, then Lea will guess something, so he just claims the opposite (in $\{0,1\}$). If $i \in L$, say $i = \ell(j)$, then he picks $y_{\ell(j)}$, such that $\{y_j, y_{\ell(j)}\} = \{0,1\}$. (Note that when choosing the $\ell(j)$th coordinate, he already fixed y_j by $\ell(j) < j$.) This way, Stan will have exactly $|J|$ ones in $(y_i : i \in J \cup L)$. Therefore he picks $m - |J|$ ones from the y_k $(k \in K)$, and wins.

Now it follows immediately, that the lex standard monomials of $I(\binom{[n]}{m})$ of degree at most i are the same as the lex standard monomials of $I(\binom{[n]}{i})$. In particular, there are $\binom{n}{i}$ of them, and then there are $\binom{n}{i} - \binom{n}{i-1}$ standard monomials of degree i. This proves the lemma. □

The approach given here allows a considerable generalization of the rank formula. We present without proof a result of this type (for details, see [16]). Suppose that $0 \le m_1 < m_2 < \cdots < m_r \le d \le n - m_r$ and let p be a prime. Consider the set family $\mathcal{F} = \binom{[n]}{m_1} \cup \binom{[n]}{m_2} \cup \cdots \cup \binom{[n]}{m_r}$. Then

$$\operatorname{rank}_{\mathbb{F}_p}\left(I\left(\binom{[n]}{d}, \mathcal{F}\right)\right) = \sum_{\substack{0 \le i \le m_r \\ p \nmid n_i}} \binom{n}{i} - \binom{n}{i-1},$$

where $n_i = \gcd\left(\binom{d-i}{m_1-i}, \binom{d-i}{m_2-i}, \dots, \binom{d-i}{m_r-i}\right)$.

4.3 Applications to Modulo q ℓ-wide Families

In this subsection we give two applications of the Gröbner methods to extremal set theory. We prove upper bounds on the cardinality of a family of subsets of $[n]$ with certain restrictions: we will consider modulo q L-intersecting, L-avoiding families, and families that do not shatter large sets. We shall omit a part of the proof, but give the ideas. A detailed proof can be found in [11].

Let us consider the following family of sets. Let q, d, and ℓ be integers, such that $1 \le \ell < q$. Then the *modulo q complete ℓ-wide family* is

$$\mathcal{G} = \{G \subseteq [n] : \exists g \in \mathbb{Z} \text{ such that } d \le g < d + \ell, \text{ and } |G| \equiv g \pmod{q}\}.$$

In other words, $\mathcal{G}$ contains all subsets of $[n]$ which have cardinality modulo q in the interval $[d, d+\ell-1]$ (of length ℓ). The restrictions on the parameters ℓ and q tell us exactly that if $|G| \equiv d + \ell \pmod{q}$, then $G \notin \mathcal{G}$ (that is, $\mathcal{G}$ is in fact ℓ-wide). Subfamilies of $\mathcal{G}$ are called *modulo q ℓ-wide families.*

The following theorem will be crucial in both applications.

Theorem 5. *Let p be a prime, and q be a power of p. Denote by $H_{\mathcal{G}}(m)$ the Hilbert-function over $\mathbb{F}_p$ of a modulo q complete ℓ-wide family $\mathcal{G}$. If $0 \le m \le \frac{n+\ell}{2}$, then*

$$H_{\mathcal{G}}(m) \le \sum_{j=0}^{\infty} \sum_{k=0}^{\ell-1} \binom{n}{m-jq-k}.$$

A sketch of the proof is the following. One can obtain the lex standard monomials of $I(\mathcal{G})$ by the Lex Game method. Then also the lexicographic Gröbner basis can be constructed: for each minimal lex leading monomial $\mathbf{x}^{\mathbf{w}}$, we can exhibit a polynomial $f_{\mathbf{w}}$ in the ideal, such that $\mathrm{lm}(f_{\mathbf{w}}) = \mathbf{x}^{\mathbf{w}}$. It turns out that the lex and deglex leading monomials of these polynomials are the same. From this fact it follows that what we got is a deglex Gröbner basis as well, and that the lex and deglex standard monomials are the same. (This is the same way to compute the deglex Gröbner basis as in Proposition 2.) This is good news, since by counting the deglex standard monomials of degree at most m, we obtain the exact value of $H_{\mathcal{G}}(m)$. The formula in Theorem 5 is then a convenient upper bound of that value.

One may compare this result to Lemma 2, noting that if $q > n$ and $\ell = 1$, then the modulo q complete ℓ-wide family is just $\binom{[n]}{d}$.

4.3.1 Modulo q L-intersecting, L-avoiding Families

Let L be a subset of integers and $\mathcal{F}$ be a system of sets. We say that $\mathcal{F}$ is *modulo q L-avoiding* if $F \in \mathcal{F}$ and $f \in L$ implies $|F| \not\equiv f \pmod{q}$. We call $\mathcal{F}$ *modulo q L-intersecting* if for any two distinct sets $F_1, F_2 \in \mathcal{F}$ a congruence $|F_1 \cap F_2| \equiv f \pmod{q}$ holds for some $f \in L$.

The maximum number of sets a modulo q L-avoiding, L-intersecting set family can contain has been studied extensively, see [11] for more details. We have the following result in this direction.

We call a set $L \subseteq \{0, \dots, q-1\}$ a *modulo q interval* if it is either an interval of integers, or a union of two intervals L_1 and L_2, such that $0 \in L_1$ and $q-1 \in L_2$.

Theorem 6. *Let q be a power of a prime, L be a modulo q interval and $\mathcal{F} \subseteq 2^{[n]}$ be a modulo q L-avoiding, L-intersecting family of sets. If $|L| \le n-q+2$, then*

$$|\mathcal{F}| \le \sum_{k=|L|}^{q-1} \binom{n}{k}.$$

The following lemma is left as an exercise for the reader.

Lemma 3. *If f is an integer, q is a power of a prime p, then*

$$\binom{f-1}{q-1} \equiv \begin{cases} 0 \pmod{p}, & \text{if } f \not\equiv 0 \pmod{q} \\ 1 \pmod{p}, & \text{if } f \equiv 0 \pmod{q}. \end{cases}$$

Proof of Theorem 6. Put $\ell = q - |L|$. If L is an interval of integers, then set $d = \max L + 1$, otherwise, when L is the union of two (separate) intervals L_1, L_2 and $0 \in L_1$, set $d = \max L_1 + 1$. Denote by $\mathcal{G}$ the modulo q complete ℓ-wide family with this parameter d. Then by the definitions $\mathcal{F} \subseteq \mathcal{G}$.

For any $F \in \mathcal{F}$ we define the polynomial $\hat{f}_F(\mathbf{x}) \in \mathbb{Q}[\mathbf{x}]$ to be

$$\hat{f}_F(\mathbf{x}) = \left(\sum_{\substack{k=0 \\ k \notin L}}^{q-1} \binom{\mathbf{x} \cdot \mathbf{v}_F - k - 1}{q-1} \right) \text{ reduced by } x_i^2 - x_i \ (i \in [n]),$$

where $\mathbf{x} \cdot \mathbf{v} = \sum_{i=1}^{n} x_i v_i$ is the usual scalar product of row vectors.

We claim that $\hat{f}_F \in \mathbb{Z}[\mathbf{x}]$. Since we have reduced with $x_i^2 - x_i$, we have that $\hat{f}_F(\mathbf{x})$ is multilinear, thus $\hat{f}_F = \sum_{G \subseteq [n]} \alpha_G x_G$ with some coefficients $\alpha_G \in \mathbb{Q}$. If $\hat{f}_F \notin \mathbb{Z}[\mathbf{x}]$, then let G be a minimal set with respect to inclusion, such that $\alpha_G \notin \mathbb{Z}$. Clearly, the reduction with the polynomials $x_i^2 - x_i$ does not change the value of the original polynomial on 0-1 vectors, therefore $f_F(\mathbf{v}_G)$ is an integer. Thus substituting $\mathbf{v}_G$ we get that $f_F(\mathbf{v}_G) = \sum_{G' \subsetneq G} \alpha_{G'} + \alpha_G$, a contradiction since the coefficients $\alpha_{G'}$ are integers. We have proven that $\hat{f}_F \in \mathbb{Z}[\mathbf{x}]$.

Suppose that q is a power of a prime p and let $F' \in \mathcal{F}$ be a set. Then

$$\hat{f}_F(\mathbf{v}_{F'}) = \sum_{\substack{k=0 \\ k \notin L}}^{q-1} \binom{|F' \cap F| - k - 1}{q-1}. \tag{6}$$

If $F' \neq F$, then, as $\mathcal{F}$ is modulo q L-intersecting, $|F' \cap F| - k$ cannot be congruent to 0 modulo q for $k \notin L$. That is (by Lemma 3), if $F' \neq F$, then all terms of the sum in (6) are zero modulo p. If $F' = F$, then using that $\mathcal{F}$ is modulo q L-avoiding, we have exactly one nonzero term modulo p, which is actually congruent to 1. Write f_F for the polynomial in $\mathbb{F}_p[\mathbf{x}]$ we obtain from $\hat{f}_F$ by reducing its integer coefficients modulo p. The above argument yields

$$f_F(\mathbf{v}_{F'}) = \begin{cases} 0 & \text{if } F \neq F', \\ 1 & \text{if } F = F'. \end{cases}$$

Since the degree of $\hat{f}_F$ is at most $q-1$, the same is true for f_F as well. Using our earlier notation, this means that $f_F \in \mathbb{F}_p[\mathbf{x}]_{\leq q-1}$. We claim that the images $\overline{f}_F$ of the f_F in the quotient space $\mathbb{F}_p[\mathbf{x}]_{\leq q-1} / I(\mathcal{G})_{\leq q-1}$ are linearly independent over $\mathbb{F}_p$. Indeed, suppose that

$$\sum_{F \in \mathcal{F}} \alpha_F \overline{f}_F = 0 \tag{7}$$

for some $\alpha_F \in \mathbb{F}_p$. The elements of $\mathbb{F}_p[\mathbf{x}]/I(\mathcal{G})$ are functions on the characteristic vectors of $\mathcal{G}$. In particular (7) still holds if we substitute $\mathbf{v}_F$ for some $F \in \mathcal{F} \subseteq \mathcal{G}$. The substitution gives $\alpha_F = 0$ immediately.

To conclude, note that the number of the polynomials f_F is bounded by the dimension of $\mathbb{F}_p[\mathbf{x}]_{\leq q-1}/I(\mathcal{G})_{\leq q-1}$, that is

$$|\mathcal{F}| \leq \dim_{\mathbb{F}_p}\left(\mathbb{F}_p\left[\mathbf{x}\right]_{\leq q-1}/I(\mathcal{G})_{\leq q-1}\right) = H_{\mathcal{G}}(q-1)$$
$$\leq \sum_{j=0}^{\infty}\sum_{k=0}^{\ell-1}\binom{n}{q-1-jq-k} = \sum_{k=|L|}^{q-1}\binom{n}{k}$$

by Theorem 5 (which we are allowed to use as $|L| \leq n-q+2$ implies the assumption $q-1 \leq \frac{n+\ell}{2}$ of the theorem). □

4.3.2 Set Families which do not Shatter Large Sets

Consider a family $\mathcal{F}$ of subsets of $[n]$. We say that $\mathcal{F}$ *shatters* $M \subseteq [n]$ if

$$\{F \cap M : F \in \mathcal{F}\} = 2^M.$$

The system of sets $\mathcal{F}$ is an *ℓ-antichain* if it does not contain $\ell+1$ distinct sets $F_0, F_1, \ldots, F_\ell$ such that $F_0 \subsetneq F_1 \subsetneq \cdots \subsetneq F_\ell$.

Frankl [14] conjectured that if an ℓ-antichain $\mathcal{F}$ shatters no set of size $m+1$ for some integer $0 \leq m \leq \frac{n+\ell}{2} - 1$, then $|\mathcal{F}| \leq \sum_{k=0}^{\ell-1}\binom{n}{m-k}$ must hold.

An ℓ-wide family (which of course can be understood as a modulo q ℓ-wide family for some q large enough) is an ℓ-antichain. In their article [15], Friedl, Hegedűs and Rónyai showed that the upper bound is valid for ℓ-wide families. The next theorem is a generalization of that result, the special case follows by choosing $q > n$.

Theorem 7. *Let $\mathcal{F} \subseteq 2^{[n]}$ be a modulo q ℓ-wide family of sets, where q is a prime power. If $\mathcal{F}$ shatters no set of size $m+1$ for some integer $0 \leq m \leq \frac{n+\ell}{2}$, then*

$$|\mathcal{F}| \leq \sum_{j=0}^{\infty}\sum_{k=0}^{\ell-1}\binom{n}{m-jq-k}.$$

Proof. We first prove that if x_M is a standard monomial of any set system $\mathcal{F}$, then $\mathcal{F}$ shatters M. Suppose that $N \subseteq M$, but $N \notin \{F \cap M : F \in \mathcal{F}\}$. Let $\mathbf{v} = \mathbf{v}_N$ be the characteristic vector of N. Then the polynomial

$$\prod_{i \in M}(x_i + v_i - 1)$$

vanishes on $V_{\mathcal{F}}$ and its leading monomial is x_M, thus $x_M \in \mathrm{Lm}(I(\mathcal{F}))$. We conclude that $x_M \in \mathrm{Sm}(I(\mathcal{F}))$ implies $|M| \leq m$ for a family $\mathcal{F}$ which does not shatter any set of size $m+1$.

Recall that $\mathcal{F} \subseteq \mathcal{G}$, where $\mathcal{G}$ is a modulo q complete ℓ-wide family. This gives $\mathrm{Sm}(I(\mathcal{F})) \subseteq \mathrm{Sm}(I(\mathcal{G}))$, and so we can bound the cardinality of the standard monomials of $\mathcal{F}$ with the number of standard monomials of $\mathcal{G}$ of degree at most m. This latter is exactly $H_{\mathcal{G}}(m)$, if we consider a degree compatible term ordering. (Actually, in this case, we can take any term order, see the discussion after Theorem 5.) Therefore

$$|\mathcal{F}| = |\mathrm{Sm}\,(I(\mathcal{F}))| \leq H_{\mathcal{G}}(m),$$

and hence Theorem 5 gives the desired bound. □

The inequality in Theorem 7 is sharp. Choose $d = m - \ell + 1$ for a modulo q complete ℓ-wide family $\mathcal{G}$, and put $\mathcal{F} = \mathcal{G} \cap \binom{[n]}{\leq m}$. Then the fact that $\mathcal{F}$ does not contain any set of size $m+1$ implies that it cannot shatter any set of cardinality $m+1$. The size of $\mathcal{F}$ is precisely $\sum_{j=0}^{\infty} \sum_{k=0}^{\ell-1} \binom{n}{m-jq-k}$.

4.4 Harima's Theorem for Set Families

Here we prove an important special case of a theorem by T. Harima. It establishes a connection among the Hilbert functions of complementary set families.

Theorem 8. *Suppose $\mathcal{F} \subseteq 2^{[n]}$ and $\mathcal{G} = 2^{[n]} \setminus \mathcal{F}$ are nonempty set families. Then for their Hilbert functions we have*

$$\sum_{i=0}^{m} \binom{n}{i} = |\mathcal{G}| + H_{\mathcal{F}}(m) - H_{\mathcal{G}}(n-1-m)$$

for every $m = 0, 1, \ldots, n$.

Theorem 8 was proved by Tadahito Harima for much more general point sets. In formula (3.1.5) of [18] the result is given for two disjoint finite point sets $\mathbb{X}, \mathbb{Y} \subset \mathbf{P}^n(\mathbb{F})$ in the projective n-space over $\mathbb{F}$, instead of $V_{\mathcal{F}}$ and $V_{\mathcal{G}}$, such that $\mathbb{X} \cup \mathbb{Y}$ is a complete intersection. The formula was used in his characterization of the Hilbert functions of Artinian Gorenstein algebras with the weak Stanley property.

Here we focus on 0,1-vectors only. Our approach is based on direct computations with polynomial functions.

Proof. For a subset $M \subseteq [n]$, let M^c stand for the set $[n] \setminus M$.

We claim that a monomial x_M is a leading monomial for $I(\mathcal{F})$ if and only if x_{M^c} is a standard monomial for $I(\mathcal{G})$.

Among the monomials of the form x_M, the number of leading monomials for $I(\mathcal{F})$ is the same as the number of standard monomials for $I(\mathcal{G})$, namely $2^n - |\mathcal{F}| = |\mathcal{G}|$, hence the claim will follow if we show that $x_M \in \mathrm{Lm}(I(\mathcal{F}))$ implies $x_{M^c} \in \mathrm{Sm}(I(\mathcal{G}))$. Indeed, suppose for contradiction that we have polynomials $f \in I(\mathcal{F})$ and $g \in I(\mathcal{G})$ with leading terms x_M and x_{M^c}, respectively. Then $f \cdot g$ vanishes on $V_{2^{[n]}}$ and its leading term is $x_{[n]}$. This is impossible, because $|\mathrm{Sm}(I(2^{[n]}))| = 2^n = |\{x_{M'} : M' \subseteq [n]\}|$ implies that every multilinear monomial is a standard monomial for $V_{2^{[n]}}$.

Let $\prec$ be a degree compatible term order on $\mathbb{F}[\mathbf{x}]$. Now the number of multilinear leading monomials of degree i for $I(\mathcal{F})$ is $\binom{n}{i} - (H_{\mathcal{F}}(i) - H_{\mathcal{F}}(i-1))$.

By the claim above, this is $H_{\mathcal{G}}(n-i) - H_{\mathcal{G}}(n-i-1)$, the number of standard monomials of degree $n-i$ for $I(\mathcal{G})$. We have

$$\binom{n}{i} = H_{\mathcal{F}}(i) - H_{\mathcal{F}}(i-1) + H_{\mathcal{G}}(n-i) - H_{\mathcal{G}}(n-i-1),$$

for every $0 \le i \le n$ (we use the convention $H_{\mathcal{F}}(-1) = H_{\mathcal{G}}(-1) = 0$). By adding these up for $i = 0, \ldots, m$, we obtain

$$\sum_{i=0}^{m} \binom{n}{i} = H_{\mathcal{F}}(m) + H_{\mathcal{G}}(n) - H_{\mathcal{G}}(n-m-1).$$

The theorem follows now from $H_{\mathcal{G}}(n) = |\mathcal{G}|$. □

Theorem 8 allows us to formulate an interesting min-max relation. Let $\mathcal{F} \subset 2^{[n]}$ be a family different from $\emptyset$ and $2^{[n]}$. Let $a(\mathcal{F})$ stand for the smallest degree of a nonzero multilinear polynomial from $\mathbb{F}[\mathbf{x}]$ which vanishes on $V_{\mathcal{F}}$. We have $1 \le a(\mathcal{F}) \le n$.

Also, we define $b(\mathcal{F})$ to be the smallest integer k such that $H_{\mathcal{F}}(k) = |\mathcal{F}|$. In other words, $b(\mathcal{F})$ is the smallest degree k such that every function from $V_{\mathcal{F}}$ to $\mathbb{F}$ can be represented by a polynomial from $\mathbb{F}[\mathbf{x}]$ of degree at most k. We have $0 \le b(\mathcal{F}) \le n$.

It is easily seen that any polynomial $\chi_{\mathbf{v}} \in \mathbb{F}[\mathbf{x}]$ which is 1 on the vector $\mathbf{v} \in \{0,1\}^n$, and 0 on all other vectors from $\{0,1\}^n$ must have degree at least n. From that we readily infer that

$$a(\mathcal{F}) + b(2^{[n]} \setminus \mathcal{F}) \ge n. \tag{8}$$

Theorem 8 implies that, in fact, we have an equality here.

Corollary 4. *Let $\mathcal{F} \subset 2^{[n]}$ and $\mathcal{G} = 2^{[n]} \setminus \mathcal{F}$. Assume that both $\mathcal{F}$ and $\mathcal{G}$ are nonempty. Then we have*

$$a(\mathcal{F}) + b(\mathcal{G}) = n.$$

Proof. We apply Theorem 8 with $m = a(\mathcal{F}) - 1$. Note first, that $m \ge 0$ and $H_{\mathcal{F}}(m) = H_{2^{[n]}}(m)$, because the multilinear monomials of degree $\le m$ are linearly independent over $\mathbb{F}$, as functions on $V_{\mathcal{F}}$. Theorem 8 gives now that $H_{\mathcal{G}}(n-m-1) = |\mathcal{G}|$, hence $b(\mathcal{G}) \le n-m-1 = n - a(\mathcal{F})$. This, together with (8) proves the assertion. □

In [22] Theorem 8 is proved over more general coefficient rings, rather than fields, which include the rings $\mathbb{Z}_k = \mathbb{Z}/k\mathbb{Z}$, where k is a positive integer. An application to the (modular weak degree) complexity of Boolean functions is also given there.

Acknowledgements

Part of this work was done during the Special Semester on Gröbner Bases (February 1 – July 31, 2006), organized by RICAM, Austrian Academy of Sciences, and RISC, Johannes Kepler University, Linz, Austria. The authors are pleased to acknowledge Prof. Bruno Buchberger and the coordinators of the Special Semester for their hospitality and attention to this project.

Research supported in part by OTKA grants NK72845, NK63066 and K77476.

References

1. J. Abbott, A. Bigatti, M. Kreuzer, and L. Robbiano, Computing ideals of points, *J. Symbol. Comput.*, **30** (2000), 341–356.
2. R. P. Anstee, L. Rónyai, and A. Sali, Shattering news, *Graphs Comb.*, **18** (2002), 59–73.
3. B. Buchberger, *Ein Algorithmus zum Auffinden der Basiselemente des Restklassenringes nach einem nulldimensionalen Polynomideal*, Doctoral thesis, University of Innsbruck, 1965. *English translation:* An algorithm for finding the basis elements in the residue class ring modulo a zero dimensional polynomial ideal. J. Symbol. Comput., Special Issue on Logic, Mathematics, and Computer Science: Interactions, **41** (2006), 475–511.
4. B. Buchberger, Ein algorithmisches Kriterium fur die Lösbarkeit eines algebraischen Gleichungssystems, *Aequationes mathematicae*, **4** (1970), 374–383. *English translation*: An algorithmic criterion for the solvability of algebraic systems of equations, in B. Buchberger, F. Winkler (eds.) Gröbner Bases and Applications, London Mathematical Society Lecture Note Series, Vol. 251, pp. 535–545, Cambridge University Press, Cambridge, 1998.
5. B. Buchberger, Gröbner-Bases: An algorithmic method in polynomial ideal theory, in N. K. Bose (ed.) *Multidimensional Systems Theory – Progress, Directions and Open Problems in Multidimensional Systems Theory*, pp. 184–232, Reidel, Dordrecht, 1985. Chapter 6.
6. B. Buchberger and J. Elias, Using Gröbner bases for detecting polynomial identities: a case study on Fermat's ideal, *J. Number Theory*, **41** (1992), 272–279.
7. H. M. Möller and B. Buchberger, The construction of multivariate polynomials with preassigned zeros, in *Computer Algebra* (1982), Lecture Notes in Comput. Sci., Vol. 144, pp. 24–31, Springer, Berlin, 1982.
8. L. Cerlienco and M. Mureddu, From algebraic sets to monomial linear bases by means of combinatorial algorithms, *Discrete Math.*, **139** (1995), 73–87.
9. D. Cox, J. Little, and D. O'Shea, *Ideals, Varieties, and Algorithms*, Springer, Berlin, 1992.
10. J. B. Farr and S. Gao, Computing Gröbner bases for vanishing ideals of finite sets of points, in *Applied Algebra, Algebraic Algorithms and Error-*

correcting Codes, Lecture Notes in Comput. Sci., Vol. 3857, pp. 118–127, Springer, Berlin, 2006.
11. B. Felszeghy, G. Hegedűs, and L. Rónyai, Algebraic properties of modulo q complete ℓ-wide families. *Combinatorics, Probability and Computing*, **18** (2009), 309–333.
12. B. Felszeghy, B. Ráth, and L. Rónyai, The lex game and some applications, *J. Symbol. Comput.*, **41** (2006), 663–681.
13. B. Felszeghy, and L. Rónyai, On the lexicographic standard monomials of zero dimensional ideals, in *Proc. 10th Rhine Workshop on Computer Algebra (RWCA)*, pp. 95–105 (2006).
14. P. Frankl, Traces of antichains, *Graphs and Combinatorics*, **5** (1989), 295–299.
15. K. Friedl, G. Hegedűs, and L. Rónyai, Gröbner bases for complete ℓ-wide families, *Publ. Math. Debrecen*, **70** (2007), 271–290.
16. K. Friedl and L. Rónyai, Order shattering and Wilson's theorem, *Discrete Math.*, **270** (2003), 127–136.
17. A. M. Garsia, Pebbles and expansions in the polynomial ring, in *Polynomial Identities and Combinatorial Methods*, Lecture Notes in Pure and Appl. Math., Vol. **235**, pp. 261–285 (2003).
18. T. Harima, Characterization of Hilbert functions of Gorenstein Artin algebras with the weak Stanley property, *Proc. Amer. Math. Soc.*, **123** (1995), 3631–3638.
19. G. Hegedűs, A. Nagy, and L. Rónyai, Gröbner bases for permutations and oriented trees, *Ann. Univ. Sci. Budapest, Sectio Computatorica*, **23** (2004), 137–148.
20. G. E. Moorhouse, Approaching some problems in finite geometry through algebraic geometry, in M. Klin et al. (eds.) *Algorithmic Algebraic Combinatorics and Gröbner Bases*, pp. 285–296, Springer, Berlin, 2009 (this volume).
21. M. G. Marinari, H. M. Möller, and T. Mora, Gröbner bases of ideals defined by functionals with an application to ideals of projective points, *Appl. Algebra Engrg. Comm. Comput.*, **4** (1993), 103–145.
22. D. Pintér and L. Rónyai, On the Hilbert function of complementary set families, *Ann. Univ. Sci. Budapest, Sectio Combinatorica*, **29** (2008), 175–198.
23. R. M. Wilson, A diagonal form for the incidence matrices of t-subsets vs k-subsets, *Europ. J. Combin.*, **11** (1990), 609–615.

A Construction of Isomorphism Classes of Oriented Matroids

Ralf Gugisch

Mathematisches Institut, University of Bayreuth, 95440 Bayreuth, Germany.
ralf.gugisch@uni-bayreuth.de

Summary. We developed a computer program for generating all oriented matroids corresponding to a prescribed underlying matroid. The generation process and its output can be controlled via a variety of possible restrictions allowing to generate specific sets of oriented matroids.

The tool was mainly intended for application in chemistry: According to an idea of A. Dreiding and A. Dress, oriented matroid generation may be used as a first step towards a conformation generation for chemical structures.

We describe the main ideas of the generation algorithm as well as an application to the conformation analysis of cyclohexane. A combination with the Gröbner base approach for conformation analysis going back to P. Hazebroek and L. Oosterhoff is suggested.

Key words: Generation, Oriented matroid, Chirotope, Affine point configuration, Order type, Conformation, Cyclohexane

1 Motivation: Conformation Spaces in Chemistry

As a motivation we want to consider the chemical compound cyclohexane C_6H_{12}. It consists of six carbon atoms (C) and 12 hydrogen atoms (H), each carbon atom is attached to two hydrogen atoms and the carbon atoms form a 6-ring, see Fig. 1. As usual in organic chemistry and for sake of simplicity, the hydrogen atoms are not taken into consideration any further.

The atoms of chemical compounds often appear in more than one possible geometric arrangement, called *conformations.* In Fig. 2 you can see three different conformations of cyclohexane, namely the chair form and two twisted forms (which are mirror images of each other). However, the number of reasonable conformations is mostly very big. The *conformation space* of cyclohexane consists of infinitely many conformations similar to or somewhere "in between" the shown ones.

M. Klin et al. (eds.), Algorithmic Algebraic Combinatorics and Gröbner Bases,
DOI 10.1007/978-3-642-01960-9_7, © Springer-Verlag Berlin Heidelberg 2009

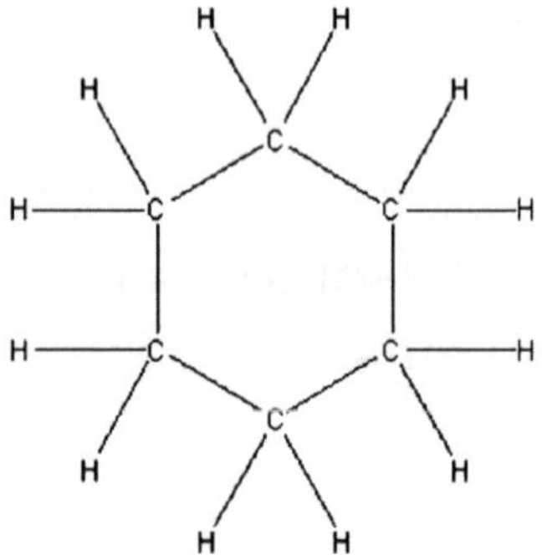

Fig. 1. The chemical compound cyclohexane

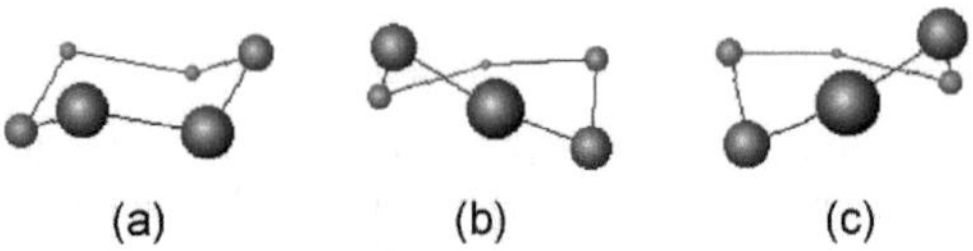

Fig. 2. Conformations of cyclohexane: The chair form (**a**) and two twisted forms (**b**, **c**)

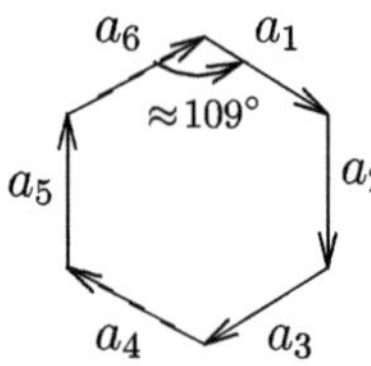

Fig. 3. The Gröbner base approach for analyzing the conformation space of cyclohexane

The conformation space of cyclohexane was analyzed using techniques which may be regarded as predecessors of Gröbner bases by Hazebroek and Oosterhoff [15]. Since then, this example is often used for demonstrating the use of Gröbner bases, e.g. in [16] or in [33]. Assuming standard uniform bond length 1 and bond angles $\alpha = \operatorname{acos}(-\frac{1}{3}) \approx 109°$ (this is the angle between the middlepoint of a regular tetrahedron and two of its vertices respective the carbon atom and two of its four neighbor atoms), one derives a system of polynomial equations from the cyclic structure of the carbon atoms in the following way: Let $a_1, \ldots, a_6 \in \mathbf{R}^3$ be the bond-vectors in a given conformation of cyclohexane, i.e. the vectors between two adjacent carbon atoms respectively, oriented in cyclic way, see Fig. 3. The fixed bond lengths and angles lead to quadratic equations using the norm and the inner product of bond vectors:

$$\|a_i\| = a_{i,1}^2 + a_{i,2}^2 + a_{i,3}^2 = 1,$$

for $i = 1, \ldots, 6$, and

$$\langle a_i, a_j \rangle = a_{i,1}a_{j,1} + a_{i,2}a_{j,2} + a_{i,3}a_{j,3} = \cos\alpha = -\frac{1}{3},$$

where i and j refer to consecutive vectors. Further, the cyclic structure is encoded by the following vectorial equation:

$$a_1 + \cdots + a_6 = 0,$$

expanding to three equations in the coordinates. Finally, the following equations fix the position of the structure in space:

$$a_{1,1} = 0, \qquad a_{1,2} = 0, \qquad a_{2,1} = 0.$$

This algebraic system of linear and quadratic equations can be solved easily using Gröbner bases. A more symmetric approach solves a substituted system of linear equations using the variables $S_{ij} := \langle a_i, a_j \rangle$ and some additional polynomial equations coming from Gram determinants in order to enforce three-dimensional solutions. It turns out that the solutions depend on only three independent variables S_{13}, S_{35} and S_{51}, and that the set of triples $(S_{13}, S_{35}, S_{51}) \in \mathbf{R}^3$ leading to feasible conformations decomposes into two connected components: a singular point leading to the chair form, and a closed curve leading to flexible conformations containing both twisted forms. Thus under the assumption that the bond lengths and bond angles are preserved, the chair form and the two twisted forms lie in two different connected components of the conformation space.

There is an extension [21] of this approach to cycloheptane C_7H_{14}. However, to the knowledge of the author, larger molecular compounds like hydrocarbonic nine- or ten-rings, or more complex structures containing a double bond were not successfully analyzed with this approach, yet, though posted as interesting problem in [23].

In this paper, we want to demonstrate a complementary approach using combinatorial tools. We will not get as detailed information about the structure of the conformation space as the connected components. Instead, we just try to provide a proper set of sample conformations helping to get an overview over the whole space. For practical chemical purposes as the elucidation of structure-property relationships this is quite convenient: One often considers a chemical compound as a mixture of different conformations. Of course, the sample set should be of reasonable size and reasonably distributed over the conformation space.

Thus, we are faced with the problem of *conformation generation,* whereby we need to keep in mind, that we do not generate all conformations rather than a (more or less) proper set of sample conformations.

One possibility to get such a sample set is to divide the conformation space into proper equivalence classes and to take a representant out of each of the classes. The idealistically best classification would be into "watersheds" of a proper energy function, as mentioned in [6]: Two points belong to the same equivalence class if and only if a steepest descents path on the energy function starting in either of the two points ends in the same sink. However, energy functions of molecular structures are very complex having a lot of different sinks and small watersheds, such that both calculation of all sinks and calculation of watersheds appear not to be feasible.

A much coarser division into equivalence classes very common in practice is the one into configurational *stereoisomers*. (Two configurational stereoisomers are distinguished by different configurations at stereocenters, i.e. at carbon atoms having – sloppily circumscribed – four different neighbors.) Generation of all configurational stereoisomers of a chemical compound may be done by a computer very efficiently using Nourse's algorithm [24–26]. Cyclohexane, for example has no stereocenters, i.e. there is only one configurational stereoisomer of cyclohexane. It turns out that the distinction of stereoisomers is too coarse for the purpose of conformation analysis. Commercial conformation generators as TRIPOS' Confort or academic ones like Frog [20] additionally generate conformations which are distinguished by standard torsion angles of butane substructures (i.e. four consecutively connected carbon atoms). For discussion of further similar approaches, see [29].

As long as we consider only three standard classes for the classified torsion angles, namely around the standard torsion angles $+60°$, $-60°$ and $180°$, we can unify and generalize the two classification aspects mentioned above using the ideas of Dreiding and Dress [7, 8], see also [31]: Certain *oriented matroids*, which are in strong relation to the *orientation function* of the atoms (as points in space), serve well for classification and description of conformations. This approach and the necessary mathematical structures are described in Sect. 2.

According to the sketched ideas, we promote the following strategy for conformation generation in two steps:

1. Combinatorial level:
   Generate all oriented matroids which are feasible (by means of simple combinatorial tests) for a chemical conformation.
2. Geometric level:
   Try to find a feasible conformation as (affine) realization for each of these oriented matroids.
   (Note that not for each oriented matroid there exist affine realizations. Though the problem of deciding whether a chirotope is affinely realizable and of finding a realization is known to be NP-hard, there exist algorithms which are suitable for small sizes [30, 28].)

In [14] the computer program `origen` was presented, a generator of oriented matroids serving for step one. In Sect. 3, we describe the features of `origen` and sketch some algorithmic ideas thereof. Note that there does not yet exist a sophisticated and adjusted solution for the second step of finding feasible conformations.

Nevertheless we are able to demonstrate the conformation generation according to the promoted strategy at hand of the example cyclohexane in Sect. 4.

2 Oriented Matroids, Chirotopes and Affine Point Configurations

One occurrence of oriented matroids [3, 4] is in connection with studies on sets of points in d-dimensional euclidean space [13, 17]. To any sequence of $d+1$ affinely independent points is assigned an orientation (positive or negative). One can determine the orientation of the points $p_1, \ldots, p_{d+1}$ having coordinate vectors $p_i = (p_{i,1}, \ldots, p_{i,d})$ by calculating the sign of the determinant:

$$\mathrm{ori}(p_1, \ldots, p_{d+1}) = \det \begin{pmatrix} 1 & \cdots & 1 \\ p_{1,1} & \cdots & p_{d+1,1} \\ \vdots & & \vdots \\ p_{1,d} & \cdots & p_{d+1,d} \end{pmatrix}.$$

For example, in two-dimensional plane, a sequence of three non-collinear points is oriented either anticlockwise (positive) or clockwise (negative).

In three-dimensional space, the orientation of four points p_1, p_2, p_3, and p_4 in arbitrary position can be determined by the common "right-hand rule": Identifying the point p_1 with the wrist of the right hand, the four points have positive orientation if and only if it is possible to point with thumb, index and middlefinger in direction to the points p_2, p_3 and p_4 without overstretching the middle finger.

By this concept of orientation, we assign to any sequence of n points spanning the whole d-dimensional euclidean space an *orientation function* $\chi : n^{d+1} \to \{0, \pm 1\}$, where the function value 0 means, that the corresponding $(d+1)$-tuple of points lies in a hyperplane. Here and below n denotes the n-element set $\{1, 2, \ldots, n\}$. Obviously, the function χ is alternating. We can write χ as sequence of its function values at the ordered $(d+1)$-tuples, using the reverse lexical order for listing the tuples.

Example 1. Let us consider the following six points in space (see Fig. 4).

$$p^{(1)} = (4,3,2), \quad p^{(2)} = (5,2,1), \quad p^{(3)} = (4,1,2),$$
$$p^{(4)} = (2,1,2), \quad p^{(5)} = (1,2,3), \quad p^{(6)} = (2,3,2).$$

The orientation function as sequence of signs for each quadruple of points is:

1234	1235	1245	1345	2345	1236	1246	1346	2346	1256	1356	2356	1456	2456	3456
+	+	0	−	−	+	+	0	−	+	+	0	+	+	+

$$\chi = ++0\ --++0\ -++0\ +++$$

For example, the function value $\chi(1,2,3,4) = +1$ means, that the four points $p^{(1)}$, $p^{(2)}$, $p^{(3)}$ and $p^{(4)}$ are positively oriented, while $\chi(1,2,4,5) = 0$ denotes, that the points $p^{(1)}$, $p^{(2)}$, $p^{(4)}$ and $p^{(5)}$ are coplanar.

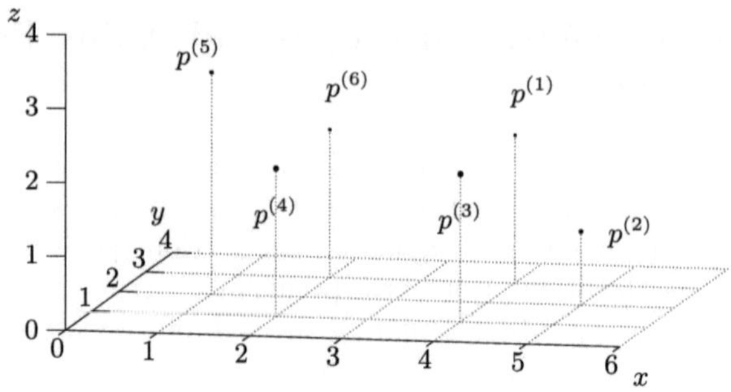

Fig. 4. Six points in space

As determinant functions, the orientation functions fulfill the binary Grassmann-Plücker relations: Let $k := d+1$. For any $\mathbf{a}, \mathbf{b} \in n^k$ the following holds:

$$\chi(\mathbf{a}) \cdot \chi(\mathbf{b}) = +1 \quad \Longrightarrow$$
$$\exists i \in \{1, \dots, k\} : \chi(b_i, a_2, \dots, a_k) \cdot \chi(b_1, \dots, \underset{\substack{\uparrow \\ i\text{th position}}}{a_1}, \dots, b_k) = +1. \quad \text{(GP)}$$

(When calling tuples $\mathbf{a}$ with $\chi(\mathbf{a}) \neq 0$ *bases*, we can interpret the Grassmann-Plücker relations as an oriented version of the base exchange axiom known from linear algebra.)

In general, alternating, non-trivial (i.e. not constantly zero) functions $\chi : n^k \to \{0, \pm 1\}$ fulfilling (GP) are called *chirotopes* of rank k. Chirotopes imply a huge amount of further structure which is embraced in the concept of *oriented matroids*. For the purpose of this work, it is enough to know that the oriented matroids are in one-to-one correspondence to the pairs $\{\chi, -\chi\}$ of chirotopes. From each oriented matroid, we get an unoriented matroid, the *underlying matroid*, by taking the domain of the chirotope as bases. The oriented matroid is called *uniform*, if the chirotope has no zero function values (i.e. if the underlying matroid is uniform).

The orientation function of a sequence of n spanning points in d-dimensional euclidean space is a chirotope of rank $k = d + 1$ and thus realizes an oriented matroid. The corresponding oriented matroid is uniform if and only if the sequence of points is in general position, i.e. if each $(d + 1)$-subset is affinely independent. (If the points are not spanning, then the orientation function is the zero function.)

Note that not each chirotope is an orientation function. We call oriented matroids *affinely realizable*, if a corresponding chirotope is the orientation function of a sequence of points in euclidean space.

We call spanning sequences of points in d-dimensional space combinatorially equivalent, or of same *signed configuration*, if they have the same orientation function. Further, we call spanning sets of points combinatorially equivalent, if there exist numberings of the sets, such that the correspond-

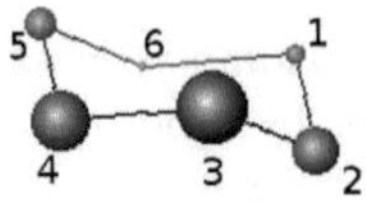

Fig. 5. The chair form of cyclohexane, with numbered atoms

ing sequences are combinatorially equivalent. The corresponding equivalence classes of sets of points we call *signed affine point configurations.*

Let χ be the orientation function of a spanning sequence of points. Then the mirror image of the point sequence has $-\chi$ as orientation function (i.e. all function values of χ are negated). We call a point set *achiral*, if the point set and its mirrored point set belong to the same signed configuration. Otherwise the point set is *chiral.* The two signed configurations of a chiral point set and its mirrored point set are called *enantiomeric* to each other. The equivalence classes of point sets arising by identifying chiral signed point configurations with their enantiomeric configurations, respectively, we call *unsigned affine point configurations.* (Unfortunately, in literature the term (affine) point configuration is used both for signed as well as for unsigned point configurations. Thus we decided to explicitly distinguish both definitions. Unsigned point configurations are often called the *order type* of a point set, too. See also the discussion of isomorphism aspects of oriented matroids below.)

A lot of geometrical applications like convex hulls or triangulations can be realized by solely considering the purely combinatorial information of affine point configurations, and so it is no surprise that these structures (or equivalent ones) were studied intensively in literature, especially in 2-dimensional case; see for example [13, 17, 18].

Example 2. If we number the atoms of a chemical structure, then each conformation corresponds to a sequence of points in space and thus to a signed point configuration. For example, if we number the atoms of the chair form of cyclohexane as shown in Fig. 5, then the atoms have the same signed configuration as the points in Fig. 4. Thus the chirotope χ from Example 1 is assigned to the chair form of cyclohexane.

2.1 Isomorphism

Reorderings of a sequence of points in euclidean space lead to different orientation functions. For example, if we exchange the labels at the points $p^{(5)}$ and $p^{(6)}$ in Example 1, we get the chirotope $\chi' := \chi \circ (5,6)$:

	1234	1235	1245	1345	2345	1236	1246	1346	2346	1256	1356	2356	1456	2456	3456
$\chi' =$	+	+	+	0	−	+	0	−	−	−	−	0	−	−	−

If the order of the sequence of points is not important, i.e. when considering *sets* of points, we need to identify the corresponding chirotopes for all reorderings. Two chirotopes are *isomorphic with respect to relabeling*, if they can be obtained from each other by a permutation of n:

$$\chi' \cong \chi \iff \exists \pi \in S_n : \forall \mathbf{a} \in n^k : \chi'(a_1, \ldots, a_n) = \chi(\pi(a_1), \ldots, \pi(a_n)).$$

It is not trivial to recognize whether two chirotopes are isomorphic with respect to relabeling, or not. In [14], an algorithm generating a canonic form for each isomorphism class of chirotopes is described. This algorithm exploits similar ideas as the iterated refinement algorithm of B. McKay for graphs and digraphs [22]. In general, one may consider chirotopes as generalizations of simple digraphs with no (anti-)parallel edges, as the adjacency matrix of such a digraph is a mapping $n^2 \to \{0, \pm 1\}$, and the generalization is towards mappings $n^k \to \{0, \pm 1\}$.

There are further relevant isomorphisms for chirotopes: Considering the mirror-image of a set of points leads to the negative chirotope $-\chi$, i.e. each function value inverts its sign. In other geometric applications of chirotopes as polytopes and zonotopes, *reorientations* are playing a role, too. A reorientation at label i changes the signs for all those tuples containing i. For a more detailed explanation of reorientation symmetry, refer to the standard literature on oriented matroids [3, 4].

Thus when generating chirotopes, three kinds of isomorphy may be considered:

- relabeling (i.e. permutations of the labels),
- negation (i.e. identifying χ and $-\chi$),
- reorientation (has no obvious interpretation for affine point configurations).

Note that generation of oriented matroids is equivalent to generation of chirotopes up to negation.

For chemical compounds, any permutation of labels acts on the molecular graph, too. Thus relabellings of the atoms of a chemical structure lead to isomorphic conformations if and only if the corresponding permutation is a graph automorphism of the molecular graph. In other words, when we are interested in isomorphic conformations, we need to consider relabellings of the chirotope under action of the graph automorphism group only.

2.2 Radon Partitions and Oriented Circuits

An important structure connected to oriented matroids and thus induced by the chirotope are the (*oriented*) *circuits*: A pair (C^+, C^-) of two disjoint subsets of n is called a circuit, if there exists an ordered $(k+1)$-tuple $(c_1, \ldots, c_{k+1})$ and an $\epsilon \in \{\pm 1\}$, such that $C^+ \cup C^- \subseteq \{c_1, \ldots, c_{k+1}\}$ and for each $i = 1, \ldots, k+1$:

$$\chi(c_1, \ldots, c_{i-1}, c_{i+1}, \ldots, c_{k+1}) = \begin{cases} +\epsilon \cdot (-1)^i & \text{if } c_i \in C^+, \\ -\epsilon \cdot (-1)^i & \text{if } c_i \in C^-, \\ 0 & \text{else.} \end{cases}$$

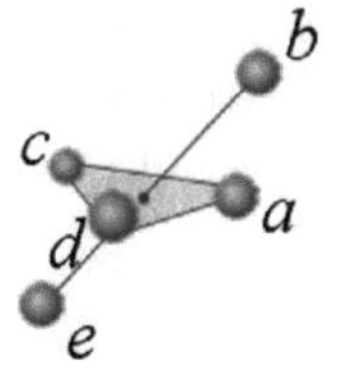

Fig. 6. A minimal Radon partition $\{+a, -b, +c, +d, -e\}$

The circuits of an oriented matroid can be obtained from the chirotope by checking all ordered $(k+1)$-tuples. We write circuits as signed sets, i.e. we attach the appropriate sign to the elements of C^+ resp. C^- and write it as a set of signed elements. For example,

$$\{+1, -3, +4, -6\}$$

is a circuit of the chirotope given in Example 1.

In an affine point set, two disjoint subsets of points whose convex hulls intersect are called a *Radon partition.* The circuits of the assigned oriented matroid denote exactly the minimal Radon partitions.

Example 3. Consider the five points a, b, c, d and e in three-dimensional space shown in Fig. 6. The convex hulls of $\{a, c, d\}$, a triangle, and of $\{b, e\}$, a line, do intersect. The pair of sets is minimal with this property, i.e. deleting one point from either set would lead to non-intersecting convex hulls. Thus, $\{+a, -b, +c, +d, -e\}$ is a minimal Radon partition.

Note that an affinely realizable oriented matroid cannot have positive circuits, i.e. circuits (C^+, C^-) with $C^- = \emptyset$. In other words, absence of positive circuits is a necessary criterion for affine realizability. Chirotopes having no positive circuits are often called *acyclic.* In addition, Radon partitions, respective (oriented) circuits, serve as useful language to formulate necessary conditions for chemical feasible conformations. See Sect. 4.

2.3 Partial Chirotopes

When considering conformations of molecular structures, the orientations of quadruples of atoms are not all of same importance. For example, the orientation of four non-connected, distant atoms is of minor importance for conformational classification, while the orientations of the four neighbor atoms of the stereocenters classify the stereoisomers.

In order to take this into consideration, the concept of *partial chirotopes* is useful. A partial chirotope is a mapping from a subset of n^k to $\{0, \pm 1\}$ being alternating and fulfilling (GP) where applicable. A partial chirotope is *extendable,* if one can extend it to a chirotope. Note that testing extendability for partial chirotopes is, though NP-complete [32], manageable for small n

(e.g. using `origen`; already [6] presented a generation algorithm suitable for this task). One can interpret the extendable partial chirotopes as equivalence classes of chirotopes, namely the sets of extensions.

3 The Generator `origen`

An obvious strategy for generation of affine signed point configurations over n points is to generate all isomorphism classes of chirotopes up to relabeling over n points and to find affine realizations in a second step. For unsigned point configurations, one can generate chirotopes up to relabeling and negation – or oriented matroids in one of its other representations. (Recall that oriented matroids are in one-to-one correspondence to sets $\{\chi, -\chi\}$ of chirotopes and their negatives.) The latter was done by Aichholzer, Aurenhammer and Krasser in the two-dimensional case [1, 2] as well as by Finschi and Fukuda for arbitrary dimension [10, 11]. The generator of uniform oriented matroids by Bokowski and de Oliveira [5] is applicable to this task, too.

However, generating full catalogs of point configurations is limited to small numbers of points. In Table 1, you see the numbers of three-dimensional unsigned point configurations (including degenerated cases) for $n \leq 8$ points, as computed in [10] and approved by `origen`. It does not make much sense to continue storing full catalogs for larger n's. We want to exploit restrictions already during the generation process and specifically compute configurations e.g. for different chemical compounds.

Generation of partial chirotopes in the sense of partial two- or three-dimensional signed point configurations was discussed in [34] in a chemical context. Generation up to relabeling with respect to an automorphism group of some additional structure (the molecular graph) was considered, too.

We developed the generator `origen` generating all oriented matroids (in form of chirotopes) corresponding to a prescribed set of parameters and restrictions. The oriented matroids are represented as chirotopes. The generation process and its output can be controlled via a variety of possible parameters and restrictions:

- The parameters n and k specify size (number of points) and rank ($k = d+1$) of the chirotopes.
- Properties as simple or cosimple underlying matroids, acyclic chirotopes (i.e. no positive circuits) can be specified.
- The domain $D \subseteq n^k$ of the chirotope (i.e. the underlying matroid) as well as single orientations can be prescribed.

Table 1. Unsigned 3D point configurations for small numbers of points (from [10])

Number of points	4	5	6	7	8
Point configurations	1	5	55	5083	10 775 236

- There is the possibility to specify a list of forbidden circuits.
- You can specify a subset $R \subseteq n^k$ of relevant k-tuples. Only extendable partial chirotopes on R are generated. (Actually, `origen` generates a transversal of extensions.)
- Different kinds of isomorphisms for chirotopes can be combined arbitrarily: relabeling, negation and reorientation.
- For relabeling isomorphy, the acting group can be restricted to a subgroup G of the symmetric group S_n.
- A group A of relabeling automorphisms can be prescribed, such that each generated solution has A as subgroup of its automorphism group.

The generator is available in internet, see

http://www.mathe2.uni-bayreuth.de/ralfg/origen.php

3.1 Notes on the Generation Algorithm

An important well-known fact for efficient validation of chirotopes during generation process is, that we do not need to test all n^{2k} Grassmann-Plücker relations rather than only the ones belonging to a pair $\mathbf{a}, \mathbf{b}$ of k-tuples differing in exactly two points (the so-called three-term Grassmann-Plücker relations). Thus having an alternating, non-trivial function $\chi : n^k \to \{0, \pm 1\}$, it suffices to test the following $\binom{n}{k+2} \cdot \binom{k+2}{4}$ conditions (see [14]):

For each $(k+2)$-subset X of n, and for each 4-subset $\{a_1, a_2, b_1, b_2\}$ thereof, with $a_1 < a_2 < b_1 < b_2$, let $\mathbf{x}$ be the ordered $(k-2)$-sequence of $X \setminus \{a_1, a_2, b_1, b_2\}$. Using the abbreviations[1]

$$\begin{array}{lll} \mathbf{a} := (\mathbf{x}, a_1, a_2) & \mathbf{a}' := (\mathbf{x}, b_1, a_2) & \mathbf{a}'' := (\mathbf{x}, b_2, a_2) \\ \mathbf{b} := (\mathbf{x}, b_1, b_2) & \mathbf{b}' := (\mathbf{x}, a_1, b_2) & \mathbf{b}'' := (\mathbf{x}, b_1, a_1) \end{array}$$

and

$$s_1 := \chi(\mathbf{a}) \cdot \chi(\mathbf{b}) \qquad s_2 := \chi(\mathbf{a}') \cdot \chi(\mathbf{b}') \qquad s_3 := \chi(\mathbf{a}'') \cdot \chi(\mathbf{b}''),$$

we need to test

$$(s_1 = 0 \wedge (s_2 = -s_3)) \vee (s_1 \neq 0 \wedge (s_1 = s_2 \vee s_1 = s_3)) \qquad \text{(GP3}'\text{)}$$

The generation is split into two main levels. In the first main level, all underlying matroids satisfying the given restrictions are generated (i.e. the possible domains for the chirotopes), and in the second main level, orientations are distributed to a given underlying matroid. This splitting is beneficial according to the homomorphism principle [19].

[1] For a $(k-2)$-tuple $\mathbf{x} \in n^{k-2}$ and $i, j \in n$ we write $(\mathbf{a}, i, j)$ as the elongation of $\mathbf{x}$ to a k-tuple.

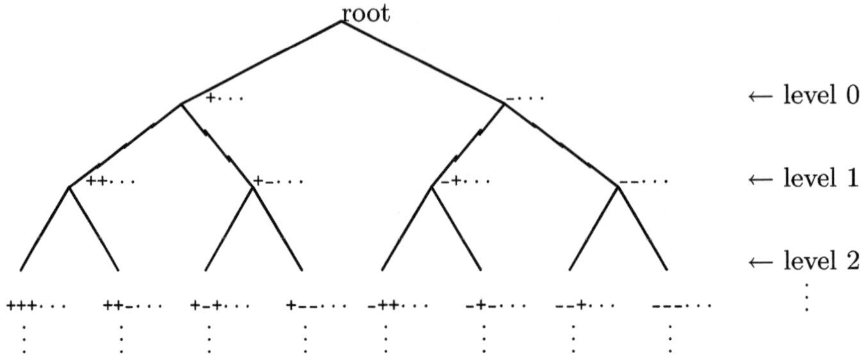

Fig. 7. Scheme of the backtrack generating algorithm

In both main levels, we generate mappings on n^k: The underlying matroid is represented as symmetric function $n^k \to \{0,1\}$ specifying the bases, and the chirotope itself is an alternating function $n^k \to \{0,\pm 1\}$ as described above. Thus in both cases it suffices to specify the function values at increasing k-tuples, and we implement both steps as backtrack algorithms of depth $\binom{n}{k}$, see Fig. 7. Principally, we generate this way all alternating functions $\chi : n^k \to \{0,\pm 1\}$ in a global backtrack algorithm of depth $2 \cdot \binom{n}{k}$. By pruning branches of this tree due to several kinds of tests on each level, we ensure, that only chirotopes with the requested properties are reached at the last level.

In either main generation step, denote the root of the (local) backtrack tree as level -1. Then, on level i, the function value $\chi(\mathbf{a}^{(i)})$ for the ith ordered k-tuple $\mathbf{a}^{(i)} = (a_1^{(i)}, \ldots, a_k^{(i)}) \in n^k$ (using the reverse lexical order on k-tuples) is specified. In case of the matroid generation (first main level) there is the choice between 0 (no base) and 1 (base), in the second generation step for given underlying matroid there is a choice for bases only, namely to choose their orientations (-1 or $+1$). The following rules lead to prunings:

- If a value $\chi(\mathbf{a})$ is prescribed due to the generation input, then the alternative choice is skipped.
- If a group $A \leq S_n$ of automorphisms is prescribed, then we calculate the orbits on the k-subsets in advance. After the first function value of an k-tuple corresponding to an orbit is specified, the remaining values ensue.
- On level $\binom{n'}{k}$, $n' \leq n$, we completed a mapping on n'^k. Test if the restricted mapping on n'^k is canonic.

The last point guarantees that each chirotope is generated only once up to considered symmetry aspects. The idea behind is the principle of *orderly generation* [9, 27]. It works only if the definition of the canonic form is compatible with the generation strategy in the following sense: For a canonic chirotope $\chi : n^k \to \{0,\pm 1\}$, each restriction to n'^k, $n' \leq n$ is canonic as well as the restricted underlying matroids are canonic.

We use the minimal lexicographic representation of the base function as canonic form for the underlying matroids. A chirotope is canonic by definition, if the underlying matroid is in canonic form, and if the orientations are lexicographically maximally distributed among all isomorphic chirotopes with canonic underlying matroid. This definition meets the described requirements for orderly generation and thus allows to do the denoted early canonicity tests during backtrack generation.

During the distribution of orientations (second main level), the following additional tests are noteworthy:

- When generating up to negation, we can restrict the first otherwise freely choosable base orientation to $+1$. This is a necessary condition for the canonic form up to negation.
- Similarly in the case when we generate up to reorientation: We fix up to n orientations to $+1$. The chosen k-tuples are the first ones affected by an reorientation of the points $1 \leq i \leq n$, respectively. (These tuples are computed once for each underlying matroid.)
- Assure, that we generate chirotopes only: For all ordered $(k+2)$-tuples $(x, y, a_1^{(i)}, \ldots, a_k^{(i)})$ with $x < y < a_1^{(i)}$, test the corresponding three-term Grassmann-Plücker relations (GP3$'$).

Example 4. In Table 2 we demonstrate the splitting of the generation process into two main levels with the example of all simple oriented matroids of rank 4 over 7 points. On main level one, the listed 49 simple matroids are generated. For each of them, on main level two, all oriented matroids with given underlying matroid are generated.

Remark that the main part of computation time is needed for generating the uniform oriented matroids. This is because we need to consider the whole S_7 as operating group, thus the canonicity tests are quite expensive. For the other candidates of underlying matroids, we can restrict to smaller operating groups on the second main level, such that canonicity tests get easier.

Also note that there are non-orientable matroids, i.e. matroids where there exists no corresponding oriented matroid.

3.2 Comparison

Using `origen`, we recomputed published results of L. Finschi from [10]. We approved the information given there in Tables 6.3 and 6.4 in page 141 concerning isomorphism classes of simple oriented matroids up to relabeling, reorientation and negation as well as Tables 7.1 and 7.2 in page 151 concerning simple acyclic oriented matroids up to relabeling and negation alias abstract unsigned affine point configurations (including degenerated cases).

In Table 6.6 in page 143, Finschi gives some computation times for the generated structures of Table 6.3. We compared them with the computation times of `origen`. Note that `origen` is designed for handling several specific

Table 2. Generation of simple oriented matroids of rank 4 over 7 points (up to relabeling, negation and reorientation)

Underlying matroid																																			Group order	Oriented matroids	Computing time
1234	1235	1245	1345	2345	1236	1246	1346	2346	1256	1356	2356	1456	2456	3456	1237	1247	1347	2347	1257	1357	2357	1457	2457	3457	1267	1367	2367	1467	2467	3467	1567	2567	3567	4567			
×	×	×	×	×	×	×	×	×	×	×	×	×	×	×	×	×	×	×	×	×	×	×	×	×	×	×	×	×	×	×	×	×	×	×	5040	11	0.676 s
0	×	×	×	×	×	×	×	×	×	×	×	×	×	×	×	×	×	×	×	×	×	×	×	×	×	×	×	×	×	×	×	×	×	×	144	27	0.084 s
0	×	×	×	×	×	×	×	×	×	×	×	×	×	×	×	×	×	×	×	×	×	×	×	×	×	×	×	×	×	×	0	×	×	×	72	7	0.048 s
0	×	×	×	×	×	×	×	×	0	×	×	×	×	×	×	×	×	×	×	×	×	×	×	×	×	×	×	×	×	×	×	×	×	×	16	35	0.068 s
0	×	×	×	×	×	×	×	×	0	×	×	×	×	×	×	×	×	×	×	×	×	×	×	0	×	×	×	×	×	×	×	×	×	×	8	9	0.020 s
0	×	×	×	×	×	×	×	×	0	×	×	×	×	×	×	×	×	×	×	0	×	×	×	×	×	×	×	×	×	×	×	×	×	×	6	14	0.024 s
0	×	×	×	×	×	×	×	×	0	×	×	×	×	×	×	×	×	×	×	0	×	×	×	×	×	×	×	×	0	×	×	×	×	×	4	4	0.024 s
0	×	×	×	×	×	×	×	×	0	×	×	×	×	×	×	×	×	×	×	0	×	×	×	×	×	×	×	0	×	×	×	×	×	×	24	2	0.016 s
0	×	×	×	×	×	×	×	×	0	×	×	×	×	0	×	×	×	×	×	×	×	×	×	×	×	×	×	×	×	×	×	×	×	×	48	6	0.032 s
0	×	×	×	×	×	×	×	×	0	×	×	×	×	0	×	×	×	×	×	0	×	×	×	×	×	×	×	×	×	×	×	×	×	×	6	5	0.024 s
0	×	×	×	×	×	×	×	×	0	×	×	×	×	0	×	×	×	×	×	0	×	×	×	×	×	×	×	×	0	×	×	×	×	×	12	1	0.016 s
0	×	×	×	×	×	×	×	×	0	×	×	×	×	0	×	×	×	×	×	0	×	×	0	×	×	×	×	×	×	×	×	×	×	×	8	1	0.016 s
0	×	×	×	×	×	×	×	×	0	×	×	×	×	0	×	×	×	×	×	0	×	×	0	×	×	×	0	×	×	×	×	×	×	×	24	1	0.016 s
0	×	×	×	×	×	×	×	×	0	×	×	×	×	0	×	×	×	×	×	0	×	×	0	×	×	×	0	0	×	×	×	×	×	×	168	0	0.008 s
0	0	×	×	×	0	×	×	×	×	×	×	×	×	×	0	×	×	×	×	×	×	×	×	×	×	×	×	×	×	×	×	×	×	×	144	4	0.024 s
0	0	×	×	×	0	×	×	×	×	×	×	×	×	×	0	×	×	×	×	×	×	×	×	×	×	×	×	×	×	×	×	×	×	0	144	1	0.020 s
0	0	×	×	×	0	×	×	×	×	×	×	0	×	×	0	×	×	×	×	×	×	×	×	×	×	×	×	×	×	×	×	×	×	×	12	6	0.016 s
0	0	×	×	×	0	×	×	×	×	×	×	0	×	×	0	×	×	×	×	×	×	×	0	×	×	×	×	×	×	×	×	×	×	×	4	4	0.016 s
0	0	×	×	×	0	×	×	×	×	×	×	0	×	×	0	×	×	×	×	×	×	×	0	×	×	×	×	×	×	0	×	×	×	×	6	1	0.012 s
0	0	×	×	×	0	×	×	×	×	×	×	0	0	0	0	×	×	×	×	×	×	×	×	×	×	×	×	×	×	×	×	×	×	0	72	1	0.008 s
0	0	0	0	0	×	×	×	×	×	×	×	×	×	×	×	×	×	×	×	×	×	×	×	×	×	×	×	×	×	×	×	×	×	×	240	6	0.056 s
0	0	0	0	0	×	×	×	×	×	×	×	×	×	×	×	×	×	×	×	×	×	×	×	×	0	×	×	×	×	×	×	×	×	×	24	6	0.036 s
0	0	0	0	0	×	×	×	×	×	×	×	×	×	×	×	×	×	×	×	×	×	×	×	×	0	×	×	×	×	0	×	×	×	×	16	3	0.056 s
0	0	0	0	0	×	×	×	×	×	×	×	×	×	×	×	×	×	×	×	×	×	×	×	×	0	0	×	0	×	×	0	×	×	×	48	1	0.016 s
0	0	0	0	0	0	×	×	×	×	×	×	×	×	×	0	×	×	×	×	×	×	×	×	×	×	×	×	×	×	×	×	×	×	×	24	6	0.020 s
0	0	0	0	0	0	×	×	×	×	×	×	×	×	×	0	×	×	×	×	×	×	×	×	×	×	×	×	×	×	×	×	×	×	0	24	2	0.020 s
0	0	0	0	0	0	×	×	×	×	×	×	×	×	×	0	×	×	×	×	×	×	×	×	×	×	×	×	0	×	×	×	×	×	×	4	6	0.024 s
0	0	0	0	0	0	×	×	×	×	×	×	×	×	×	0	×	×	×	×	×	×	×	×	×	×	×	×	0	×	×	×	0	×	×	4	2	0.024 s
0	0	0	0	0	0	×	×	×	×	×	×	×	×	×	0	×	×	×	×	×	×	×	×	×	×	×	×	0	0	0	×	×	×	0	12	1	0.012 s
0	0	0	0	0	0	×	×	×	×	×	×	×	×	×	0	×	×	×	×	×	×	×	×	×	0	0	0	×	×	×	×	×	×	×	48	2	0.008 s
0	0	0	0	0	0	×	×	×	×	×	×	×	×	×	0	×	×	×	×	×	×	×	×	×	0	0	0	×	×	×	×	×	×	0	48	1	0.008 s
0	0	0	0	0	0	×	×	×	×	×	×	0	×	×	0	×	×	×	×	×	×	0	×	×	×	×	×	×	×	×	×	×	×	×	16	3	0.012 s
0	0	0	0	0	0	×	×	×	×	×	×	0	×	×	0	×	×	×	×	×	×	0	×	×	×	×	×	×	0	×	×	×	×	×	4	2	0.016 s
0	0	0	0	0	0	×	×	×	×	×	×	0	×	×	0	×	×	×	×	×	×	0	×	×	×	×	×	×	0	×	×	×	0	×	8	1	0.012 s
0	0	0	0	0	0	×	×	×	×	×	×	0	×	×	0	×	×	×	×	×	×	0	×	×	0	0	0	×	×	×	×	×	×	×	8	2	0.008 s
0	0	0	0	0	0	×	×	×	×	×	×	0	×	×	0	×	×	×	×	×	×	0	×	×	0	0	0	×	0	×	×	0	×	×	8	1	0.012 s
0	0	0	0	0	0	×	×	×	×	×	×	0	×	×	0	×	×	×	×	×	×	0	×	×	0	0	0	0	×	×	0	×	×	0	48	1	0.004 s
0	0	0	0	0	0	0	0	0	×	×	×	×	×	×	0	0	0	0	×	×	×	×	×	×	×	×	×	×	×	×	×	×	×	×	144	1	0.020 s
0	0	0	0	0	0	0	0	0	×	×	×	×	×	×	0	0	0	0	×	×	×	×	×	×	×	×	×	×	×	×	0	×	×	×	36	1	0.016 s
0	0	0	0	0	0	0	0	0	×	×	×	×	×	×	0	0	0	0	×	×	×	×	×	×	×	×	×	×	×	×	0	0	0	0	144	1	0.012 s
0	0	0	0	0	0	0	0	0	0	0	0	0	0	0	×	×	×	×	×	×	×	×	×	×	×	×	×	×	×	×	×	×	×	×	720	4	0.040 s
0	0	0	0	0	0	0	0	0	0	0	0	0	0	0	0	×	×	×	×	×	×	×	×	×	×	×	×	×	×	×	×	×	×	×	36	3	0.016 s
0	0	0	0	0	0	0	0	0	0	0	0	0	0	0	0	×	×	×	×	×	×	×	×	×	×	×	×	×	×	×	×	×	×	0	72	1	0.012 s
0	0	0	0	0	0	0	0	0	0	0	0	0	0	0	0	×	×	×	×	×	×	0	×	×	×	×	×	×	×	×	×	×	×	×	8	3	0.024 s
0	0	0	0	0	0	0	0	0	0	0	0	0	0	0	0	×	×	×	×	×	×	0	×	×	×	×	×	×	0	×	×	×	×	×	6	2	0.032 s
0	0	0	0	0	0	0	0	0	0	0	0	0	0	0	0	×	×	×	×	×	×	0	×	×	×	×	×	×	0	×	×	×	0	×	24	1	0.020 s
0	0	0	0	0	0	0	0	0	0	0	0	0	0	0	0	0	0	0	×	×	×	×	×	×	×	×	×	×	×	×	×	×	×	×	48	1	0.008 s
0	0	0	0	0	0	0	0	0	0	0	0	0	0	0	0	0	0	0	×	×	×	×	×	×	×	×	×	×	×	×	0	×	×	×	12	1	0.020 s
0	0	0	0	0	0	0	0	0	0	0	0	0	0	0	0	0	0	0	0	0	0	0	0	0	×	×	×	×	×	×	×	×	×	×	240	1	0.008 s
																																			Sum:	206	

parameters as for example an acting symmetry group G which may differ from S_n. This ensues a lot of overhead in implementation, which is not needed for the generation tasks in question. On the other hand, as our runs were about six years later than Finschi's, our system (a Dual Core AMD Opteron (tm)

Table 3. Number of isomorphism classes of oriented matroids. The entry for $k = 7, n = 10$ is new vs. Table 6.3 in [10]

$n =$	1	2	3	4	5	6	7	8	9	10
$k = 1$	1									
$k = 2$		1	1	1	1	1	1	1	1	1
$k = 3$			1	2	4	17	143	4890	461 053	95 052 532
$k = 4$				1	3	12	206	181 472		
$k = 5$					1	4	25	6 029		
$k = 6$						1	5	50	508 321	
$k = 7$							1	6	91	*99 875 033*
$k = 8$								1	7	164
$k = 9$									1	8
$k = 10$										1

Table 4. CPU times needed for computing the results from Table 3

	Finschi's generator				`origen`			
$n =$	7	8	9	10	7	8	9	10
$k = 1$								
$k = 2$	–	–	–	–	–	–	–	–
$k = 3$	3 s	2.2 min	3.6 h	≈1700 h	0.6 s	13 s	13 min	33 h
$k = 4$	10 s	4.1 h			0.6 s*	17 min		
$k = 5$	2 s	48.3 min			0 s*	14 s*		
$k = 6$	–	26 s	≈240 h		–	0.2 s*	13.8 min*	
$k = 7$	–	–	9.9 min		–	–	0.6 s*	36 h*
$k = 8$		–	–	4.8 h		–	–	1.5 s*
$k = 9$			–	–			–	–
$k = 10$				–				–

Table 5. Relative time speedups for $k = 3$

$n =$	7	8	9	10
Ratio of comp. times	5	10	17	52

Processor 265, 1000 MHz, running on a 64 bit system) is most probably much faster than the one Finschi used (a Sun Sparc Ultra-60, 360 MHz). Thus only relative comparisons of CPU times do make sense.

In Table 3, we list for given rank k and number n of points the number of simple oriented matroids up to reorientation, negation and relabeling, as given in Table 6.3 in [10], and as approved by `origen`. We added one number for the parameter set $k = 7$ and $n = 10$, which we were able to compute in less than two days. The CPU times given by Finschi as well as the ones needed by `origen` are shown in Table 4. For the sake of fairness, we need to mention, that we used a well known trick to compute some of the parameter sets: In the cases marked with a*, it was easier to compute the dual cosimple oriented matroids of rank $k' = n - k$ instead of direct computation. Finschi noted, that this is possible, but he did not make profit of this trick.

We made a relative comparison of computation times for $k = 3$, as there are four comparable times available in one row. In Table 5, we show the factor

$$\frac{\text{computation time by Finschi's generator}}{\text{computation time by } \texttt{origen}}$$

indicating a relative increase of computation speed with growing number n of points.

Klin et al. computed numbers of non-isomorphic binary (i.e. non-degenerated) signed abstract 2D-configurations of n points, see Table 6 in [17]. The listed numbers for $n = 3, 4, 5, 6$ and 7 points, namely $1, 2, 3, 20$ and 242 2D-configurations, respectively, coincide with the numbers of relabeling classes of acyclic uniform chirotopes of rank $k = 3$, as calculated by `origen`. However, a discrepancy for $n = 8$ remains, where Klin et al. reported 6406 abstract 2D-configurations, whereas `origen` constructed 6405 relabeling classes of acyclic uniform chirotopes.

4 Application in Chemical Conformation Analysis: The Example Cyclohexane

Reconsider the chemical compound cyclohexane C_6H_{12}. We number the atoms as indicated in Fig. 5.

We generate chirotopes of rank $k = 4$ over $n = 6$ points. The affinely realizable ones thereof correspond to the three-dimensional point configurations of 6 atoms.

- The molecular graph has as automorphism group the dihedral group D_6 with 12 elements. Thus, we generate chirotopes up to relabeling under the D_6.
- Assuming that any 4 atoms are affinely independent, we restrict to uniform chirotopes. Note that this is not a great restriction, as one can enforce general position of the atoms by infinitesimal small movements.
- Generating all isomorphism classes of uniform chirotopes over 6 points with D_6 as acting group results in *386* solutions. The first 5 are listed below:

1234	1235	1245	1345	2345	1236	1246	1346	2346	1256	1356	2356	1456	2456	3456
+	+	+	+	+	+	+	+	+	+	+	+	+	+	+
+	+	+	+	+	+	+	+	+	+	+	+	+	+	−
+	+	+	+	+	+	+	+	+	+	+	+	+	−	−
+	+	+	+	+	+	+	+	+	+	+	+	−	+	+
+	+	+	+	+	+	+	+	+	+	+	+	−	−	+

⋮

Fig. 8. Two kinds of forbidden oriented circuits for molecular conformations (using bond-information from the molecular graph)

- We prescribe a list of forbidden circuits reflecting mathematical and chemical knowledge.
  - positive circuits are not affinely realizable.
  - Three consecutively connected atoms have a triangle as convex hull. In general, we can exclude conformations where a bond intersects with such a triangle, as the steric energy of this conformation would be very high. Thus we can exclude all chirotopes containing a circuit of the form $(+a, +b, +c, -d, -e)$ where atoms a, b, and c are consecutively connected as well as atoms d and e, see Fig. 8(a).
  - Similarly, we can exclude chirotopes having a circuit of the form $(+a, +b, +c, +d, -e)$, where atom a is connected to atoms b, c and d, see Fig. 8(b).

  By excluding the forbidden circuits

$$\begin{array}{lll}(+1,-3,-4,-5,+6), & (+1,-3,-4,+5,+6), & (+1,-2,-3,-4,+6),\\ (+1,-2,-3,+5,+6), & (+1,+2,-4,-5,-6), & (+1,+2,-4,-5,+6),\\ (+1,+2,-3,-4,-5), & (+1,+2,-3,-4,+6), & (+1,+2,+3,-5,-6),\\ (+1,+2,+3,-4,-5), & (+2,+3,-4,-5,-6), & (+2,+3,+4,-5,-6),\end{array}$$

  we reduce the number of solutions to *162*.
- Focusing on a set R of relevant quadruples of atoms, we consider partial chirotopes only. For example we could take as R the following quadruples of atoms:
  - the four neighbors of each stereocenter; These quadruples classify the stereoisomers;
  - four consecutively connected atoms.

  Using the relevant quadruples

$$R = \{1234, 2345, 1236, 1256, 1456, 3456\},$$

  we obtain *13* partial chirotopes shown in Table 6.

Note that thus far we only considered combinatorial aspects. The remaining part is of geometrical nature, and there exist no sophisticated algorithms adjusted to the chemical situation, yet. Nevertheless using constrained optimization with the aid of the software package sqp by Gerdts [12], we can demonstrate, that in principle, the expected conformations are generated by the promoted twofolded strategy:

Table 6. Partial chirotopes generated for cyclohexane

1234	1235	1245	1345	2345	1236	1246	1346	2346	1256	1356	2356	1456	2456	3456
+				+	+				+			+		+
+				+	+				+			+		−
+				+	+				+			−		+
+				+	+				+			−		−
+				+	+				−			+		+
+				+	+				−			+		−
+				+	+				−			−		+
+				+	−				+			−		+
+				−	+				+			+		+
+				−	+				+			+		−
+				−	+				−			+		−
+				−	+				−			−		−
−				−	+				−			+		−

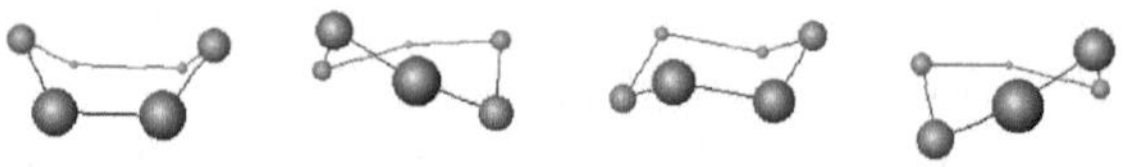

+++++-   ++++--   +-++++   +-+---

Fig. 9. Four conformations found by constrained optimization

Table 7. Number of partial chirotopes generated for medium sized cyclic hydrocarbons

Structure	C_6H_{12}	C_7H_{14}	C_8H_{16}	C_9H_{18}	$C_{10}H_{20}$
Partial chirotopes	13	18	30	46	78
Comp. time	0 s	0 s	0.2 s	4.9 s	62 s

- For only *4* of the given partial chirotopes, we found a stable realization, see Fig. 9.

Note that the first of the conformations, known as boat form, is indeed an unstable conformation having zero gradient (i.e. a saddle point of the energy function). Though it is interesting that this was found by our algorithms, its unstability can easily be recognized by considering the Hesse matrix of the energy function.

The remaining three conformations represent exactly the usual classification of all possible conformations of cyclohexane (compare Fig. 9 with Fig. 2), namely the chair form (the third conformation) as well as two twisted forms (second and fourth conformation).

Repeating the generation up to negation leads to *3* orientation patterns, reflecting the fact, that two of the found conformations (the two twisted forms) are mirror images of each other.

We extended the combinatorial part of our example to medium-sized molecular structures, too. Table 7 shows numbers of partial chirotopes generated correspondingly for cyclic hydrocarbons with 6 to 10 carbon atoms together with the computation times needed by `origen`.

5 Conclusions and Outlook

We presented `origen`, a generator of oriented matroids. Generation of oriented matroids can be seen as a first combinatorial step towards the generation of conformations for chemical compounds.

The second step, finding chemically feasible affine realizations of the generated oriented matroids, is still an open task. There exist general algorithms finding affine realizations of oriented matroids for small sizes [30, 28]. However the found realizations are often far away from being chemically feasible conformations. For more sophisticated approaches meeting the chemical demands, distances and maybe angles of the atoms should be taken into consideration. Thus distance geometry [6] as well as the Gröbner bases approach by P. Hazebroek and L. Oosterhoff [15] could be of great benefit, here.

A combination with the Gröbner base approach could work out as follows: In Sect. 1 we shortly sketched how Hazebroek and Oosterhoff formulated the structural properties of cyclohexane as polynomial equalities. It should not be a big problem to automatically build up systems of polynomial equations for arbitrary chemical structures, using tabularized standard bond lengths and bond angles for various bond and atom types. The prescribed orientations of a given partial chirotope actually are signs of determinants and thus directly lead to polynomial inequalities. Thus we can analyze an extended system of polynomial equations and inequalities corresponding to a conformation space with prescribed stereo-information. Depending on how much stereo-information is prescribed, the solution space could be manageable small. Still, the open question is how difficult it will be to solve the resulting systems of polynomial equations and inequalities.

Acknowledgments

This work has been conducted during the Special Semester on Gröbner Bases, February 1 – July 31, 2006, organized by RICAM, Austrian Academy of Sciences, and RISC, Johannes Kepler University, Linz, Austria. We especially want to thank B. Buchberger for creating this nice opportunity for discussion and further for the possibility of presentation in form of this paper. Special thank goes to M. Klin for detailed and motivating discussions.

References

1. O. Aichholzer, F. Aurenhammer, and H. Krasser, Enumerating order types for small point sets with applications, *Order*, **19** (2002), 265–281.
2. O. Aichholzer and H. Krasser, The point set order type data base: A collection of applications and results, in *Proc. 13th Annual Canadian Conference on Computational Geometry CCCG 2001*, pp. 17–20, 2001.
3. A. Björner, M. Las Vergnas, B. Sturmfels, N. White, and G. M. Ziegler, *Oriented Matroids*, 2nd edn., Cambridge University Press, Cambridge, 2000.
4. J. Bokowski, *Computational Oriented Matroids*, Cambridge University Press, Cambridge, 2006.
5. J. Bokowski and A. Guedes de Oliveira, On the generation of oriented matroids, *Discrete Comput. Geom.*, **24** (2000), 197–208.
6. G. M. Crippen and T. F. Havel, *Distance Geometry and Molecular Conformation*, Research Studies Press, Taunton, 1988.
7. A. Dreiding and K. Wirth, The multiplex a classification of finite ordered point sets in oriented d-dimensional space, *Commun. Math. Comput. Chem.*, **8** (1980), 341–352.
8. A. Dress, A. Dreiding, and H. Haegi, Classification of mobile molecules by category theory, *Stud. Phys. Theor. Chem.*, **23** (1983), 39–58.
9. I. A. Faradžev, Constructive enumeration of combinatorial objects, *Problèmes Combinatoires et Théorie des Graphes*, **260** (1978), 131–135. Colloq. Internat. CNRS, University of Orsay, Orsay 1976.
10. L. Finschi, *A Graph Theoretical Approach for Reconstruction and Generation of Oriented Matroids*, PhD thesis, ETH Zürich, 2001.
11. L. Finschi and K. Fukuda, Generation of oriented matroids, *Discrete Comput. Geom.*, **27** (2002), 117–136.
12. M. Gerdts, SQP V1.1 – Ein Fortran77-Programmpaket zur Lösung von nichtlinearen, restringierten Optimierungsproblemen, Universität Bayreuth, Bayreuth, 2004.
13. J. E. Goodman and R. Pollack, The complexity of point configurations, *Discrete Appl. Math.* (2), **31** (1991), 167–180.
14. R. Gugisch, *Konstruktion von Isomorphieklassen orientierter Matroide*, PhD thesis, University of Bayreuth, 2005.
15. P. Hazebroek and J. Oosterhoff, The isomers of cyclohexane, *Discrete Faraday Soc.*, **10** (1951), 87–93.
16. A. Heck, *Introduction to MAPLE*, 3rd edition, Springer, New York, 2003.
17. M. H. Klin, S. S. Tratch, and N. S. Zefirov, 2D-configurations and clique-cyclic orientations of the graphs $L(K_p)$, *Rep. Mol. Theory*, **1** (1990), 149–163.
18. D. E. Knuth, *Axioms and Hulls*, Lecture Notes in Computer Science, Vol. 606, Springer, Berlin, 1992.
19. R. Laue, Construction of combinatorial objects – a tutorial, *Bayreuther Mathematische Schriften*, **43** (1993), 53–96.

20. T. B. Leite, D. Gomes, M. A. Miteva, J. Chomilier, B. O. Villoutreix, and P. Tufféry, Frog: a free online drug 3d conformation generator, *Nucl. Acids Res.*, **35**(Web Server issue) (2007), W568–W572.
21. A. H. M. Levelt, *The Cycloheptane Molecule, a Challenge to Computer Algebra*; Invited Lecture, ISSAC'97, http://www.math.ru.nl/~ahml/engels.pdf, 1997.
22. B. D. McKay, Practical graph isomorphism, *Congr. Numer.*, **30** (1981), 45–87.
23. M. Minimair and M. P. Barnett, Solving polynomial equations for chemical problems using Gröbner bases, *Mol. Phys.* (23–24), **102** (2004), 2521–2535.
24. J. G. Nourse, The configuration symmetry group and its application to stereoisomer generation, specification, and enumeration, *J. Am. Chem. Soc.*, **101** (1979), 1210–1215.
25. J. G. Nourse, R. E. Carhart, D. H. Smith, and C. Djerassi, Exhaustive generation of stereoisomers for structure elucidation, *J. Am. Chem. Soc.*, **101** (1979), 1216–1223.
26. J. G. Nourse, D. H. Smith, R. E. Carhart, and C. Djerassi, Computer-assisted elucidation of molecular structure with stereochemistry, *J. Am. Chem. Soc.*, **102** (1980), 6289–6295.
27. R. C. Read, Everyone a winner, *Ann. Discrete Math.*, **2** (1978), 107–120.
28. J. Richter-Gebert, Two interesting oriented matroids, *Doc. Math.*, **1** (1996), 137–148.
29. M. Saunders, K. N. Houk, Y.-D. Wu, W. C. Still, J. Lipton, G. Chang, and W. C. Gouda, Conformations of cycloheptadecane. A comparison of methods for conformational searching, *J. Am. Chem. Soc.*, **112** (1990), 1419–1427.
30. P. Shor, Stretchability of pseudolines is np-hard, in P. Gritzmann and B. Strumfels (eds.) *Applied Geometry and Discrete Mathematics – The "Victor-Klee Festschrift"*, DIMACS Series in Discrete Mathematics and Theoretical Computer Science, Vol. 4, pp. 531–554, AMS, Providence, 1991.
31. S. S. Tratch, Mathematical models in stereochemistry. I. Combinatorial characteristics of composition, connection, and configuration of organic molecules, *Russian J. Org. Chem.* (9), **31** (1995), 1189–1217.
32. F. Tschirschnitz, *Testing Extendability for Partial Chirotopes is np-Complete*, http://citeseer.ist.psu.edu/449661.html.
33. J. von zur Gathen and J. Gerhard, *Modern Computer Algebra*, Cambridge University Press, Cambridge, 1999.
34. N. S. Zefirov and S. S. Tratch, Some notes on Randić–Razinger's approach to characterization of molecular shapes, *J. Chem. Inf. Comput. Sci.*, **37** (1997), 900–912.

Algorithmic Approach to Non-symmetric 3-class Association Schemes

Leif K. Jørgensen

Department of Mathematical Sciences, Aalborg University, Fr. Bajers Vej 7, 9220 Aalborg, Denmark. leif@math.aau.dk

Summary. There are 24 feasible parameter sets for a primitive non-symmetric association schemes with 3 classes and at most 100 vertices. Using computer search, we prove non-existence for three feasible parameter sets. Ten cases are still open.

In the imprimitive case, we survey the known results including some constructions of infinite families of schemes. In the smallest case that has been open up to now, we use computer search to find new schemes. These schemes are equivalent to "skew" Bush-type Hadamard matrices of order 36. We also consider directed graphs that satisfy only some of the conditions required for a non-symmetric association scheme with 3 classes.

Key words: Association schemes, Orientation of strongly regular graphs, Computer search

1 Introduction

The theory of association schemes was for a long time concentrated on the investigation of the symmetric association schemes generated by distance regular graphs. In this context the symmetric association schemes with two classes are exactly the schemes generated by strongly regular graphs.

More opportunities appear as soon as we are dealing with at least three classes. A good survey of symmetric association schemes with three classes was provided by van Dam [6].

In this paper we consider non-symmetric association schemes with three classes. From each such association scheme, a symmetric association scheme with two classes can be obtained by merging the non-symmetric relations. Feasibility conditions for the existence of these association schemes have previously been considered by Bannai and Song [2], Song [37] and by Goldbach and Claasen [13].

In this paper we make an attempt of a more systematic investigation of non-symmetric 3-class association schemes with a relatively small number of

M. Klin et al. (eds.), Algorithmic Algebraic Combinatorics and Gröbner Bases, DOI 10.1007/978-3-642-01960-9_8, © Springer-Verlag Berlin Heidelberg 2009

vertices. In the primitive case we generate all feasible parameter sets with at most 100 vertices. There are 24 such parameter sets. We review known results and prove non-existence results for three parameter sets, while 10 cases still remain open.

We also briefly consider normally regular digraphs (in the sense of [25]) as a generalization of non-symmetric 3-class association schemes.

For the imprimitive case we start from a consideration of doubly regular (m, r)-team tournament in the sense of [27]. In [27] we distinguish three possible types of such directed graphs. A graph of type 3 cannot be a relation of an association scheme, however we do not know if any graph of this type exists. Types 1 and 2 indeed correspond to imprimitive non-symmetric 3-class association schemes. Graphs of type 1 are easily reduced to doubly regular tournaments in the sense of [35]. Thus we concentrate on graphs of type 2 and the corresponding association schemes. In particular we consider a subtype of type 2 which has links to Bush-type Hadamard matrices.

Here we investigate the smallest open case of order 36. We find four such association schemes by computer search, but we leave open the problem of complete enumeration of all association schemes with this set of parameters. We expect that there may be a large number of such schemes – probably all with small automorphism groups. N. Ito [19] has proved that they cannot have an automorphism group of rank 4.

It should be stressed that our approach is strictly algorithmic, essentially depending on the use of computers. The computer is used already on the initial stage of the generation of all feasible sets of parameters of primitive schemes.

For computer-aided constructive enumeration of all association schemes with a given set of parameters we use two different approaches. The first approach makes use of a complete catalog of strongly regular graphs with the parameters that would be obtained by merging the non-symmetric relations. In the cases considered in this paper we use the complete catalog of strongly regular graphs with parameters $(45, 12, 3, 3)$ found by Coolsaet, Degraer and Spence [5] and the classical result of Hoffman and Singleton [16] about the uniqueness of the strongly regular graph with parameters $(50, 7, 0, 1)$. The second approach is an orderly generation algorithm in the spirit of Faradžev [8] and Read [34].

These techniques are used to exclude existence of three feasible parameter sets for primitive association schemes. The second technique was also used to find the above mentioned imprimitive association schemes of order 36. In that case the search space is huge and it was not possible to complete the full search. But we successfully used some ad hoc tricks in order to catch in the whole search space a few lucky directions leading to a construction of the desired combinatorial objects.

We hope that the results presented in this paper may help to promote further approaches towards constructive enumeration of association schemes.

2 Preliminaries

Let X be a finite set ($|X| = v$) and let $\{R_0, R_1, \ldots, R_d\}$ be a partition of $X \times X$. Then we say that $\mathcal{X} = (X, \{R_0, R_1, \ldots, R_d\})$ is an *association scheme* with *d classes* if the following conditions are satisfied

- $R_0 = \{(x,x) \mid x \in X\}$,
- for each i, $R_i^{\mathrm{t}} := \{(x,y) \mid (y,x) \in R_i\} = R_{i'}$, for some i',
- and for each triple (i,j,h), $i,j,h \in \{0,\ldots,d\}$ there exists a so-called *intersection number* p_{ij}^h such that for all $x,y \in X$ with $(x,y) \in R_h$ there are exactly p_{ij}^h elements $z \in X$ so that $(x,z) \in R_i$ and $(z,y) \in R_j$.

For $i > 0$ the relation R_i can be viewed as the edge set of the (undirected or directed) graph (X, R_i). We will frequently identify this graph with the relation R_i.

If $i = i'$ for all i then $\mathcal{X}$ is said to be *symmetric*, otherwise it is *non-symmetric*. If the graphs $R_1, \ldots, R_d$ all are connected then we say that $\mathcal{X}$ is *primitive*, otherwise it is *imprimitive*.

A relation (say R_1) of a symmetric association scheme with two classes is a strongly regular graph with parameters (v,k,a,c), where $v = |X|, k = p_{11}^0, a = p_{11}^1, c = p_{11}^2$. And conversely, if R_1 is a strongly regular graph and R_2 is the complementary graph of R_1, then R_1 and R_2 form a symmetric association scheme with two classes.

A relation of a non-symmetric association scheme with two classes is called a *doubly regular tournament*. Reid and Brown [35] proved that there exists a doubly regular tournament with n vertices if and only if there exists a skew Hadamard matrix of order $n+1$. Thus a necessary condition is that $n \equiv 3 \mod 4$.

In this paper we consider non-symmetric association schemes with $d = 3$ classes. We will assume that the relations are enumerated so that R_1 and R_2 are non-symmetric, $R_2 = R_1^{\mathrm{t}}$, and R_3 is a symmetric relation. In this case the association scheme is determined uniquely by relation R_1.

If A denotes the adjacency matrix of the relation R_1 then the adjacency matrices of R_0, R_2 and R_3 are I, A^{t} and $J - I - A - A^{\mathrm{t}}$, respectively. The *Bose-Mesner algebra* of $\mathcal{X}$ is the matrix algebra $\mathcal{A}$ spanned by these four matrices, see Bannai and Ito [1].

Higman [15] proved that an association scheme with $d \leq 4$ has a commutative Bose-Mesner algebra, which means that $p_{ij}^h = p_{ji}^h$, for all i,j,h.

Thus multiplication in the Bose-Mesner algebra is determined by the following equations.

$$AJ = JA = \kappa J, \tag{1}$$

$$AA^{\mathrm{t}} = \kappa I + \lambda(A + A^{\mathrm{t}}) + \mu(J - I - A - A^{\mathrm{t}}), \tag{2}$$

$$A^{\mathrm{t}}A = \kappa I + \lambda(A + A^{\mathrm{t}}) + \mu(J - I - A - A^{\mathrm{t}}), \tag{3}$$

$$A^2 = \alpha A + \beta A^{\mathrm{t}} + \gamma(J - I - A - A^{\mathrm{t}}), \tag{4}$$

where $\kappa = p^0_{12}$, $\lambda = p^1_{12} = p^1_{21} = p^2_{12} = p^2_{21}$, $\mu = p^3_{12} = p^3_{21}$, $\alpha = p^1_{11}$, $\beta = p^2_{11}$ and $\gamma = p^3_{11}$.

We note that $\alpha = \lambda$. This is seen by counting in two ways the pairs (y, z) so that $(x, y), (x, z), (y, z) \in R_1$, for a fixed vertex x.

Since $\mathcal{A}$ is commutative and consists of normal matrices, the matrices of $\mathcal{A}$ have a common diagonalization, i.e., $\mathcal{A}$ has a basis $\{E_0, E_1, E_2, E_3\}$ of orthogonal projections.

Since a non-symmetric association scheme $\mathcal{X}$ with 3 classes is commutative, the symmetrization $(X, \{R_0, R_1 \cup R_2, R_3\})$ is also an association scheme, thus R_3 is a strongly regular graph and R_1 and R_2 are orientations of a strongly regular graph. In fact $R_1 \cup R_2$ is a strongly regular graph with parameters

$$\begin{aligned}(v, k, a, c) &= (v, 2p^0_{12}, p^1_{11} + p^1_{12} + p^1_{21} + p^1_{22}, p^3_{11} + p^3_{12} + p^3_{21} + p^3_{22}) \\ &= (v, 2p^0_{12}, 3p^1_{12} + p^1_{22}, 2(p^3_{11} + p^3_{12})). \qquad (5)\end{aligned}$$

In [25], we prove the following.

Lemma 1. *If A is the adjacency matrix of a regular directed graph (i.e., (1) is satisfied), then (2) and (3) are equivalent.*

(This is also an alternative proof of the commutativity of the Bose-Mesner algebra $\mathcal{A}$ in this particular case.) A directed graph whose adjacency matrix satisfies these equations is called a *normally regular digraph.* The eigenvalues of a normally regular digraph have the following property.

Theorem 1. (See [25].) *If the adjacency matrix A of a regular directed graph satisfies (2) then an eigenvalue $\theta \neq k$ lies on the circle in the complex plane with center $\lambda - \mu$ and radius $\sqrt{k - \mu + (\lambda - \mu)^2}$ and $\theta + \overline{\theta}$ is an eigenvalue of $A + A^t$.*

If A satisfies all Eqs. 1, 2, 3 and 4 then it has four eigenvalues κ, and say ρ, σ and $\overline{\sigma}$ with multiplicities 1, m_1, m_2 and m_2, respectively, and the eigenvalues of $A + A^t$ are 2κ, 2ρ, and $\sigma + \overline{\sigma}$ with multiplicities 1, m_1, and $2m_2$.

For parameters v and p^h_{ij}, $i, j, h \in \{0, 1, 2, 3\}$ the parameters of $R_1 \cup R_2$ can be computed from (5). Using standard formulas, the spectrum of $R_1 \cup R_2$ can then be computed. From this it is possible to compute eigenvalues and multiplicities of R_1 (e.g. using Theorem 1). For an arbitrary set of intersection numbers, using expressions for multiplicities, one may get non-integer values, thus excluding this parameter set.

Definition 1. *We say that v and p^h_{ij}, $i, j, h \in \{0, 1, 2, 3\}$ form a feasible parameter set for a non-symmetric association scheme with three classes if they are non-negative integers and the multiplicities of the (four) eigenvalues computed from these intersection numbers are positive integers.*

Bannai and Song proved that the spectrum of A can be computed from the spectrum of $A + A^t$. (We note that if the eigenvalues of $A + A^t$ are $2\kappa, r, s$ then either r or s can be split in two complex eigenvalues, if their multiplicities are even.)

Lemma 2. (See Bannai and Song [2].) *Suppose A is an adjacency matrix of a non-symmetric relation R_1 of a 3-class association scheme. If s is the eigenvalue (of multiplicity $2m$) of $A + A^t$ that is split in two complex eigenvalues σ and $\overline{\sigma}$ (i.e., $s = \sigma + \overline{\sigma}$) then $\sigma = \frac{1}{2}(s + i\sqrt{v\kappa/m})$.*

It is well known that the intersection numbers can be computed from the spectrum of A.

The *Hadamard product* of matrices $B = (b_{ij})$ and $C = (c_{ij})$ is the matrix $B \circ C = (b_{ij}c_{ij})$. Since $\{I, A, A^t, J - A - A^t - I\}$ is a basis of $\mathcal{A}$, it follows by considering the Hadamard product of these matrices that $\mathcal{A}$ is closed under the Hadamard product. In particular there exist numbers q_{ij}^h, for $i, j, h \in \{0, 1, 2, 3\}$, so that $E_i \circ E_j = \frac{1}{v}\sum_h q_{ij}^h E_h$. These numbers are called *Krein parameters*. It is known that each Krein parameter is a non-negative real number, see Bannai and Ito [1]. Since the Krein parameters can be computed from the spectrum of A, this can be used to prove non-existence for some feasible parameter sets.

Neumaier [33] found another way to exclude feasible parameter sets. Let m_i be the rank of E_i, for $i \in \{0, 1, 2, 3\}$. (Thus $m_0, \ldots, m_3$ are the multiplicities of eigenvalues.)

Theorem 2. (See [33].) *The following inequalities are satisfied for a commutative association scheme.*

$$\sum_{h:q_{ii}^h>0} m_h \leq \frac{1}{2}m_i(m_i+1), \quad \text{for } i = 0, \ldots, d,$$

$$\sum_{h:q_{ij}^h>0} m_h \leq m_i m_j, \quad \text{for } i, j = 0, \ldots, d, \ i \neq j.$$

3 Primitive Association Schemes with Three Classes

We now use a computer to generate a list of all feasible parameter sets for primitive association schemes with three classes and $|X| \leq 100$. For each feasible parameter set (v, k, a, c) of a strongly regular graph we investigate the feasible parameters of non-symmetric association schemes with three classes such that $R_1 \cup R_2$ has parameters (v, k, a, c). It follows from (5) that we need only consider parameters where k and c are even. It is also useful to know that the eigenvalues of $R_1 \cup R_2$ are integers. This follows from the next lemma.

Lemma 3. (See Goldbach and Claasen [13].) *There is no non-symmetric association scheme with three classes so that $R_1 \cup R_2$ has parameters $(4c + 1, 2c, c - 1, c)$.*

In Goldbach and Claasens's terminology they proved non-existence if the strongly regular graph is *pseudo-cyclic*, i.e., the two non-trivial eigenvalues have the same multiplicities. It is well-known that this is equivalent to having

Table 1. A list of all feasible parameter sets for primitive non-symmetric 3-class association schemes with at most 100 vertices. The second column is the parameters of the strongly regular graph $R_1 \cup R_2$. The third column gives information on the number of strongly regular graphs with these parameters. (In fact in some cases where we write ≥ 1, there are several known strongly regular graphs, e.g. with parameters $(100, 44, 18, 20)$, see [26].) These numbers are from [3, 5, 32]. In column six "NO" means that we prove non-existence of the association scheme in this paper and "no" means that non-existence follows from general results or it was proved in other papers

No.	Parameters for $R_1 \cup R_2$	No. of SRGs	p_{12}^1	p_{12}^3	Exists	Reference
1	$(16, 10, 6, 6)$	1	1	2	no	Goldbach and Claasen [12]
2	$(21, 10, 3, 6)$	1	1	1	no	Enomoto and Mena [7]
3	$(36, 14, 4, 6)$	180	0	2	yes	Goldbach and Claasen [11]
4	$(36, 20, 10, 12)$	32548	3	2	NO	Theorem 5
5	$(45, 32, 22, 24)$	78	6	4	NO	Theorem 4
6	$(50, 42, 35, 36)$	1	8	12	NO	Theorem 3
7	$(57, 42, 31, 30)$	0	7	9	no	Wilbrink and Brouwer [38]
8	$(64, 28, 12, 12)$	≥ 1	4	2	yes	Enomoto and Mena [7]
9	$(64, 36, 20, 20)$	≥ 1	4	6	yes	Jørgensen [28]
10	$(64, 42, 26, 30)$	≥ 1	7	6	?	
11	$(64, 42, 30, 22)$	0	7	6	no	Absolute bound
12	$(81, 50, 31, 30)$	≥ 1	9	5	?	
13	$(85, 64, 48, 48)$	≥ 1	13	8	?	
14	$(85, 70, 57, 60)$	?	13	20	?	
15	$(96, 38, 10, 18)$	?	3	4	?	
16	$(96, 50, 22, 30)$	?	3	10	no	Neumaier
17	$(96, 60, 38, 36)$	?	11	6	no	Krein
18	$(96, 76, 60, 60)$	≥ 1	16	10	?	
19	$(100, 44, 18, 20)$	≥ 1	3	6	?	
20	$(100, 54, 28, 30)$	≥ 1	8	6	?	
21	$(100, 66, 39, 52)$	0	10	12	no	Absolute bound
22	$(100, 66, 41, 48)$	≥ 1	8	16	no	Neumaier
23	$(100, 66, 44, 42)$	?	10	12	?	
24	$(100, 72, 50, 56)$	≥ 1	13	12	?	

parameters $(4c + 1, 2c, c - 1, c)$, see [3]. (Only in this pseudo-cyclic case it is possible that a strongly regular graph has irrational eigenvalues.)

The resulting list of feasible parameter sets is presented in Table 1.

The association scheme with parameter set no. 3 was constructed by Iwasaki [23] and independently by Ivanov, Klin and Faradžev [22], see also [9]. Later Goldbach and Claasen [11] proved that it is the unique association scheme with these parameters. The association scheme with parameter set no. 8 was constructed by Enomoto and Mena [7]. Liebler and Mena [30] showed this scheme belongs to an infinite family of association schemes. These schemes have order $4s^4$ where s is a power of 2. Four association schemes with parameter set no. 9 were constructed in [28].

These are the only known primitive non-symmetric association schemes with three classes.

In parameter sets nos. 7, 11 and 21 it is known that the strongly regular graph does not exist, see Brouwer [3]. Thus the 3-class association scheme does not exist in these cases.

In parameter set no. 17 some of the Krein parameters are negative. Thus this case is excluded. The multiplicities of eigenvalues for parameter sets nos. 16 and 22 do not satisfy Neumaier's condition.

We will now use computers to prove non-existence of association schemes with parameter sets nos. 4, 5 and 6. We use two different techniques.

For parameter sets nos. 5 and 6, the computation is based on the classification of all strongly regular graphs which have the same parameters as the strongly regular graph R_3 (assuming that $(X, \{R_0, R_1, R_2, R_3\})$ is the required association scheme). For a given graph R_3 we try to construct R_1 by orienting the complement of R_3. We first consider orientation of edges of the complement of R_3 incident with a fixed vertex x. We let $N^+(x)$ denote the out-neighbors of x in R_1 and we let $N_2(x)$ denote the vertices at distance 2 from x in R_3. Then a candidate for $N^+(x)$ consists of exactly half of the vertices in $N_2(x)$. But it also has some other properties. Let $x_1, \dots, x_k$ be the neighbors of x in R_3 and S_i denote the set of neighbors of x_i in $N_2(x)$, for $i = 1, \dots, k$. Then a candidate for $N^+(x)$ must satisfy $|S_i \cap N^+(x)| = p_{13}^3 = \frac{1}{2}|S_i|$. It also satisfies that the subgraph of R_3 induced by $N^+(x)$ is regular of degree p_{13}^1 and the subgraph induced by $N_2(x) \setminus N^+(x)$ is regular of degree $p_{23}^2 = p_{13}^1$. When we have computed the list of all candidates for $N^+(x)$ for every vertex x, we try if it is possible to combine these orientations in such a way that for any two vertices x and y the orientation of the edges incident with x and the orientation of the edges incident with y should agree on the orientation of the edge $\{x, y\}$ if x and y are non-adjacent in R_3, and they should satisfy that for all i, j the number of vertices z so that $(x, z) \in R_i$ and $(z, y) \in R_j$ is exactly p_{ij}^h where $(x, y) \in R_h$.

For parameters no. 6, R_3 is a strongly regular graph with parameters $(50, 7, 0, 1)$, i.e., it is the Hoffman-Singleton graph, see [16]. This case can be excluded by investigating possible orientations of the complement of the Hoffman-Singleton graph.

Theorem 3. *There is no non-symmetric association scheme with three classes where R_3 is the Hoffman-Singleton graph.*

Proof. Suppose that there exists a non-symmetric association scheme with three classes where R_3 is the Hoffman-Singleton graph. When applying the method described above we may use that the Hoffman-Singleton graph has a large group of automorphisms. Computations using this group are done in GAP [10] with GRAPE [36] and nauty [31]. Other computations are done in a C-program.

Let x be a vertex and let $x_1, \ldots, x_7$ be the neighbors of x in R_3. Let S_i be the set of neighbors of x_i other than x, for $i = 1, \ldots, 7$. Let $N^+(x)$ be the set out-neighbors of x in R_1. Then $N^+(x)$ is a set of 21 vertices in the set $N_2(x) = S_1 \cup \ldots \cup S_7$ of vertices at distance 2 from x, and $|S_i \cap N^+(x)| = p_{13}^3 = 3$, for $i = 1, \ldots, 7$. The subgraph of R_3 spanned by $N^+(x)$ is regular of degree $p_{13}^1 = 4$. The complement of $N^+(x)$ in $S_1 \cup \ldots \cup S_7$ is the set of in-neighbors of x in R_1 and this set also spans a 4-regular subgraph of R_3.

A computer enumeration shows that there are exactly 1140 subsets of $N_2(x)$ with the properties required for $N^+(x)$. These 1140 subsets form three orbits under the action of the subgroup of the automorphism group of the Hoffman-Singleton graph stabilizing the vertex x.

Thus we need only consider three possibilities for $N^+(x)$, but then we must consider all 1140 candidates $N^+(y)$ for any other vertex y. It turns out that we only need to consider orientations of edges incident with $x, x_1, \ldots, x_5$. These edges must by oriented such that $|N^+(x) \cap N^+(x_i)| = p_{12}^3 = 12$, as $(x, x_i) \in R_3$, $|N^+(x_i) \cap N^+(x_j)| = p_{12}^1 = p_{12}^2 = 8$, as $(x_i, x_j) \notin R_3$, and such that $x_j \in N^+(x_i)$ if and only if $x_i \notin N^+(x_j)$.

A computer search shows that there are no orientations of all edges incident with x, x_1, x_2, x_3, x_4 and x_5 that satisfy these conditions. Thus the required association scheme does not exist. □

For parameters no. 5, R_3 is a strongly regular graph with parameters $(v, k, a, c) = (45, 12, 3, 3)$.

Coolsaet, Degraer and Spence [5], have shown that there are exactly 78 strongly regular graphs with these parameters. Thus the method from the previous theorem can be applied to each of these 78 graphs.

Theorem 4. *There is no primitive non-symmetric association scheme with three classes with parameter set no. 5.*

Proof. Suppose that there exists such an association scheme. Let x be a vertex and let $x_1, \ldots, x_{12}$ be the neighbors of x in R_3. Let S_i be the set of neighbors of x_i at distance 2 from x, $|S_i| = k - a - 1 = 8$, for $i = 1, \ldots, 12$. Let $N^+(x)$ be the set out-neighbors of x in R_1. Then $N^+(x)$ is a set of 16 vertices in the set $N_2(x) := S_1 \cup \ldots \cup S_{12}$, and $|S_i \cap N^+(x)| = p_{13}^3 = 4$, for $i = 1, \ldots, 12$. The subgraph of R_3 spanned by $N^+(x)$ is regular of degree $p_{13}^1 = 3$.

The computer search shows that if N is a set with $|S_i \cap N| = 4$, for $i = 1, \ldots, 12$, and in which every vertex has degree at most 3 then N is 3-regular and the subgraph of R_3 spanned by $N_2(x) \setminus N$ is also 3-regular.

The number of such sets N depend on the graph and the vertex x. The largest number of sets is 396, which appear in the graph with a rank 3 automorphism group.

44 of the 78 candidates for R_3 can be excluded because, for at least one vertex x, there is no such set N.

For each of the other 34 graphs we find by computer search a set W of at most 8 vertices so that there is no combination of orientations of edges in the complement of R_3 incident with w, for each $w \in W$ that satisfies the required properties. (This search took 45 minutes on a 2.4 GHz PC.)

Thus an association scheme with parameter set no. 5 does not exist. □

For parameter set no. 4 (and for one case of imprimitive association schemes, see Sect. 4) we use a different computer search technique. This does not depend on characterization of strongly regular graphs.

We use an orderly generation algorithm (see Faradžev [8] or Read [34]) to search for the matrix $B = 3A_3 + 2A_2 + A_1$, where A_1, A_2, A_3 are adjacency matrices of the relations R_1, R_2, R_3 of the required association scheme. Recall that for $i \in \{1,2,3\}$ we define $i' \in \{1,2,3\}$ so that $R_i^{\mathrm{t}} = R_{i'}$. In our usual enumeration of relations this means that $1' = 2$, $2' = 1$ and $3' = 3$, but in the first application of the algorithm (Theorem 5) we use a different enumeration (where $1' = 1$, $2' = 3$ and $3' = 2$).

We want the vertices to be labeled with numbers $1, \ldots, v$ so that the matrix B is in maximal form, i.e., the sequence obtained by reading the entries of the first row followed by the entries of the second row, etc., is as large as possible (in the lexicographic order) among all labellings of the vertices.

Suppose that the first $r-1$ rows of the matrix $B = (b_{ij})$ has been filled in. We then investigate all possible ways to fill in row r with 0 on the diagonal entry, $p^0_{11'}$ entries with 1's, $p^0_{22'}$ entries with 2's, and $p^0_{33'}$ entries with 3's in such a way that

- the first $r-1$ entries are in accordance with the entries of column r of the previous rows,
- for each $x < r$ the number of columns s, so that $b_{xs} = i$ and $b_{rs} = j'$ is exactly p^h_{ij}, where $b_{xr} = h$,
- the matrix is still in maximal form.

For each possible way to fill row r we repeat the procedure for row $r+1$.

Theorem 5. *There is no primitive non-symmetric association scheme with three classes with parameter set no. 4.*

Proof. As described above, we search for the matrix $B = 3A_3 + 2A_2 + A_1$.

In turns out that with the maximality condition on the matrix and for this particular parameter set it is convenient to enumerate the relations so that R_1 is symmetric and $R_2^{\mathrm{t}} = R_3$. Thus the first row of B should consist of one 0 followed by $p^0_{33'} = 10$ entries with 3's followed by $p^0_{22'} = 10$ entries with 2's and finally $p^0_{11'} = 15$ entries with 1's.

When using the algorithm described above we find that the number of ways to fill in the first r rows is 1, 1, 100, 24161, 205671, 1116571, 52650, 39, 0, …, 0, for $r = 1, \ldots, 36$. Thus the required association scheme does not exist. (This search took 81 minutes on a 2.4 GHz PC.) □

4 Imprimitive Association Schemes with Three Classes

4.1 General Results

If R_3 is connected but R_1 and R_2 are disconnected then each connected component of R_1 is a doubly regular tournament on $2p_{12}^0 + 1$ vertices. Thus the study of these schemes reduces to the study of doubly regular tournaments.

We will thus assume that R_1 and R_2 are connected and R_3 is disconnected. Then R_3 consists of m copies of a complete graph on r vertices, for some constants m and r. We denote this graph by $m \circ K_r$. Then R_1 is an orientation of the complement $\overline{m \circ K_r}$. The vertex set of $\overline{m \circ K_r}$ is partitioned in m independent sets of size r, denoted by $V_1, \ldots, V_m$.

In [27] we introduce the following family of graphs that do not necessarily satisfy all the conditions on a relation of a non-symmetric association scheme with three classes. We say that a directed graph is a *doubly regular* (m,r)-*team tournament* if it is an orientation of $\overline{m \circ K_r}$ with adjacency matrix A satisfying (1) and (4) in Sect. 2.

In [27] we give a combinatorial proof of the following, i.e., we do not use eigenvalues.

Theorem 6. (See Jørgensen, Jones, Klin and Song [27].) *Every doubly regular* (m,r)*-team tournament is of one of the following types.*

1. *For every pair* i, j *either all the edges between* V_i *and* V_j *are directed from* V_i *to* V_j*, or they are all directed from* V_j *to* V_i*. The graph with vertices* $v_1, \ldots, v_m$ *and an edge directed from* v_i *to* v_j *if edges are directed from* V_i *to* V_j *is a doubly regular tournament.*
2. *For every vertex* $x \in V_i$*, exactly half of the vertices in* V_j $(j \neq i)$ *are out-neighbors of* x*, and* $\alpha = \beta = \frac{(m-2)r}{4}$*, and* $\gamma = \frac{(m-1)r^2}{4(r-1)}$.
3. *For every pair* $\{i, j\}$ *either* V_i *is partitioned in two sets* V_i' *and* V_i'' *of size* $\frac{r}{2}$ *so that all edges between* V_i *and* V_j *are directed from* V_i' *to* V_j *and from* V_j *to* V_i''*, or similarly with* i *and* j *interchanged. The parameters are* $\alpha = \frac{(m-1)r}{4} - \frac{3r}{8}$, $\beta = \frac{(m-1)r}{4} + \frac{r}{8}$, $\gamma = \frac{(m-1)r^2}{8(r-1)}$.

A graph of type 3 cannot be a relation of an association scheme. In this case 8 divides r and $4(r-1)$ divides $m-1$. We do not know if any graph of this type exists.

Every graph of type 1 or type 2 is a relation of a non-symmetric association scheme with 3 classes. The results for these types where first proved by Goldbach and Claasen [14].

Clearly, the graph of type 1 exists if and only if a doubly regular tournament of order m exists. Thus in the remaining part this section we will only consider graphs of type 2.

4.2 Association Schemes of Type 2

We first show that a graph of type 2 is a relation of a non-symmetric association scheme with 3 classes. This is done by proving that (2) and (3) are satisfied.

Lemma 4. *Let A be the adjacency matrix of a doubly regular (m,r)-team tournament of type 2. Then A satisfies (2) and (3) with*

- $\lambda = \alpha = \frac{(m-2)r}{4}$ *and*
- $\mu = \frac{(m-1)r(r-2)}{4(r-1)}$.

In particular if $m = r$ then $\lambda = \mu = \frac{m(m-2)}{4}$.

Proof. Let $x \in V_i$ and $y \in V_j$, $i \neq j$, and suppose that there is an edge directed from x to y. Then x has $\kappa - \frac{r}{2}$ out-neighbors outside $V_i \cup V_j$, α of these are in-neighbors of y and the remaining $\kappa - \frac{r}{2} - \alpha$ are out-neighbors of y. Thus $\lambda = \kappa - \frac{r}{2} - \alpha = \frac{(m-2)r}{4}$, since $\kappa = \frac{(m-1)r}{2}$.

Similarly, for $x, y \in V_i$, we get $\mu = \kappa - \gamma = \frac{(m-1)r(r-2)}{4(r-1)}$. Thus (2) is satisfied. Equation 3 can be proved in a similar way, or by applying Lemma 1. □

Since the parameters of a graph of type 2 are integers, it follows that r is even and $r-1$ divides $m-1$. Using eigenvalues, it can be shown that m is even, see [27] or Goldbach and Claasen [14].

Existence in the case $r = 2$ is equivalent to existence of a doubly regular tournament of order $m-1$.

Theorem 7. (See [27].) *If there exists a doubly regular $(m,2)$-team tournament Γ then 4 divides m and the out-neighbors of a vertex in Γ span a doubly regular tournament of order $m-1$.*
Conversely, for every doubly regular tournament T of order $m-1$, there exists a doubly regular $(m,2)$-team tournament Γ, such that for some vertex x in Γ the out-neighbors of x span a subgraph isomorphic to T.

In [28] we found 40 association schemes with $(r,m) = (4,16)$. No other schemes with $4 \leq r < m$, where $r-1$ divides $m-1$ are known.

We will now consider the case $m = r$. By Lemma 4, the directed graph is then a normally regular digraph with $\mu = \lambda$. Such digraphs are also known as doubly regular asymmetric digraphs. These graphs were introduced and studied in a series of papers by N. Ito [18–21] and also studied by Ionin and Kharaghani [17].

The first non-trivial case of an association scheme of type 2 with $m = r$ is for $m = 4$. In this case there exist two non-isomorphic schemes. One of these schemes has an automorphism group of rank 4, i.e., the group acts transitively on the vertices and the stabilizer of a vertex x has four orbits: $\{x\}$, the set of out-neighbors of x, the set of in-neighbors of x and the set of vertices not adjacent to x. Any doubly regular asymmetric digraph with automorphism group of rank 4 is a relation of a non-symmetric association scheme with 3 classes. Ito [19] has proved that a primitive non-symmetric 3-class association scheme with $\mu = \lambda$ does not satisfy the feasibility condition. Thus a doubly regular asymmetric digraph with automorphism group of rank 4 is a relation of an imprimitive non-symmetric 3-class association scheme of type 2 with $m = r$ (as $\mu = \lambda$). In this case Ito [19] has proved that $m = r$ is a power of 2. He also claims to have proved that the only possibility is $m = 4$. But the proof of this does not seem to be correct and in fact Ito in his paper gives an example of a vertex transitive scheme with $m = r = 8$. According to computations in GAP [10] using share package GRAPE [36] with nauty [31] the automorphism group of this scheme has rank 4.

We will now consider the links between such association schemes and a special case of some well-known structures.

Definition 2. *An Hadamard matrix H of order n is an $n \times n$ matrix in which every entry is either 1 or -1 and $HH^t = nI$.*

An Hadamard matrix H of order m^2 is said to be Bush-type if H is block matrix with $m \times m$ blocks H_{ij} of size $m \times m$ such that $H_{ii} = J_m$ and $H_{ij}J_m = J_m H_{ij} = 0$, for $i \neq j$.

Theorem 8. *An imprimitive 3-class association scheme of type 2 and with $r = m$ is equivalent to a Bush-type Hadamard matrix of order m^2 with the property that $H_{ij} = -H^t_{ji}$, for all pairs i, j with $i \neq j$.*

Proof. Let A be an adjacency matrix of relation R_1, for some imprimitive 3-class association scheme of type 2 and with $r = m$. We may assume that vertices are enumerated such that the vertices in V_i corresponds to columns/rows $mi - m + 1, \ldots, mi$. Let $H = J_{m^2} - 2A$. Then H is partitioned in blocks H_{ij} of size $m \times m$ corresponding to the partition of vertices in sets $V_1, \ldots, V_m$. Clearly $H_{ii} = J_m$ and since a vertex in V_i has exactly $\frac{m}{2}$ out-neighbors and $\frac{m}{2}$ in-neighbors in V_j, $H_{ij}J_m = J_m H_{ij} = 0$.

From (1) and (2) we get (since $\kappa = \frac{m(m-1)}{2}$ and $\mu = \lambda = \frac{m(m-2)}{4}$)

$$HH^{\mathrm{t}} = (J_{m^2} - 2A)(J_{m^2} - 2A^{\mathrm{t}}) = (m^2 - 4\kappa)J_{m^2} + 4(\kappa I + \mu(J_{m^2} - I)) = m^2 I.$$

Thus H is an Hadamard matrix.

Conversely, suppose that H is a Bush-type Hadamard matrix which is skew in the sense that $H_{ij} = -H^{\mathrm{t}}_{ji}$, for $i \neq j$.

Let $A = \frac{1}{2}(J - H)$, where $J = J_{m^2}$. Then A is a $\{0, 1\}$ matrix. Since H is Bush-type it has exactly $m + (m-1)\frac{m}{2}$ entries equal to 1 and $(m-1)\frac{m}{2}$ entries equal to -1 in each row. Thus $HJ = mJ$ and the transposed equation is $JH^{\mathrm{t}} = mJ$. Similarly $JH = mJ$. Thus $AJ = JA = \frac{m(m-1)}{2} J$ and

$$AA^{\mathrm{t}} = \frac{1}{4}(J - H)(J - H^{\mathrm{t}}) = \frac{m(m-2)}{4} J + \frac{m^2}{4} I.$$

We see that (1) and (2) are satisfied. Equation 3 can be proved in a similar way, or by applying Lemma 1.

Let K denote the block diagonal matrix with diagonal blocks equal to J_m. Then the Bush-type property of H implies that $HK = mK$ and the skew property of H implies that $H + H^t = 2K$. Thus $H^2 = H(2K - H^t) = 2mK - m^2 I$, and so

$$A^2 = \frac{1}{4}(J - H)^2 = \frac{1}{4}(m(m-2)J + 2mK - m^2 I).$$

Since $J - I - A - A^t = K - I$, it follows that (4) is satisfied with $\alpha = \beta = \frac{m(m-2)}{4}$ and $\gamma = \frac{m^2}{4}$. □

Kharaghani [29] proved that if there exists an Hadamard matrix of order m then there exists a Bush-type Hadamard matrix of order m^2.

Ionin and Kharaghani [17] modified this construction and proved that if there exists an Hadamard matrix of order m then there exists a Bush-type Hadamard matrix of order m^2, which has the skew property required in Theorem 8.

Thus in many cases with $m = r$ a multiple of 4, an association scheme can be constructed.

The case with $m = r$ congruent to 2 modulo 4 seems to be more difficult and no general constructions are known. But in the special case $m = r = 6$ we may apply the orderly generation algorithm described before Theorem 5.

The number of ways to fill the first s rows is 1, 1, 4, 12, 8, 6, 29077, 76216458, for $s = 1, 2, \ldots, 8$. (Note that the first six rows correspond to a connected component of the undirected graph R_3.) We estimate that a complete search through all 76 million ways to fill the first 8 rows would take several years. But we guessed (especially because there are no such schemes with a rank 4 group) that if a scheme exists then there are many schemes and so a partial search may lead to a least one scheme.

Probably starting a complete search and let the computer run until a scheme is discovered is not an optimal strategy. Instead, we chose 2405 cases randomly among all ways to fill 8 rows. This search gave 47 ways to fill 13 rows but no ways to fill 14 rows. The idea is now to do a complete search in the "neighbourhood" of the most successful 8-row matrices, where the neighborhood of an 8-row matrix is the set of all 8-row matrices with which it has the first 7 rows in common. This search lead to two association schemes. A repetition (with another set of randomly chosen 8-row matrices) gave two other schemes.

Thus we have:

Theorem 9. *There exist at least four imprimitive non-symmetric 3-class association schemes of type 2 with $m = r = 6$.*
Each of these four schemes has a trivial automorphism group.

Table 2. Matrix of a 3-scheme with $m = r = 6$

0 0 0 0 0 0	1 1 1 0 0 0	1 1 1 0 0 0	1 1 1 0 0 0	1 1 1 0 0 0	1 1 1 0 0 0
0 0 0 0 0 0	1 1 0 1 0 0	1 0 0 1 1 0	1 0 0 1 1 0	1 1 0 1 0 0	0 0 0 1 1 1
0 0 0 0 0 0	1 0 0 0 1 1	1 0 0 1 0 1	0 1 0 1 0 1	0 0 1 1 1 0	1 1 0 1 0 0
0 0 0 0 0 0	0 1 1 0 1 0	0 1 0 0 1 1	1 0 0 0 1 1	0 0 0 1 1 1	1 0 1 0 1 0
0 0 0 0 0 0	0 0 0 1 1 1	0 1 1 0 1 0	0 0 1 0 1 1	1 0 1 0 0 1	0 1 0 1 0 1
0 0 0 0 0 0	0 0 1 1 0 1	0 0 1 1 0 1	0 1 1 1 0 0	0 1 0 0 1 1	0 0 1 0 1 1
0 0 0 1 1 1	0 0 0 0 0 0	1 1 0 1 0 0	0 1 0 0 1 1	1 0 1 0 1 0	0 0 1 0 1 1
0 0 1 0 1 1	0 0 0 0 0 0	1 0 0 0 1 1	0 1 1 0 1 0	0 1 0 1 0 1	0 1 1 1 0 0
0 1 1 0 1 0	0 0 0 0 0 0	0 1 1 0 0 1	1 0 0 1 0 1	0 1 1 1 0 0	0 1 0 0 1 1
1 0 1 1 0 0	0 0 0 0 0 0	0 0 1 1 0 1	1 0 0 1 1 0	1 0 1 0 0 1	1 0 1 1 0 0
1 1 0 0 0 1	0 0 0 0 0 0	0 1 1 0 1 0	0 1 1 1 0 0	1 0 0 1 1 0	1 0 0 1 1 0
1 1 0 1 0 0	0 0 0 0 0 0	1 0 0 1 1 0	1 0 1 0 0 1	0 1 0 0 1 1	1 1 0 0 0 1
0 0 0 1 1 1	0 0 1 1 1 0	0 0 0 0 0 0	1 1 0 1 0 0	1 0 0 1 0 1	1 1 0 0 0 1
0 1 1 0 0 1	0 1 0 1 0 1	0 0 0 0 0 0	0 0 0 1 1 1	1 1 0 0 1 0	1 1 1 0 0 0
0 1 1 1 0 0	1 1 0 0 0 1	0 0 0 0 0 0	0 1 1 0 1 0	0 0 1 1 0 1	1 0 0 0 1 1
1 0 0 1 1 0	0 1 1 0 1 0	0 0 0 0 0 0	0 0 1 1 1 0	0 1 1 0 1 0	0 1 0 1 1 0
1 0 1 0 0 1	1 0 1 1 0 0	0 0 0 0 0 0	1 0 1 0 0 1	0 0 1 1 1 0	0 0 1 1 0 1
1 1 0 0 1 0	1 0 0 0 1 1	0 0 0 0 0 0	1 1 0 0 0 1	1 1 0 0 0 1	0 0 1 1 1 0
0 0 1 0 1 1	1 1 0 0 1 0	0 1 1 1 0 0	0 0 0 0 0 0	0 1 0 0 1 1	1 0 0 1 0 1
0 1 0 1 1 0	0 0 1 1 0 1	0 1 0 1 1 0	0 0 0 0 0 0	0 1 1 1 0 0	1 0 1 1 0 0
0 1 1 1 0 0	1 0 1 1 0 0	1 1 0 0 0 1	0 0 0 0 0 0	1 0 0 0 1 1	0 1 0 1 1 0
1 0 0 1 1 0	1 1 0 0 0 1	0 0 1 0 1 1	0 0 0 0 0 0	1 0 0 1 1 0	0 1 1 0 0 1
1 0 1 0 0 1	0 0 1 0 1 1	1 0 0 0 1 1	0 0 0 0 0 0	1 1 1 0 0 0	1 0 0 0 1 1
1 1 0 0 0 1	0 1 0 1 1 0	1 0 1 1 0 0	0 0 0 0 0 0	0 0 1 1 0 1	0 1 1 0 1 0
0 0 1 1 0 1	0 1 1 0 0 1	0 0 1 1 1 0	1 1 0 0 0 1	0 0 0 0 0 0	0 1 0 1 1 0
0 0 1 1 1 0	1 0 0 1 1 0	1 0 1 0 1 0	0 0 1 1 0 1	0 0 0 0 0 0	1 0 1 0 1 0
0 1 0 1 0 1	0 1 0 0 1 1	1 1 0 0 0 1	1 0 1 1 0 0	0 0 0 0 0 0	0 0 1 1 0 1
1 0 0 0 1 1	1 0 0 1 0 1	0 1 0 1 0 1	1 0 1 0 1 0	0 0 0 0 0 0	1 1 0 0 1 0
1 1 0 0 1 0	0 1 1 1 0 0	1 0 1 0 0 1	0 1 0 0 1 1	0 0 0 0 0 0	1 0 0 1 0 1
1 1 1 0 0 0	1 0 1 0 1 0	0 1 0 1 1 0	0 1 0 1 1 0	0 0 0 0 0 0	0 1 1 0 0 1
0 1 0 0 1 1	1 1 1 0 0 0	0 0 0 1 1 1	0 0 1 1 0 1	1 0 1 0 0 1	0 0 0 0 0 0
0 1 0 1 0 1	1 0 0 1 1 0	0 0 1 0 1 1	1 1 0 0 1 0	0 1 1 0 1 0	0 0 0 0 0 0
0 1 1 0 1 0	0 0 1 0 1 1	1 0 1 1 0 0	1 0 1 0 1 0	1 0 0 1 1 0	0 0 0 0 0 0
1 0 0 1 0 1	1 0 1 0 0 1	1 1 1 0 0 0	0 0 0 1 1 1	0 1 0 1 0 1	0 0 0 0 0 0
1 0 1 0 1 0	0 1 0 1 0 1	1 1 0 0 1 0	1 1 0 1 0 0	0 0 1 0 1 1	0 0 0 0 0 0
1 0 1 1 0 0	0 1 0 1 1 0	0 1 0 1 0 1	0 1 1 0 0 1	1 1 0 1 0 0	0 0 0 0 0 0

The computation of automorphism groups can be done in GAP [10] using share package GRAPE [36] with nauty [31].

The adjacency matrix of R_1 is listed in Table 2 for one of these four schemes. Note that we have reordered rows and columns so that the imprimitive structure is clear. The matrix is not in maximal form in this ordering (even with the 3's and 2's that have been replaced by 0's).

A Bush-type Hadamard matrix of order 36 was first constructed by Janko [24]. But a "skew" Bush-type Hadamard matrix was not previously known. Bussemaker, Haemers and Spence [4] proved that a symmetric Bush-type Hadamard matrix of order 36 does not exist.

5 Concluding Remarks

We have seen in Sect. 3 that very few primitive non-symmetric 3-class association schemes are known. In fact (except for the first 9 cases) the problem of existence is still open for the majority of feasible parameter sets. We do not expect that the orderly generation algorithm described in Sect. 3 can be applied to the remaining open cases in the primitive case. However, the other technique using information about the strongly regular graph obtained by merging the non-symmetric relations may still be used in some particular cases. It would also be very useful to develop new computer aided search methods or even some computer free methods. It could also be interesting to get information about existence of association schemes with a given group of automorphisms.

In the imprimitive case the situation is quite different. Here we have many constructions, especially because of the connection to Hadamard matrices. The most interesting open problem in the imprimitive case is whether there exist association schemes of type 2 with $4 \leq r < m$, other than $(r, m) = (4, 16)$. The smallest feasible case is $r = 4,\ m = 10$ with order 40. We tried to attack this problem with the orderly generation algorithm, but it seems that the search space is too large. However, it may be that the algorithm can be improved so that this problem can be solved. But it seems that it is easier to solve the still open problem of complete enumeration of association schemes in the case $m = r = 6$.

Acknowledgments

The author wishes to thank Prof. Mikhail Klin for some very helpful discussions and suggestions on this paper and the research reported in it.

The author is pleased to acknowledge Prof. Bruno Buchberger and the coordinators of the Special Semester on Gröbner Bases (February 1 – July 31, 2006), organized by RICAM, Austrian Academy of Sciences, and RISC, Johannes Kepler University, Linz Austria for their attention to this project.

References

1. E. Bannai and T. Ito, *Algebraic Combinatorics. I*, Benjamin-Cumming, Menlo Park, 1984.

2. E. Bannai and S.-Y. Song, Character tables of fission schemes and fusion schemes, *Europ. J. Combin.*, **14** (1993), 385–396.
3. A. E. Brouwer, Strongly regular graphs, in C. J. Colbourn and J. H. Dinitz (eds.) *The CRC Handbook of Combinatorial Designs*, pp. 667–685, CRC Press, Boca Raton, 1996.
4. F. C. Bussemaker, W. Haemers, and E. Spence, The search for pseudo orthogonal Latin squares of order six, *Designs Codes Cryptogr.*, **21** (2000), 77–82.
5. K. Coolsaet, J. Degraer, and E. Spence, The strongly regular (45, 12, 3, 3) graphs, *Electron. J. Combin.* (1), **13** (2006), Research Paper 32, 9 pp.
6. E. R. van Dam, Three-class association schemes, *J. Algebraic Combin.*, **10** (1999), 69–107.
7. H. Enomoto and R. A. Mena, Distance-regular digraphs of girth 4, *J. Combin. Theory, Ser. B*, **43** (1987), 293–302.
8. I. A. Faradžev, Constructive enumeration of combinatorial objects, in *Problèmes combinatoires et théorie des graphes.* Colloq. Internat. CNRS, Vol. 260, pp. 131–135, CNRS, Paris, 1978.
9. I. A. Faradžev, M. H. Klin, and M. E. Muzichuk, Cellular rings and groups of automorphisms of graphs, in I. A. Faradžev, A. A. Ivanov, M. H. Klin, and A. J. Woldar (eds.) *Investigations in Algebraic Theory of Combinatorial Objects*, pp. 1–152, Kluwer Academic, Dordrecht, 1994.
10. The GAP Group, *GAP – Groups, Algorithms, and Programming, Version 4.4.9*; 2006, http://www.gap-system.org.
11. R. W. Goldbach and H. L. Claasen, A primitive non-symmetric 3-class association scheme on 36 elements with $p_{11}^1 = 0$ exists and is unique, *Europ. J. Combin.*, **15** (1994), 519–524.
12. R. W. Goldbach and H. L. Claasen, On splitting the Clebsch graph, *Indag. Math. (N.S.)*, **5** (1994), 285–290.
13. R. W. Goldbach and H. L. Claasen, Feasibility conditions for non-symmetric 3-class association schemes, *Discrete Math.*, **159** (1996), 111–118.
14. R. W. Goldbach and H. L. Claasen, The structure of imprimitive non-symmetric 3-class association schemes, *Europ. J. Combin.*, **17** (1996), 23–37.
15. D. G. Higman, Coherent configurations, *Geom. Dedicata*, **4** (1975), 1–32.
16. A. J. Hoffman and R. R. Singleton, On Moore graphs with diameters 2 and 3, *IBM J. Res. Develop.*, **4** (1960), 497–504.
17. Y. J. Ionin and H. Kharaghani, Doubly regular digraphs and symmetric designs, *J. Combin. Theory, Ser. A*, **101** (2003), 35–48.
18. N. Ito, Doubly regular asymmetric digraphs, *Discrete Math.*, **72** (1988), 181–185.
19. N. Ito, Automorphism groups of DRADs, in *Group Theory*, Singapore, 1987, pp. 151–170, de Gruyter, Berlin, 1989.

20. N. Ito, Doubly regular asymmetric digraphs with rank 5 automorphism groups, in *Groups–Korea 1988*. Lecture Notes in Math., Vol. 1398, pp. 94–99, Springer, Berlin, 1989.
21. N. Ito, On spectra of doubly regular asymmetric digraphs of RH-type, *Graphs Combin.*, **5** (1989), 229–234.
22. A. A. Ivanov, M. H. Klin, and I. A. Faradžev, *The Primitive Representations of the Non-abelian Simple Groups of Order less than* 10^6. *Part 2* (Russian), Preprint, Moscow, Institute for System Studies, 1984, 76 pp.
23. S. Iwasaki, A characterization of $\mathrm{PSU}(3,3^2)$ as a permutation group of rank 4, *Hokkaido Math. J.*, **2** (1973), 231–235.
24. Z. Janko, The existence of a Bush-type Hadamard matrix of order 36 and two new infinite classes of symmetric designs, *J. Combin. Theory, Ser. A*, **95** (2001), 360–364.
25. L. K. Jørgensen, Normally regular digraphs, preprint R-94-2023, Institute for Electronic Systems, Aalborg University, 1994, 44 pp.
26. L. K. Jørgensen and M. H. Klin, Switching of edges in strongly regular graphs. I. A family of partial difference sets on 100 vertices, *Electron. J. Combin.*, **10** (2003), Research Paper 17, 31 pp.
27. L. K. Jørgensen, G. A. Jones, M. H. Klin, and S. Y. Song, Normally Regular Digraphs, Association Schemes and Related Combinatorial Structures (in preparation).
28. L. K. Jørgensen, *Schur Rings and Non-symmetric Association Schemes on 64 Vertices*, Preprint R-2008-17, Department of Mathematical Sciences, Aalborg University, 2008, 16 pp.
29. H. Kharaghani, On the twin designs with the Ionin-type parameters, *Electron. J. Combin.*, **7** (2000), Research Paper 1, 11 pp.
30. R. A. Liebler and R. A. Mena, Certain Distance regular digraphs and related rings of characteristic 4, *J. Combin. Theory, Ser. B*, **47** (1988), 111–123.
31. B. D. McKay, *Nauty User's Guide (Version 1.5)*, Technical report TR-CS-90-02, Australian National University, Computer Science Department, 1990.
32. B. D. McKay and E. Spence, Classification of regular two-graphs on 36 and 38 vertices, *Australas. J. Combin.*, **24** (2001), 293–300.
33. A. Neumaier, New inequalities for the parameters of an association scheme, in *Combinatorics and Graph Theory*, Calcutta, 1980. Lecture Notes in Math., Vol. 885, pp. 365–367, Springer, Berlin, 1981.
34. R. C. Read, Every one a winner or How to avoid isomorphism search when cataloguing combinatorial configurations, *Ann. Discrete Math.*, **2** (1978), 107–120.
35. K. B. Reid and E. Brown, Doubly regular tournaments are equivalent to skew Hadamard matrices, *J. Combin. Theory, Ser. A*, **12** (1972), 332–338.
36. L. H. Soicher, GRAPE: a system for computing with graphs and groups, in L. Finkelstein and W. M. Kantor (eds.) *Groups and Computation*, DIMACS

Series in Discrete Mathematics and Theoretical Computer Science, Vol. 11, pp. 287–291, AMS, Providence, 1993.

37. S.-Y. Song, Class 3 association schemes whose symmetrization have two classes, *J. Combin. Theory, Ser. A*, **70** (1995), 1–29.
38. H. A. Wilbrink and A. E. Brouwer, A $(57, 14, 1)$ strongly regular graph does not exist, *Indag. Math.*, **45** (1983), 117–121.

Sets of Type (d_1, d_2) in Projective Hjelmslev Planes over Galois Rings

Axel Kohnert

Mathematisches Institut, University of Bayreuth, 95440 Bayreuth, Germany.
axel.kohnert@uni-bayreuth.de

Summary. In this paper we construct sets of type (d_1, d_2) in the projective Hjelmslev plane. For computational purposes we restrict ourself to planes over $\mathbb{Z}_{p^s}$ with p a prime and $s > 1$, but the method is described over general Galois rings. The existence of sets of type (d_1, d_2) is equivalent to the existence of a solution of a Diophantine system of linear equations. To construct these sets we prescribe automorphisms, which allows to reduce the Diophantine system to a feasible size. At least two of the newly constructed sets are 'good' u-arcs. The size of one of them is close to the known upper bound.

Key words: Projective Hjelmslev plane, Two-weight codes, Arcs

1 Introduction and Motivation

The projective Hjelmslev plane over a Galois ring is a generalization of the projective plane over a finite field $GF(q)$ with field size q a power of a prime p. Similar to the finite field case the Galois ring $GR(p^s, p^{sm})$ is defined for positive integers s, m as the ring $\mathbb{Z}_{p^s}[x]/(h)$ where h is a monic polynomial of degree m over $\mathbb{Z}_{p^s}$ which is irreducible over $\mathbb{Z}_p$. For different choices of the polynomial h, the resulting Galois rings are isomorphic.

Two limiting special cases of Galois rings are the finite fields $GF(q)$ which are isomorphic to $GR(p, q)$ and the modular integers $\mathbb{Z}_{p^s}$ which are isomorphic to $GR(p^s, p^s)$. Basic facts about Galois rings can be found in [21]. For computational purposes we will restrict us to $\mathbb{Z}_{p^s}$ in this paper.

To construct the projective Hjelmslev plane we define the points as the free rank 1 submodules of $GR(p^s, p^{sm})^3$. The lines are the free submodules of rank 2. The incidence is given by set inclusion. In general this construction works for the larger class of chain rings R, the corresponding projective Hjelmslev plane is denoted by $PHG(2, R)$. In this paper the ring R is always a Galois ring. Much more on projective Hjelmslev planes can be found in the work of Honold, Landjev and their coworkers [12, 13, 17, 16]. A useful tool

DOI 10.1007/978-3-642-01960-9_9, © Springer-Verlag Berlin Heidelberg 2009

is the homomorphism $\phi : \mathbb{Z}_{p^{s+1}} \rightarrow \mathbb{Z}_{p^s}$ which maps an representing element from $\mathbb{Z}_{p^{s+1}}$ to its remainder modulo p^s. ϕ can be extended to a mapping $\hat{\phi} : PHG(2, GR(p^{s+1}, p^{(s+1)m})) \rightarrow PHG(2, GR(p^s, p^{sm}))$. This function maps points to points and lines to lines. It allows to define a neighborhood of a point (or a line) in $PHG(2, GR(p^{s+1}, p^{(s+1)m}))$ as the set of points (or lines) having the same image under $\hat{\phi}$.

For two nonnegative integers d_1 and d_2 a set C of type (d_1, d_2) (also called two-intersection set) in a projective Hjelmslev plane is a set of points such that every line of the plane contains either d_1 or d_2 points of C. We always assume $d_1 < d_2$. In the case of a finite field the problem of sets of type (d_1, d_2) has been studied in a large number of papers (e.g. [9–11, 18–20]). They also study the more general case of point sets in $PG(k, q)$ with two intersection numbers with respect to the hyperplanes.

The interest in such point sets in the projective plane comes from the fact that they include hyperovals, some maximal arcs, unitals and Baer subplanes [11]. In the general case of the projective space $PG(k, q)$ with $k > 2$ the sets of type (d_1, d_2) have been also studied in the equivalent language of linear codes. Then these point sets are two-weight codes. For a survey see [8]. In coding theory one is interested in a high minimum distance for a fixed length n of the code, this corresponds to a point set with n points and intersection numbers as low as possible. There are cases where the best (for coding theory) point sets are such of type (d_1, d_2). More on the connection between linear codes and projective geometry can be found in [1, 3].

Also in the case of a projective Hjelmslev plane over a Galois ring there are links to coding theory. There are famous codes like the Nordstrom-Robinson-Code which are 'better' than the linear codes which are connected to $PG(k, q)$. These better codes are $\mathbb{Z}_4$-linear codes. To describe these $\mathbb{Z}_4$-linear codes using projective geometry we need the projective Hjelmslev geometry. Now the hope is to find more good codes studying $PHG(k, GR(p^s, p^{sm}))$ in general.

2 Parameters

There are several relations connecting the two parameters of the set C of type (d_1, d_2) to the number of lines and points in $PHG(2, GR(p^s, p^{sm}))$. These will allow to restrict the search to the cases of feasible pairs of parameters. We denote by t_1 and t_2 the number of lines intersecting with d_1 points respectively d_2 points from the set C. For a projective Hjelmslev plane over $GR(p^s, p^{sm})$ with point set P and line set L we know with $q := p^m$:

Lemma 1. ([16] Fact 1.)

1. $|L| = |P| = (q^2 + q + 1)q^{2(s-1)}$
2. Each line (point) is incident with $(q + 1)q^{s-1}$ *points (lines).*

The first equations show that the numbers of lines and points in a Hjelmslev plane over a Galois ring is a multiple of the number of points and lines in $PG(2,q)$. It is possible to get the Hjelmslev plane by substituting one point of $PG(2,q)$ by $q^{2(s-1)}$ points building a neighborhood in $PHG(2, GR(p^s, p^{sm}))$. Using the lemma above we can derive the following relations with $c = |C|$:

1. $t_1 + t_2 = (q^2 + q + 1)q^{2(s-1)}$
2. $d_1t_1 + d_2t_2 = c(q + 1)q^{s-1}$

These two equations give restrictions on possible values of d_1 and d_2 as t_1 and t_2 have to be integral numbers. In the case $s = 1$ (i.e. finite field) there is a third relation, which we get by counting the number of pairs of different points in C in two ways:

3. $d_1(d_1 - 1)t_1 + d_2(d_2 - 1)t_2 = c(c - 1)$

The right hand is the number of different pairs in C. The left hand side we get when we look at the unique line corresponding to the pair of points. In t_1 cases this is a line having intersection number d_1. Counting the possible pairs in C corresponding to this line we get the first summand. This last equation cannot easily be generalized to an s greater than 1 as the number of lines through a pair of different points from C depends then on the neighbor relation between the two points. There may be more than one line through two points, which changes relation 3 into an inequality.

In general it is possible to construct new sets of type (d_1, d_2) over $PHG(2, GR(p^{s+1}, p^{(s+1)m}))$ using two-intersection sets in $PHG(2, GR(p^s, p^{sm}))$. A useful starting point for these recursive constructions are the single points and complete lines in $PG(2,p)$ which is isomorphic to $PHG(2, GR(p,p))$ or a single point in an arbitrary projective Hjelmslev plane.

Lemma 2 (Recursive construction). *Let S be a set of type (d_1, d_2) over $GR(p^s, p^{sm})$, then there is a set of type (pd_1, pd_2) over $GR(p^{s+1}, p^{(s+1)m})$.*

Proof. The key is the function $\hat{\phi}$: $PHG(2, GR(p^{s+1}, p^{(s+1)m})) \rightarrow PHG(2, GR(p^s, p^{sm}))$. It maps two-intersection sets to two-intersection sets. Each element in S is replaced by the p^2 elements of the complete neighborhood (the preimages under $\hat{\phi}$) in $PHG(2, GR(p^{s+1}, p^{(s+1)m}))$. The t_1 lines intersecting in d_1 points are mapped to p^2t_1 lines intersecting in pd_1 points, and the t_2 lines intersecting in d_2 points are mapped to p^2t_2 lines intersecting in pd_2 points. □

Example 1. Take a line in the Fano plane $PG(2,2) = PHG(2, GR(2,2))$. This is a set of type $(1,3)$ with $t_1 = 6$ and $t_2 = 1$ and order 3. From this we construct a set of type $(2,6)$ in $PHG(2, GR(4,4))$ with $t_1 = 24$ and $t_2 = 4$ and order 12.

3 Constrution of Sets of Type (d_1, d_2)

The set P of points and the set L of lines of a projective Hjelmslev plane $PHG(2, GR(p^s, p^{sm}))$ define an incidence system. Denote by M the corresponding incidence matrix. The rows are labeled by the lines, the columns are labeled by the points, then we have for a point p and a line l:

$$M_{l,p} := \begin{cases} 1 & \text{if } p \text{ is incident with } l, \\ 0 & \text{otherwise.} \end{cases}$$

It is then possible to state the existence of a (d_1, d_2) using a Diophantine system of equations:

Theorem 1. *There is a set of type (d_1, d_2) in $PHG(2, GR(p^s, p^{sm}))$ if and only if there is a 0/1-solution $x = (x_1, \ldots, x_{|P|})^T$ of the following system of equations*

$$Mx = \begin{pmatrix} d_1 \text{ or } d_2 \\ \vdots \\ d_1 \text{ or } d_2 \end{pmatrix}.$$

This set of equation has to be read as follows: A solution x has the property that the product of a single with x is d_1 or d_2. As we want to solve this Diophantine system using some standard method we restate it as a linear equation as follows. Denote by D the matrix of the same size as M with $(d_1 - d_2)$ on the diagonal and 0 elsewhere. We denote by $(M|D)$ the block matrix built from the incidence matrix M and the matrix D :

$$(M|D) := \begin{pmatrix} m_{1,1} & m_{1,2} & & m_{1,|P|} & d_1 - d_2 & 0 & \ldots & 0 & 0 \\ m_{2,1} & m_{2,2} & & m_{2,|P|} & 0 & d_1 - d_2 & \ldots & 0 & 0 \\ \vdots & \vdots & \ddots & \vdots & \vdots & \vdots & \ddots & \vdots & \vdots \\ & & \ddots & & 0 & 0 & \ldots & d_1 - d_2 & 0 \\ m_{|L|,1} & m_{|L|,2} & & m_{|L|,|P|} & 0 & 0 & \ldots & 0 & d_1 - d_2 \end{pmatrix}.$$

Corollary 1. *There is a set of type (d_1, d_2) in $PHG(2, GR(p^s, p^{sm}))$ if and only if there is a 0/1-solution $x|y = (x_1, \ldots, x_{|P|}, y_1, \ldots, y_{|L|})^T$ of the following system of equalities*

$$(M|D)(x|y) = \begin{pmatrix} d_1 \\ \vdots \\ d_1 \end{pmatrix}.$$

Given the solution it is possible to read off if a line l intersects with d_1 points. This is the case if $y_l = 0$, or with d_2 points, in this case $y_l = 1$.

The size of this problem is given by the number of points. In general this number is growing too fast. To handle also larger cases we apply the following method. We no longer look for an arbitrary set of type (d_1, d_2). We are now only interested in a set which has a prescribed group $G < PGL(2, GR(p^s, p^{sm}))$ of automorphisms. An automorphism φ of a point set $C = \{c_1, \ldots, c_n\}$ is an element from $PGL(2, GR(p^s, p^{sm})$ such that $C = \{\varphi(c_1), \ldots, \varphi(c_n)\}$.

The main advantage of this method is that the size of the system of equations is much smaller, it will only have the size equal to the number of orbits of G on the points of $PHG(2, GR(p^s, p^{sm}))$. We can summarize this construction as a two-step process:

- In a first step the solution of a construction problem is described as a solution of a Diophantine system of linear equations.
- In a second step the size of the linear system is reduced by prescribing automorphisms.

This construction method is a general approach that works for many discrete structures as designs [2, 15], q-analogs of designs [6], arcs in projective geometries [7] or linear codes [1, 4, 5]. The general method is as follows: The matrix M is reduced by adding up columns (labeled by the points of $PHG(2, GR(p^s, p^{sm}))$ corresponding to the orbits of G. Now because of the relation

$$p \in l \quad \Longleftrightarrow \quad \varphi(p) \in \varphi(l) \tag{1}$$

for any point p and line l and any automorphism $\varphi \in G$ the rows corresponding to lines from a orbit of G are equal, therefore these are removed from the system of equations and we get a square matrix denoted by M^G. The number of orbits on the points and the number of orbits on the lines is equal, as we can label the lines by the orthogonal point and then act with the transposed matrix. We denote by $\omega_1, \ldots, \omega_s$ the orbits on the points and by $\Omega_1, \ldots, \Omega_s$ the orbits on the lines. For an entry of M^G we have:

$$M^G_{\Omega_i, \omega_j} = |\{p \in \omega_j : p \in l\}|$$

where l is a representative of the line orbit Ω_i. Because of property (1) this definition is independent of the representative. Now we can restate the above theorem in a version with the condensed matrix M^G :

Theorem 2. *Let G be a subgroup of $PGL(2, GR(p^s, p^{sm}))$. There is a set of type (d_1, d_2) in $PHG(2, GR(p^s, p^{sm}))$ whose group of automorphisms contains G as a subgroup if, and only if, there is a 0/1-solution $x = (x_1, \ldots, x_{|s|})^T$ of the following system of equations:*

$$M^G x = \begin{pmatrix} d_1 \ or \ d_2 \\ \\ \vdots \\ \\ d_1 \ or \ d_2 \end{pmatrix}.$$

To solve this using a computer we transform it like in the above corollary into a Diophantine system of linear equations and using the slack variables we get the information which lines intersect in d_1 points and which one in d_2 points.

4 Example

We describe the construction of the set of type $(2,5)$ over $\mathbb{Z}_9$ with 39 points, which is a very good 5-arc as explained in the following section with results. $PHG(2,\mathbb{Z}_9)$ has 117 points, therefore the Diophantine system of equations which is to be solved would have 234 variables and 117 equations. We prescribe a group G of automorphisms generated by a single element:

$$G := \left\langle \begin{pmatrix} 7 & 1 & 0 \\ 4 & 8 & 4 \\ 5 & 3 & 8 \end{pmatrix} \right\rangle .$$

This group has 9 orbits, each of size 13. In fact this group is a lifted version (i.e. a preimage under ϕ) of the Singer cycle in $PGL(2,3)$. $PHG(2,\mathbb{Z}_9)$ can be constructed from $PG(2,3)$ by substituting each point in $PG(2,3)$ by 9 'lifted' points of $PHG(2,\mathbb{Z}_9)$. Each orbit now contains for each point of $PG(2,3)$ one lifted point. The condensed matrix M^G is a 9×9 matrix:

$$M^G = \begin{pmatrix} 0 & 3 & 2 & 2 & 1 & 1 & 0 & 1 & 2 \\ 0 & 0 & 2 & 2 & 1 & 1 & 3 & 1 & 2 \\ 1 & 1 & 0 & 3 & 2 & 2 & 1 & 2 & 0 \\ 2 & 2 & 1 & 1 & 0 & 0 & 2 & 3 & 1 \\ 1 & 1 & 0 & 0 & 2 & 2 & 1 & 2 & 3 \\ 3 & 0 & 2 & 2 & 1 & 1 & 0 & 1 & 2 \\ 1 & 1 & 3 & 0 & 2 & 2 & 1 & 2 & 0 \\ 2 & 2 & 1 & 1 & 3 & 0 & 2 & 0 & 1 \\ 2 & 2 & 1 & 1 & 0 & 3 & 2 & 0 & 1 \end{pmatrix} .$$

The solution $x = (1,0,0,0,1,1,0,0,0)$ of the equation from Theorem 2 corresponds to the set of type $(2,5)$ with 39 points built from three orbits. From

$$M^G x^T = \begin{pmatrix} 2 \\ 2 \\ 5 \\ 2 \\ 5 \\ 5 \\ 5 \\ 5 \\ 5 \end{pmatrix}$$

we read off which line orbits have intersection size 2 and which one size 5.

5 Results

In this section we give results for projective Hjelmslev planes over the Galois rings isomorphic to $\mathbb{Z}_4, \mathbb{Z}_8, \mathbb{Z}_9, \mathbb{Z}_{16}, \mathbb{Z}_{25}, \mathbb{Z}_{27}$. As the complement of a set of type (d_1, d_2) is again a set with only two intersection numbers, we list only those sets C where $|C|$ is at most half of the points. In Table 1 we list the parameters (d_1, d_2) of two-intersection sets we constructed with the method described. By t_1 and t_2 we denote the number of lines having intersection numbers d_1 and d_2. We denote by * in the second column if this set can not be constructed using the recursive method from 2. We do not list the trivial set consisting of one point. This list is not complete, as we only construct a two-intersection set C if we first choose a group G such that there is a C with this group of automorphism, and secondly the resulting Diophantine system is small enough to be solved. So it may happen that further parameters (d_1, d_2) are possible and for pairs (d_1, d_2) already in the list there may be other sets, with different groups of automorphisms.

These results are also interesting if you look for arcs. There are at least two cases where we found improvements against previously known values for the maximal size of u-arcs. More on arcs in projective Hjelmslev planes can be found in [12]. The construction of u-arcs over Galois rings will also be covered in a forthcoming paper with M. Kiermaier. Some first results can be found in the proceedings of the 2007 conference on optimal codes [14].

The most interesting set is the 39-set of type $(2, 5)$ in $\mathbb{Z}_9$. This is a 5-arc just one point below the upper-bound of 40 points. It improves the previously known 5-arc with 31 points. The other improvement is the 310-set of type $(9, 14)$ in $\mathbb{Z}_{25}$. The paper by Landjev and Honold only cover the cases with $s = 2$. We didn't find tables for $\mathbb{Z}_8$ and $\mathbb{Z}_{16}$.

Table 1. Sets of type (d_1, d_2)

R	$\|C\|$	d_1	d_2	t_1	t_2
$\mathbb{Z}_4$	4	0	2	16	12
	6*	0	2	10	18
	7*	0	2	7	21
	12	2	6	24	4
	14*	2	4	14	14
$\mathbb{Z}_8$	4	0	2	88	24
	6*	0	2	76	36
	8*	0	2	64	48
	16	0	4	64	48
	24	0	4	40	72
	28*	2	6	84	28
	28	0	4	28	84
	32*	2	6	72	40
	36*	3	7	88	24
	36*	2	6	60	52
	44*	2	6	36	76
	48*	2	6	24	88
	48*	4	8	80	32
	48	4	12	96	16
	52*	3	7	40	72
	52*	4	8	68	44
	56	4	8	56	56
$\mathbb{Z}_9$	9	0	3	81	36
	30*	2	5	75	42
	36	3	12	108	9
	39*	3	6	78	39
	39*	2	5	39	78
	42*	3	6	66	51

R	$\|C\|$	d_1	d_2	t_1	t_2
$\mathbb{Z}_{16}$	4	0	2	400	48
	6*	0	2	376	72
	8*	0	2	352	96
	12*	0	2	304	144
	16	0	4	352	96
	24	0	4	304	144
	28*	0	4	280	168
	32	0	4	256	192
	40*	0	4	208	240
	64	0	8	256	192
	96	0	8	160	288
	112	4	12	336	112
	112	0	8	112	336
	128	4	12	288	160
	144	4	12	240	208
	144	6	14	352	96
	176	4	12	144	304
	192	8	24	384	64
	192	8	16	320	128
	192	4	12	96	352
	208	6	14	160	288
	208	8	16	272	176
	224	10	14	224	224
$\mathbb{Z}_{25}$	25	0	5	625	150
	155*	5	10	620	155
	310*	9	14	310	465
	310*	10	15	465	310

Acknowledgements

The author is pleased to acknowledge Prof. Bruno Buchberger and the coordinators of the Special Semester on Gröbner Bases (February 1 – July 31, 2006), organized by RICAM, Austrian Academy of Sciences, and RISC, Johannes Kepler University, Linz Austria for their interest to this project.

References

1. A. Betten, M. Braun, H. Fripertinger, A. Kerber, A. Kohnert, and A. Wassermann, *Error-Correcting Linear Codes. Classification by Isometry and Applications. With CD-ROM*, Algorithms and Computation in Mathematics, Vol. 18, Springer, Berlin, 2006, xxix, 798 pp.

2. A. Betten, A. Kerber, A. Kohnert, R. Laue, and A. Wassermann, The discovery of simple 7-designs with automorphism group PΓL(2, 32), in G. Cohen et al. (eds.) *Applied Algebra, Algebraic Algorithms and Error-Correcting Codes, 11th International Symposium, AAECC-11, Paris, France, July 17–22*, Lect. Notes Comput. Sci., Vol. 948, pp. 131–145, Springer, Berlin, 1995.
3. J. Bierbrauer, *Introduction to Coding Theory*, Discrete Mathematics and Its Applications, CRC Press, Boca Raton, 2005, xxiii, 390 pp.
4. M. Braun, Construction of linear codes with large minimum distance, *IEEE Trans. Inform. Theory* (8), **50** (2004), 1687–1691.
5. M. Braun, A. Kohnert, and A. Wassermann, Optimal linear codes from matrix groups, *IEEE Trans. Inform. Theory*, **12** (2005), 4247–4251.
6. M. Braun, Some new designs over finite fields, *Bayreuther Math. Schr.*, **74** (2005), 58–68.
7. M. Braun, A. Kohnert, and A. Wassermann, Construction of (n, r)-arcs in $PG(2, q)$, *Innov. Incidence Geom.*, **1** (2005), 133–141.
8. R. Calderbank and W. M. Kantor, The geometry of two-weight codes, *Bull. Lond. Math. Soc.*, **18** (1986), 97–122.
9. M. de Finis, On k-sets of type (m, n) in projective planes of square order. Finite geometries and designs, *Lond. Math. Soc. Lect. Note Ser.*, **49** (1981), 98–103. Proc. 2nd Isle of Thorns Conf. 1980.
10. M. de Finis, On k-sets in PG$(3, q)$ of type (m, n) with respect to planes, *Ars Comb.*, **21** (1986), 119–136.
11. J. W. P. Hirschfeld, *Projective Geometries over Finite Fields*, 2nd ed., Oxford Mathematical Monographs, Clarendon, Oxford, 1998, xiv, 555 pp.
12. T. Honold and I. Landjev, On arcs in projective Hjelmslev planes, *Discrete Math.* (1–3), **231** (2001), 265–278.
13. T. Honold and I. Landjev, On maximal arcs in projective Hjelmslev planes over chain rings of even characteristic, *Finite Fields Appl.* (2), **11** (2005), 292–304.
14. M. Kiermaier and A. Kohnert, New arcs in projective Hjelmslev planes over Galois rings, in *Fifth International Workshop on Optimal Codes and Related Topics*, Balchik, pp. 112–119, 2007.
15. E. S. Kramer and D. M. Mesner, t-Designs on hypergraphs, *Discrete Math.*, **15** (1976), 263–296.
16. I. Landjev, On blocking sets in projective Hjelmslev planes, *Adv. Math. Commun.* (1), **1** (2007), 65–81.
17. I. Landjev and T. Honold, Arcs in projective Hjelmslev planes, *Discrete Math. Appl.* (1), **11** (2001), 53–70.
18. T. Penttila and G. F. Royle, Sets of type (m, n) in the affine and projective planes of order nine, *Designs Codes Cryptogr.* (3), **6** (1995), 229–245.

19. G. Tallini, Some new results on sets of type (m, n) in projective planes, *J. Geom.*, **29** (1987), 191–199.
20. M. Tallini Scafati, The k-sets of type (m, n) in a Galois space $S_{r,q}(r \geq 2)$, in *Colloq. Int. Teorie Comb.*, Roma 1973, Tomo II, pp. 459–463, 1976.
21. Z.-X. Wan, *Lectures on Finite Fields and Galois Rings*, World Scientific, River Edge, 2003, x, 342 pp.

A Construction of Designs from $PSL(2, q)$ and $PGL(2, q)$, $q = 1$ mod 6, on $q + 2$ Points

Izumi Miyamoto

Department of Computer Science and Media Engineering,
University of Yamanashi, Kofu 400-8511, Japan. imiyamoto@yamanashi.ac.jp

Summary. Let $G = PSL(2, q)$ or $PGL(2, q)$. We consider the action of G on the projective line together with one additional point, which is fixed by G. Assume $q \equiv 1$ mod 6 and set $\lambda_q = \frac{1}{24}(q-1)(q-3)(q-5)$. We construct 3-$(q+2, \frac{1}{2}(q-1), \lambda_q)$ designs admitting $PSL(2, q)$ as their automorphisms, if $q \equiv 3$ mod 4. We also construct 3-$(q + 2, \frac{1}{2}(q - 1), 2\lambda_q)$ designs admitting $PGL(2, q)$ as their automorphisms. These designs may not be simple.

Key words: Block design, Superscheme, Permutation group, Homogeneous group

1 Introduction

For $q = 3$ mod 4, $PSL(2, q)$ is 3-homogeneous on the $q + 1$ points of the projective line. So a union of certain orbits of $PSL(2, q)$ acting on the k-subsets of the projective line forms a 3-design. Most of such designs are classified in [1]. In the present paper we consider the action of $PSL(2, q)$ on the projective line together with one additional point, which is fixed by $PSL(2, q)$. We will construct 3-designs from unions of certain orbits of $PSL(2, q)$ acting on $\frac{1}{2}(q - 1)$ points, when $q = 1$ mod 3. The author considered a combinatorial approach to doubly transitive permutation groups as transitive extensions using some superschemes [8–10]. Superschemes are introduced in [6, 11]. Although $PSL(2, q)$ does not extend to a triply transitive group on $q + 2$ points, we obtained some superschemes on $q + 2$ points which have properties similar to those of triply transitive groups. Such a superscheme gives a certain number of subsets X_j of distinct 4-tuples of $q + 2$ points. The subsets form a partition of the distinct 4-tuples. In a superscheme we consider projections $\pi_1(x, y, z, w) = (y, z, w)$, similarly, $\pi_2, \ldots, \pi_4$. In the above superschemes the projection π_i, $i = 1, 2, 3, 4$, maps every subset X_j to the entirety of distinct triples, and $|\pi_i^{-1}(x, y, z)|$ is constant for any distinct triple (x, y, z). So we can consider an orbit-like set $\{w | (x, y, z, w) \in X_j\}$ for any fixed three distinct

DOI 10.1007/978-3-642-01960-9_10,

points x, y and z. The present research is motivated by the expectation that an orbit-like set may form a 3-design, since $PSL(2,q)$ is 3-homogeneous if $q = 3 \bmod 4$. As a result we can construct 3-designs even if there do not exist superschemes like those arising from triply transitive groups.

The subgroups of $PSL(2,q)$ are well known. Readers may refer to [3, 5]. The subgroups of $G = PSL(2,q)$ or $PGL(2,q)$ may also be found in [1, 2], together with results of 3-designs constructed from certain orbits of G acting on k-subsets. Data of the fixed points of subgroups are also listed in [1, 2]. Some data for t-designs, $t \geq 3$, appear in [7].

2 Definitions and Notations

Let t, v, k and λ be positive integers satisfying that $t \leq k \leq v$ and $\lambda > 0$. Let X be a set of points, and denote by $\binom{X}{k}$ the set of all k-subsets of X. Let B be a specified collection from $\binom{X}{k}$. A k-subset in B will be called a block. We allow the possibility that B be a multi-set, i.e., B may contain repeated blocks. Then a pair (X, B) is called a t-(v,k,λ) design if every t-subset of X is contained in exactly λ blocks. If B contains no repeated blocks, then the design is called simple. Note that we may describe a design simply by indicating its set B of blocks.

Let q be a prime power, and let $\mathbf{P}$ be the union of the Galois field $GF(q)$ and $\{\infty\}$. For $a, b, c, d \in GF(q)$, a mapping g on $\mathbf{P}$ is defined by

$$g : x \mapsto \frac{ax+b}{cx+d},$$

if $ad - bc \neq 0$, $x \neq -d/c$, and $g(\infty) = a/c$, $g(-d/c) = \infty$, if $c \neq 0$, and $g(\infty) = \infty$, if $c = 0$. Then the set of all such mappings becomes a permutation group, denoted $PGL(2,q)$, and we refer to the set $\mathbf{P}$ as the projective line. We use the GAP system [4] to compute permutation groups in our experiments. In the GAP library, a permutation group always acts on the point set $\{1, 2, \ldots, n\}$. Thus we shall denote the projective line $\mathbf{P} = \{1, 2, \ldots, q+1\}$ below. Let $PSL(2,q)$ be the subgroup of $PGL(2,q)$ consisting of those mappings which satisfy $ad - bc = 1$.

Let G be a permutation group on a set X. For points $x, y, \ldots, z$ of X the pointwise stabilizer of $x, y, \ldots, z$ in G is the subgroup of G consisting of all the elements g such that $x^g = x, y^g = y, \ldots, z^g = z$. For a subset Y of X the setwise stabilizer of Y in G is the subgroup consisting of all elements g such that $y^g \in Y$ for all $y \in Y$. We define the induced action of G on the set of s-subsets of X by $\{x_1, x_2, \ldots, x_s\}^g = \{x_1^g, x_2^g, \ldots, x_s^g\}$. We call G s-homogeneous if for any pair of s-subsets there exists an element g which moves one to the other. Then $PGL(2,q)$ is 3-homogeneous on $\mathbf{P}$, and $PSL(2,q)$ is 3-homogeneous on $\mathbf{P}$ if $q = 3 \bmod 4$.

3 Construction of Designs

Let $G = PSL(2,q)$ or $PGL(2,q)$ acting on the $q+1$ points of the projective line $\mathbf{P} = \{1,2,\ldots,q+1\}$. Then G is of order $\frac{1}{2}g_0(q+1)q(q-1)$, where $g_0 = 1$ or 2 according to whether $G = PSL(2,q)$ or $PGL(2,q)$. We consider an additional point, denoted $q+2$. If $G = PSL(2,q)$, we assume that $q = 3 \bmod 4$, in which case G is 3-homogeneous. Suppose furthermore that $q = 1 \bmod 3$. Let H be a subgroup of G of order 3. Since every nonidentity element of G has at most two fixed points, H has two fixed points. Suppose that H fixes the points 1 and 2. Let b_1 be the union of some $\frac{1}{6}(q-1)$ orbits of H of size 3, and let b_2 be the union of $\{1,2,q+2\}$ and some $\frac{1}{6}(q-7)$ orbits of H of size 3. Then both b_1 and b_2 consist of $\frac{1}{2}(q-1)$ points. Let G_1 be the setwise stabilizer in G of b_1, and let G_2 be that of b_2. Let $3g_1$ and $3g_2$ be the orders of G_1 and G_2, respectively. Let K be the stabilizer of the points 1 and 2 in G. Then K is cyclic of order $\frac{1}{2}g_0(q-1)$, and the subgroup of K of order $\frac{1}{2}(q-1)$ has two orbits of length $\frac{1}{2}(q-1)$. Let b_3 be one of these orbits, and let G_3 be the stabilizer in G of b_3. Then the order of G_3 is $\frac{1}{2}g_0(q-1)$. The blocks of our design are generated from the sets b_1, b_2 and b_3 under the action of G. Let B_i denote the set of blocks generated from b_i via the action of G on the $\frac{1}{2}(q-1)$-subsets of $\mathbf{P}\cup\{q+2\}$. Thus $|B_i| = \frac{1}{6}(g_0/g_i)(q+1)q(q-1)$ for $i = 1,2$ and $|B_3| = (\frac{1}{2}g_0(q+1)q(q-1))/(\frac{1}{2}g_0(q-1)) = (q+1)q$. For our design, the blocks generated from b_i are repeated g_i times. Such a set of blocks is denoted by g_iB_i. Set $\lambda_q = \frac{1}{24}(q-1)(q-3)(q-5)$. Then we have the following theorems.

Theorem 1. *Suppose that $G = PSL(2,q)$, where $q = 3 \bmod 4$ and $q = 1 \bmod 6$. Then the block set $g_1B_1 \cup g_2B_2 \cup B_3$ forms a 3-$(q+2, \frac{1}{2}(q-1), \lambda_q)$ design. If $g_1 = g_2 = 1$ and $q > 7$, then the design is simple.*

Theorem 2. *Suppose that $G = PGL(2,q)$, where $q = 1 \bmod 6$. Then the block set $g_1B_1 \cup g_2B_2 \cup 2B_3$ forms a 3-$(q+2, \frac{1}{2}(q-1), 2\lambda_q)$ design. If $g_1 = g_2 = 2$ and $q > 7$, then the block set $B_1 \cup B_2 \cup B_3$ gives a simple 3-design with $\lambda = \lambda_q$.*

Proofs. We have $|g_1B_1| = |g_2B_2| = \frac{1}{6}g_0(q+1)q(q-1)$ blocks from each of b_1 and b_2, and $|g_0B_3| = g_0(q+1)q$ blocks from b_3. Both b_1 and b_3 contain $\frac{1}{48}(q-1)(q-3)(q-5)$ distinct 3-subsets of $\{1,2,\ldots,q+1\}$, and b_2 contains $\frac{1}{48}(q-3)(q-5)(q-7)$ such subsets. Moreover, b_2 contains $\frac{1}{8}(q-3)(q-5)$ distinct 3-subsets of the form $\{x,y,q+2\}$, where $x,y \in \{1,2,\ldots,q+1\}$. Since G is 3-homogeneous on $\{1,2,\ldots,q+1\}$, a counting argument reveals that every distinct 3-subset of $\{1,2,\ldots,q+1\}$ is contained in the following number of blocks:

$$\frac{(|g_1B_1|+|g_0B_3|)\frac{1}{48}(q-1)(q-3)(q-5)+|g_2B_2|\frac{1}{48}(q-3)(q-5)(q-7)}{\frac{1}{6}(q+1)q(q-1)}$$
$$=\frac{g_0(q+1)q(q-3)(q-5)(\frac{1}{48}(\frac{1}{6}(q-1)+1)(q-1))+\frac{1}{6\cdot 48}(q-1)(q-7))}{\frac{1}{6}(q+1)q(q-1)}$$
$$=\frac{1}{48}(g_0(q-3)(q-5)\left((q-1)+6+q-7\right))$$
$$=\frac{1}{24}(g_0(q-1)(q-3)(q-5)).$$

Similarly, for $x, y \in \{1, 2, \ldots, q+1\}$ the number of blocks containing each distinct 3-subset $\{x, y, q+2\}$ is

$$\frac{\frac{1}{6\cdot 8}g_0(q+1)q(q-1)(q-3)(q-5)}{\frac{1}{2}(q+1)q}=\frac{1}{24}(g_0(q-1)(q-3)(q-5)).$$

So we have obtained a 3-design. If $g_0 = g_1 = g_2 = 1$, then we have a simple design. We also note that if $g_0 = g_1 = g_2 = 2$ then, without repetition, we also have a simple design from the above argument.

4 Experiments

We used GAP system [4] to compute our examples. Let b_1, b_2, b_3, G_1, G_2 and G_3 be as in the previous section. In Example 1 and 2 in Table 1, $G = PSL(2, 19)$ = `PrimitiveGroup(20,1)` in the GAP library. G is generated by

Table 1. Examples

Example 1. $G = PSL(2,19)$ = `PrimitiveGroup(20, 1)`	
$b_1 = \{4,6,7,10,12,13,16,18,19\}$	$\|G_1\| = 3$
$b_2 = \{1,2,5,8,11,14,17,20,21\}$	$\|G_2\| = 6$
$b_3 = \{3,5,7,9,11,13,15,17,19\}$	$\|G_3\| = 9$
Example 2. $G = PSL(2,19)$ = `PrimitiveGroup(20, 1)`	
$b_1 = \{4,5,6,10,11,12,16,17,18\}$	$\|G_1\| = 3$
$b_2 = \{1,2,5,6,11,12,17,18,21\}$	$\|G_2\| = 6$
$b_3 = \{3,5,7,9,11,13,15,17,19\}$	$\|G_3\| = 9$
Example 3. $G = PSL(2,31)$ = `PrimitiveGroup(32, 4)`	
$b_1 = \{3,4,5,6,7,8,9,11,14,17,21,26,27,29,30\}$	$\|G_1\| = 3$
$b_2 = \{1,2,3,4,5,6,9,10,11,14,21,24,26,32,33\}$	$\|G_2\| = 3$
$b_3 = \{3,6,9,10,11,13,14,20,21,23,24,25,28,31,32\}$	$\|G_3\| = 15$
Example 4. $G = PGL(2,25)$ = `PrimitiveGroup(26, 2)`	
$b_1 = \{3,4,7,8,9,13,14,16,17,21,22,23\}$	$\|G_1\| = 6$
$b_2 = \{1,2,3,4,6,7,9,11,20,21,23,27\}$	$\|G_2\| = 6$
$b_3 = \{3,4,5,7,9,10,15,18,21,23,24,26\}$	$\|G_3\| = 24$

$(3,19,17,15,13,11,9,7,5)(4,20,18,16,14,12,10,8,6)$ and $(1,2,12)(3,11,13)$ $(4,17,6)(5,14,8)(7,20,18)(10,19,16)$. We construct two 3-designs. The block b_3 is common in both of the two 3-designs. The blocks in B_2 are duplicated. Then we have two 3-$(21,9,168)$ designs. In this case, gathering all the blocks of these two designs (including repetition) gives the block set of a 4-design, namely a 4-$(21,9,112)$ design.

In Example 3, $G = PSL(2,31) =$ `PrimitiveGroup`(32,4). G is generated by $(1,2,\ldots,31)$ and $(1,32)(2,31)(3,16)(4,11)(5,24)(6,7)(8,23)(9,28)(10,25)$ $(12,15)(13,19)(14,20)(17,30)(18,21)(22,29)(26,27)$. In this example we get a simple 3-$(33,15,910)$ design.

In Example 4, $G = PGL(2,25) =$ `PrimitiveGroup`(26,2). G is generated by $(3,6,7,17,10,25,26,8,23,11,21,22,5,12,18,14,9,20,4,13,24,19,15,16)$ and (1, 19, 18, 23, 16, 17, 9, 25, 11, 3, 4, 22, 2)(5, 12, 21, 7, 10, 26, 15, 13, 24, 8, 20, 14, 6). In this example we get a simple 3-$(27,12,440)$ design.

Acknowledgments

The author is pleased to acknowledge Prof. Bruno Buchberger and the coordinators of the Special Semester on Gröbner Bases (February 1 – July 31, 2006), organized by RICAM, Austrian Academy of Sciences, and RISC, Johannes Kepler University, Linz Austria for partial support of this project.

References

1. P. J. Cameron, H. R. Maimani, G. R. Omidi, and B. Tayfeh-Rezaie, 3-Designs from $PSL(2,q)$, *Discrete Math.*, **306** (2006), 3063–3073.
2. P. J. Cameron, G. R. Omidi, and B. Tayfeh-Rezaie, 3-Designs from $PGL(2,q)$, *Electron. J. Combin.*, **13** (2006), #R50.
3. L. E. Dickson, *Linear Groups with an Exposition of the Galois Field Theory*, Dover, New York, 1958.
4. The GAP Groups, Gap – Groups, Algorithms and Programming, Version 4, Lehrstuhl D für Mathematik, Rheinisch Westfälische Technische Hochschule, Aachen, Germany and School of Mathematical and Computational Sciences, Univ. St. Andrews, Scotland, 2000.
5. B. Huppert, *Endliche gruppen I, Die Grundlehren der Mathematischen Wissenschaften, Band 134*, Springer, Berlin, 1967.
6. K. W. Johnson, and J. D. H. Smith, Characters of finite quasigroups IV: products and superschemes, *Europ. J. Combin.*, **10** (1989), 257–263.
7. G. B. Khosrovshahi and R. Laue, t-Designs with $t \geq 3$, in C. J. Colbourn, and J. H. Dinitz (eds.) *Handbook of Combinatorial Designs*, 2nd edition, CRC Press Series on Discrete Mathematics and Its Applications, pp. 79–101, CRC Press, Boca Raton, 2006.
8. I. Miyamoto, A generalization of association schemes and computation of doubly transitive groups, *RIMS Kokyuroku on Computer Algebra –*

Algorithms, Implementations and Applications, **1394** (2004), 185–189 (in Japanese).
9. I. Miyamoto, A computation of some multiply homogeneous superschemes from transitive permutation groups, in C. W. Brown (ed.) *Proceedings of the 2007 International Symposium on Symbolic and Algebraic Computation*, pp. 293–298, ACM, New York, 2007.
10. I. Miyamoto, A combinatorial approach to doubly transitive permutation groups, *Discrete Math.*, **308** (2008), 3073–3081.
11. J. D. H. Smith, Association schemes, superschemes, and relations invariant under permutation groups, *Europ. J. Combin.* (3), **15** (1994), 285–291.

Approaching Some Problems in Finite Geometry Through Algebraic Geometry

G. Eric Moorhouse

Department of Mathematics, University of Wyoming, Laramie, WY 82071, USA.
moorhous@uwyo.edu

Summary. In the study of finite geometries one often requires knowledge of the ranks of related $(0,1)$-incidence matrices. We describe some of the combinatorial questions in finite geometry for which formulas for these ranks are useful; and we describe methods from algebraic geometry that are useful in obtaining such rank formulas.

Key words: p-Rank, Polar space, Ovoid, Spread, Hilbert function

1 Motivation and Background

Here we recall the definitions of a few standard notions from finite geometry. Considering the audience for this presentation, many of whom are coding theorists, the coding-theoretic interpretations of our objects of study, and of our results, will occasionally be explicitly mentioned. For a more extensive summary description of the relevant definitions from finite geometry, see e.g. [9, 17, 32].

We denote by $P^n(\mathbb{F}_q)$ the classical projective n-space over the finite field $\mathbb{F}_q$ of order q, i.e. the incidence system formed by the subspaces of $\mathbb{F}_q^{n+1}$: the points, lines, planes, etc. being the subspaces of dimension 1, 2, 3, etc.; thus the vector $(k+1)$-subspaces of $\mathbb{F}_q^{n+1}$ are the projective k-subspaces. In particular $P^2(\mathbb{F}_q)$ denotes the classical (i.e. Desarguesian) projective plane of order q. Non-classical projective planes exist, but all projective spaces of dimension $n \geq 3$ are classical.

An *ovoid in projective 3-space* $P^3(\mathbb{F}_q)$ is a set of q^2+1 points with no three collinear. Alternatively, one may consider a linear $[n,4]$-code $\mathcal{C}$ over $\mathbb{F}_q$ such that the dual code $\mathcal{C}^\perp$ has minimum weight ≥ 4; in this case $n \leq q^2+1$, and equality holds iff a generator matrix for $\mathcal{C}$ has as its columns the points of an ovoid in $P^3(\mathbb{F}_q)$. For q odd, every ovoid is an elliptic quadric (Barlotti [2]; Panella [28]). For q even, the *known ovoids* are the elliptic quadrics and (for $q=2^{2e+1}$) the Suzuki-Tits ovoids.

M. Klin et al. (eds.), Algorithmic Algebraic Combinatorics and Gröbner Bases,
DOI 10.1007/978-3-642-01960-9_11, © Springer-Verlag Berlin Heidelberg 2009

A *spread in* $P^{2n-1}(\mathbb{F}_q)$ is a set $\mathcal{S}$ consisting of q^n+1 projective $(n-1)$-subspaces which partition the points. These exist for all $n \geq 1$ and prime powers q, and every such spread gives a plane (affine or projective) of order q^n known as a *translation plane.* This construction is responsible for most of the explicitly known finite projective planes.

An *orthogonal* (resp., *unitary*) *polar space* is the incidence system formed by the subspaces of projective space which lie on a given nondegenerate quadric (resp., Hermitian variety). A *symplectic polar space* is the incidence system formed by the totally isotropic subspaces of a projective space with respect to a nondegenerate alternating form. Let $\mathcal{P}$ be any *finite classical polar space* (i.e. a finite orthogonal, unitary or symplectic polar space). An *ovoid in* $\mathcal{P}$ is a set $\mathcal{O}$ consisting of points of $\mathcal{O}$ such that every maximal subspace of $\mathcal{P}$ contains exactly one point of $\mathcal{P}$. A *spread* in $\mathcal{P}$ is a set $\mathcal{S}$ consisting of maximal subspaces of $\mathcal{P}$, such that every point of $\mathcal{P}$ lies in exactly one member of $\mathcal{S}$. These notions of ovoid and spread are distinct from (albeit related to) the notions of ovoid and spread for projective spaces. In the polar space setting, one has a bipartite incidence graph formed by incidences between points and maximal subspaces. Whenever one has a bipartite graph with partition $A \cup B$ of the vertices (and every edge of the graph has one end in A and the other in B) then one may ask for a subset $\mathcal{O} \subseteq A$ such that every vertex in B is adjacent to exactly one member of $\mathcal{O}$; or a subset $\mathcal{S} \subseteq B$ such that every vertex in A is adjacent to exactly one member of $\mathcal{S}$. At this level of abstraction we see that ovoids and spreads are very similar notions.

Questions of existence and possible constructions of ovoids and spreads in the finite classical polar spaces are in many cases open; for an almost-current survey see [18, pp. 345–348]. Ovoids of polar spaces are most intensively studied in the orthogonal case. If $\mathcal{Q}$ is a nondegenerate quadric in $P^{2n-1}(\mathbb{F}_q)$ then $\mathcal{Q}$ is *hyperbolic* or *elliptic* according as maximal subspaces lying in the quadric have projective dimension $n-1$ or $n-2$. In the hyperbolic case an ovoid in $\mathcal{Q}$ (defined as above) is simply a set $\mathcal{O}$ consisting of $q^{n-1}+1$ points of $\mathcal{Q}$, no two on a line of the quadric. It is known [33] that ovoids do not exist in the elliptic case for $n \geq 5$; and [15] that no ovoids exist in nondegenerate quadrics of $P^{2n}(\mathbb{F}_q)$ (the *parabolic* case) for $n \geq 4$.

There are many connections between the various notions of spreads, ovoids, and other objects. The following sample of such connections is not exhaustive but is intended to hint at the central role played by these notions in finite geometry: Spreads of projective 3-space are equivalent to ovoids in the Klein quadric (i.e. the hyperbolic quadric in projective 5-space) via the Klein correspondence (the Plücker map). Ovoids and spreads in higher-dimensional spaces often give ovoids and spreads in lower-dimensional spaces, by a simple process of 'slicing'. The E_8 root lattice gives rise to numerous constructions (see [12, 23, 24, 26]) of ovoids in the hyperbolic quadric in projective 7-space, and these are in turn equivalent to spreads of the same quadric. Figure 1 depicts some of the connections arising between the finite geometric objects we have mentioned.

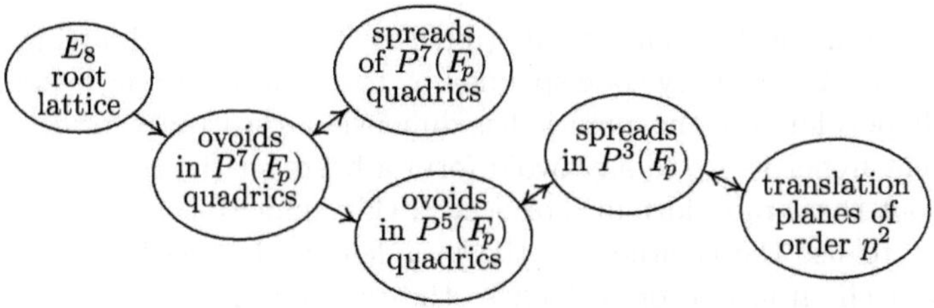

Fig. 1. Some connections between finite geometric objects

The most significant open question in this area is the question of whether there exist ovoids in nondegenerate quadrics in $P^n(\mathbb{F}_q)$ for $n > 7$. Ovoids in higher dimensions would give rise to significant numbers of ovoids in dimensions 5 and 7, which seems unlikely; yet no proof of impossibility is known. One may be tempted to mimic the construction of ovoids from the E_8 root lattice, using the Leech lattice to produce ovoids in quadrics in $P^{23}(\mathbb{F}_p)$; however this approach cannot succeed for primes $p < 59$ by virtue of Corollary 1 below. Such nonexistence results for ovoids motivated our interest in the p-rank formulas of Sect. 4.

Another motivation for our work is the desire to better understand the following striking parallel between different spaces admitting ovoids.

1.1 Ovoids in $P^3(\mathbb{F}_q)$, $q = 2^r$

Here the known ovoids belong to two infinite families, each admitting a doubly transitive subgroup of $PGL(4, q)$: the elliptic quadrics (for all q even, stabilized by $PSL(3, q)$) and the Suzuki-Tits ovoids (for $q = 2^{2e+1}$ only, admitting the Suzuki group ${}^2B_2(q)$). The binary code spanned by the (characteristic vectors of the) planes of $P^3(\mathbb{F}_q)$ has dimension $q^2{+}1$ (only for q even), and the tangent planes to any ovoid form a basis for the code.

1.2 Ovoids in Hyperbolic Quadrics in $P^7(\mathbb{F}_q)$, $q = 2^r$

Aside from one known sporadic example for $q = 8$ (Dye's ovoid; see [19]) just two infinite families of ovoids are known, each admitting a doubly transitive subgroup of $P\Omega^+(8, q)$: one family (for all q even) stabilized by $PSL(3, q)$, and the other (for $q = 2^{2e+1}$ only) admitting $PSU(3, q)$. The binary code spanned by the (characteristic vectors of the) tangent hyperplanes to the quadric has dimension $q^3{+}1$ (only for q even), and the tangent hyperplanes to any ovoid form a basis for the code.

1.3 Ovoids in Nondegenerate (Parabolic) Quadrics in $P^6(\mathbb{F}_q)$, $q = 3^r$

Here the known ovoids belong to two infinite families, each admitting a doubly transitive subgroup of $P\Omega(7, q)$: one family (for all $q = 3^r$) stabilized by

$PSU(3,q)$, and the Ree-Tits ovoids (for $q = 3^{2e+1}$ only) admitting the Ree group ${}^2G_2(q)$. The ternary code spanned by the (characteristic vectors of the) tangent hyperplanes to the quadric has dimension q^3+1 (only for $q = 3^r$), and the tangent hyperplanes to any ovoid form a basis for the code.

Without the p-rank formulas of Sect. 4 this analogy is not complete. One may hope to use the tightness of the p-rank bound in each case to classify ovoids in each situation, or to look to the recent literature on case 1.1 (for example Brown [6, 7]) in the hopes of finding techniques that may apply also to cases 1.2 and 1.3.

2 p-Ranks Related to Projective Spaces

Let A be the $(0,1)$-incidence matrix of a finite point-block incidence structure, i.e. the matrix having rows indexed by points and columns indexed by blocks, and with entries 0 and 1 corresponding to nonincident and incident point-block pairs, respectively. By the *p-rank* of the incidence structure, we mean the rank of A over a field of characteristic p. It has long been known (see [13, 20, 31]) that the symmetric design of points and hyperplanes of $P^n(\mathbb{F}_q)$ has p-rank equal to

$$\binom{p+n-1}{n}^r + 1$$

where $q = p^r$. The binomial coefficient appearing in the latter formula is in fact the coefficient of t^{p-1} in the binomial series

$$\frac{1}{(1-t)^{n+1}} = 1 + \binom{n+1}{1}t + \binom{n+2}{2}t^2 + \cdots + \binom{n+p-1}{p-1}t^{p-1} + \cdots$$

which arises as the Hilbert series for projective n-space. More explanation of the connection between p-ranks and Hilbert series is given in Sect. 3. Stronger information is in fact available: Black and List [3] give the Smith normal form of the point-hyperplane incidence matrix, although here we omit the details.

More generally, one may ask for the p-rank the design of points versus projective $(n-k)$-subspaces of $P^n(\mathbb{F}_q)$ where $q = p^r$; that is, the dimension of the linear code $\mathcal{C} = \mathcal{C}(n,k,p^r)$ spanned over $\mathbb{F}_p$ by the (characteristic vectors of the) projective subspaces of $P^n(\mathbb{F}_q)$ of codimension k. The first formula available for this is that of Hamada [16]; see also [5, p. 366]. Unfortunately the computational time required to evaluate Hamada's formula can be prohibitive, even for rather modest values of the input parameters. Fortunately however, $\dim\mathcal{C}$ can be computed quite easily using the information implicit in [1], where the structure of the code as an $\mathbb{F}_qG$-module for the group $G = PGL(n{+}1,\mathbb{F}_q)$ is given. We have

$$\dim\mathcal{C}(n,k,p^r) = 1 + \big(\text{coeff. of } t^r \text{ in } tr[(I-tM)^{-1}]\big)$$

where M is the $k \times k$ matrix with (i,j)-entry equal to the coefficient of t^{pi-j} in $(1+t+t^2+\cdots+t^{p-1})^{n+1}$. For example the following Maple™ code [21] determines the dimension of the $\mathbb{F}_5$-code spanned by lines of $P^3(\mathbb{F}_{5^r})$:

```
> with(linalg):
> p:=5: n:=3: k:=2:
> S:=simplify(((1-t^p)/(1-t))^(n+1)):
> M:=array(1..k,1..k):
> for i from 1 to k do
>   for j from 1 to k do
>     M[i,j]:=coeff(S,t,p*j-i):
>     od:
>   od:
> print(M);
```

$$\begin{bmatrix} 35 & 80 \\ 20 & 85 \end{bmatrix}$$

```
> simplify(trace(inverse(\&*()-t*M)));
```

$$-\frac{2(60t-1)}{1375t^2-120t+1}$$

```
> series(\%,t=0,6);
```

$$2+120t+11650t^2+1233000t^3+131941250t^4+14137575000t^5+O(t^6)$$

Thus the dimension of the code spanned by the lines of $P^3(\mathbb{F}_{5^r})$ is

$$120,\ 11650,\ 1233000,\ 131941250,\ 14137575000,\ \ldots$$

for $r=1,2,3,\ldots$.

Again, even stronger information is available [10] from the Smith normal form of the incidence matrix of points versus projective $(n-k)$-subspaces.

3 p-Ranks via the Hilbert Function

Consider the $\mathbb{F}_q$-linear code $\hat{\mathcal{C}}$ of length $N=(q^{n+1}-1)/(q-1)$ spanned by the (characteristic vectors of the) hyperplanes of $P^n(\mathbb{F}_q)$ where $q=p^r$. The subcode $\mathcal{C}\subset\hat{\mathcal{C}}$ spanned by the *complements* of the hyperplanes has dimension $\binom{p+n-1}{n}^r$, while $\hat{\mathcal{C}}$ itself has dimension $\binom{p+n-1}{n}^r+1$. Now let $\mathcal{V}$ be a subset of the points of $P^n(\mathbb{F}_q)$. Denote by $\hat{\mathcal{C}}_{\mathcal{V}}$ and $\mathcal{C}_{\mathcal{V}}$ the punctured codes of length $|\mathcal{V}|$ obtained by simply restricting $\hat{\mathcal{C}}$ and $\mathcal{C}$ (respectively) to the coordinate positions indexed by $\mathcal{V}$. We are interested in general methods for determining the dimensions of $\hat{\mathcal{C}}_{\mathcal{V}}$ and $\mathcal{C}_{\mathcal{V}}$. We note that $\mathcal{C}_{\mathcal{V}} \subseteq \hat{\mathcal{C}}_{\mathcal{V}}$ is a subcode of codimension at most 1, and in many cases of interest (see Theorem 3 below) the exact codimension is 1. So for now we focus attention on $\mathcal{C}_{\mathcal{V}}$. We are most interested in the case of a point set $\mathcal{V}$ arising as the set of $\mathbb{F}_q$-rational points of a projective variety. We establish notation to describe this case.

Consider the polynomial ring $R = \mathbb{F}_q[X_0, X_1, \ldots, X_n] = \bigoplus_{d\ge 0} R_d$ where R_d is the d-homogeneous part of R with respect to the standard degree grading. Let $I \subseteq R$ be a homogeneous ideal, and let $\mathcal{V}$ be the set of $\mathbb{F}_q$-rational points of projective n-space where I vanishes; thus $\mathcal{V} = \mathcal{V}(I+J)$ is the zero set of the ideal $I+J$ where J is generated by the polynomials $X_i^q X_j - X_i X_j^q$ for $0 \le i < j \le n$. Let $\mathcal{I} = \mathcal{I}(\mathcal{V}) \subseteq R$ be the ideal generated by all homogeneous polynomials vanishing on $\mathcal{V}$; this is just the radical ideal $\mathcal{I} = \sqrt{I+J}$. We denote the Hilbert function of $\mathcal{I}$ by $h_{\mathcal{I}}(d) = \dim(R_d/\mathcal{I}_d)$ where $\mathcal{I}_d = \mathcal{I} \cap R_d$. Denote by $LM(\mathcal{I})$ the set of leading monomials in $\mathcal{I}$ with respect to some fixed monomial ordering. A monomial in R is *standard* if it is *not* in $\mathcal{I}$. Then $h_{\mathcal{I}}(d)$ is the number of standard monomials of degree d. For simplicity we consider first the special case $q = p$.

Theorem 1. (See [27].) *If* $q = p$ *then* $\dim(\mathcal{C}_{\mathcal{V}}) = h_{\mathcal{I}}(p-1)$.

3.1 Computational Example

Consider the incidence system of points of the cubic surface $x^3 + y^3 + z^2 w = 0$ in $P^3(\mathbb{F}_{13})$ versus all hyperplanes of the projective space. Using Macaulay 2 [14] we compute

```
i1 : p = 13;
i2 : F = ZZ/(p);
i3 : R = F[x,y,z,w];
i4 : S = (s,t)->s^p*t-s*t^p;
i5 : J = ideal(S(x,y),S(x,z),S(x,w),S(y,z),S(y,w),S(z,w));
o5 : Ideal of R
i6 : I = ideal(x^3+y^3+z^2*w)+J;
o6 : Ideal of R
i7 : II = radical(I);
o7 : Ideal of R
i8 : hilbertSeries(I)
```

$$\texttt{o8} = \frac{1 - T^3 - 6T^{14} + 4T^{15} - T^{16} + 6T^{17} - 4T^{18} + T^{19} + 2T^{26} + \cdots + 2T^{33}}{(1-T)^4}$$

```
o8 : Divide
i9 : hilbertSeries(II)
```

$$\texttt{o9} = \frac{1 - T^3 - 2T^{10} - 4T^{11} + T^{12} + 9T^{13} - 3T^{14} + 2T^{15} - T^{16} - 2T^{17}}{(1-T)^4}$$

```
o9 : Divide
i10 : hilbertSeries(II, Order=>p)
```

$$\text{o10} = 1 + 4T + 10T^2 + 19T^3 + 31T^4 + 46T^5 + 64T^6 + \cdots + 187T^{11} + 200T^{12}$$

```
o10 : ZZ [T, MonomialOrder => RevLex, Inverses => true]}
```

From the coefficient of T^{p-1} we see that $\dim(\mathcal{C}_\mathcal{V}) = 200$, and so $\dim(\hat{\mathcal{C}}_\mathcal{V}) = 201$ by Theorem 3 below. The most time-consuming step in this example (the computation of the radical ideal) requires at most a few seconds on a typical personal computer, but in other examples this step may overwhelm the computational resources of the machine. In such cases one might try to explicitly determine the radical by other means; or it may be necessary to determine the required p-rank by Gaussian elimination. For example we check independently that the above cubic surface has 209 points, and that the 209×2380 incidence matrix of points versus hyperplanes has 13-rank equal to 201.

Now consider the general case $q = p^r$, $r \geq 1$. We define a *p-standard monomial* to be a monomial of the form $m_0\, m_1^p m_2^{p^2} \cdots$, a finite product in which each m_i is a standard monomial of degree less than p. (This definition is not standard; sorry, no pun intended!) Denote by $h_\mathcal{I}^\dagger(d)$ the number of p-standard monomials of degree d.

Theorem 2. (See [27].) $\dim(\mathcal{C}_\mathcal{V}) = h_\mathcal{I}^\dagger(q-1)$.

This requires us to count the number of monomials of the form

$$m_0 m_1^p m_2^{p^2} \cdots m_{r-1}^{p^{r-1}}$$

where each m_i is a standard monomial of degree $p-1$.

3.2 Example: Projective n-Space

Let $I = 0$, so that $\mathcal{I} = 0$ and $\mathcal{V}$ consists of all $(q^{n+1}-1)/(q-1)$ points of $P^n(\mathbb{F}_q)$. Every monomial is standard, and the p-standard monomials are those of the form $m_0\, m_1^p m_2^{p^2} \cdots m_{r-1}^{p^{r-1}}$ where each monomial m_i has degree $p-1$. There are $\binom{p+n-1}{n}$ choices for each m_i, and hence the number of p-standard monomials of degree $q-1$ is $h_\mathcal{I}^\dagger(q-1) = \binom{p+n-1}{n}^r$. This gives the well-known value for $\dim(\mathcal{C}_\mathcal{V})$; and the value $\dim(\hat{\mathcal{C}}_\mathcal{V}) = 1 + \dim(\mathcal{C}_\mathcal{V})$ may be seen as a special case of the following (for a vacuous set of $k = 0$ polynomials).

Theorem 3. *Let $f_1, \ldots, f_k \in R$ be nonconstant homogeneous polynomials of total degree $\sum_i \deg(f_i) \leq n-2$, and let $\mathcal{V}$ be the set of all points in $P^n(\mathbb{F}_q)$ where every f_i vanishes. Then* $\dim(\hat{\mathcal{C}}_\mathcal{V}/\mathcal{C}_\mathcal{V}) = 1$.

Proof. Let M be the number of vectors in $\mathbb{F}_q^{n+1}$ where all $f_1, \ldots, f_k$ vanish. Since the total degree $\sum_i \deg(f_i) < n$, the Chevalley-Warning Theorem [30, p. 5] shows that p divides M. But the homogeneity of $f_1, \ldots, f_k$ means that $q-1$ divides $M-1$, so in fact $M = mp(q-1)+q$ for some $m \geq 0$. Thus $|\mathcal{V}| = (M-1)/(q-1) = mp+1 \equiv 1 \mod p$.

Now let $h \in R_1$ be a nonzero homogeneous linear polynomial, and let M_h be the number of vectors in $\mathbb{F}_q^{n+1}$ where all $k+1$ of the polynomials $f_1, f_2, \ldots, f_k, h$ vanish. Since the total degree again satisfies $1+\sum_i \deg(f_i) \leq n-1$, the previous argument also shows that $M_h = m_h p(q-1)+q$ for some $m_h \geq 0$. Thus $|H \cap \mathcal{V}| = m_h p + 1 \equiv 1 \mod p$ where H is the hyperplane of $P^n(\mathbb{F}_q)$ consisting of all points where h vanishes.

Since $\dim(\hat{\mathcal{C}}/\mathcal{C}) = 1$, it suffices to find a nonzero linear functional $\phi : \hat{\mathcal{C}}_{\mathcal{V}} \to \mathbb{F}_p$ vanishing on $\mathcal{C}_{\mathcal{V}}$. For $v \in \hat{\mathcal{C}}_{\mathcal{V}}$, define $\phi(v)$ to be simply the sum of the coordinate entries of v. In case v is the characteristic vector of a hyperplane H (restricted to $\mathcal{V}$), we have $\phi(v) = |H \cap \mathcal{V}| \equiv 1 \mod p$. Similarly if v is the characteristic vector of the *complement* of a hyperplane H, then $\phi(v) = |\mathcal{V}| - |H \cap \mathcal{V}| \equiv 1 - 1 \equiv 0 \mod p$. Since we have considered typical generators for the codes $\hat{\mathcal{C}}_{\mathcal{V}}$ and $\mathcal{C}_{\mathcal{V}}$, our ϕ has the required properties and the conclusion follows. □

4 p-Ranks Related to Polar Spaces and Grassmannians

We survey some interesting p-rank formulas and some applications to bounds for ovoids. Each p-rank formula listed here is derived either by the approach described in Sect. 3, or from the theory of group representations. Our notation $q, p, n, R, J, \mathcal{C}_{\mathcal{V}}$, etc. is the same as in Sect. 3.

Theorem 4. (See [4].) *Let $I = (Q)$ where $Q(X_0, X_1, \ldots, X_n) \in R_2$ is an irreducible quadratic form, and let $\mathcal{Q} = \mathcal{V}(I+J)$ be the resulting quadric in $P^n(\mathbb{F}_q)$. Let $\hat{\mathcal{C}}_{\mathcal{Q}}$ be the $\mathbb{F}_q$-linear code of length $|\mathcal{Q}|$ spanned by the hyperplane sections of the quadric. Then*

$$\dim(\hat{\mathcal{C}}(\mathcal{Q})) = \left[\binom{p+n-1}{n} - \binom{p+n-3}{n}\right]^r + 1.$$

If a nondegenerate quadric $\mathcal{Q}$ in $P^n(\mathbb{F}_q)$ admits an ovoid $\mathcal{O}$ then the tangent hyperplanes to $\mathcal{Q}$ at the points of $\mathcal{O}$ span a subcode of $\hat{\mathcal{C}}_{\mathcal{Q}}$ of dimension $p^{\lfloor n/2 \rfloor r} + 1$. This gives

Corollary 1. (See [4].) *There do not exist ovoids in $\mathcal{Q}$ (using the notation of Theorem 4) if*

$$p^{\lfloor n/2 \rfloor} > \binom{p+n-1}{n} - \binom{p+n-3}{n}.$$

In particular ovoids do not exist in $\mathcal{Q}$ for $n = 9$ and $p = 2, 3$; or for $n = 11$ and $p = 2, 3, 5, 7$.

In studying finite projective planes, it is often useful to have an explicit basis for the $\mathbb{F}_p$-linear code spanned by the lines. In the case of classical (Desarguesian) planes $P^2(\mathbb{F}_p)$, this code has dimension $\binom{p+1}{2}+1$ and so it had long been speculated that an explicit basis could be formed from any conic (which has $p+1$ points). This follows also from our approach to p-ranks:

Corollary 2. (See [4].) *Let $\hat{\mathcal{C}}$ be the $\mathbb{F}_p$-linear code spanned by the lines of $P^2(\mathbb{F}_p)$, so that $\dim(\mathcal{C}) = \binom{p+1}{2}+1$. Let $\mathcal{C} \subset \hat{\mathcal{C}}$ be the subcode of dimension $\binom{p+1}{2}$ spanned by the complements of the lines. Let $\mathcal{Q}$ be any conic in the plane, so that $\mathcal{Q}$ has $p+1$ tangent lines, $\binom{p+1}{2}$ secant lines, and $\binom{p}{2}$ passant lines (i.e. lines not meeting $\mathcal{Q}$). Then the complements to the secants form a basis for $\mathcal{C}$. Moreover the tangents and the passants together form a basis for $\hat{\mathcal{C}}$.*

We remark in passing that another choice of explicit basis is found in [22].

Theorem 5. (See [25].) *Suppose $q = q_0^2 = p^r$ (r even) and let $I = (U)$ where $U \in R_{q_0+1}$ is a nondegenerate unitary form; we may choose coordinates so that $U(X_0, X_1, \dots, X_n) = X_0^{q_0+1} + X_1^{q_0+1} + \cdots + X_n^{q_0+1}$. Let $\mathcal{H} = \mathcal{V}(I+J)$ be the resulting hermitian variety in $P^n(\mathbb{F}_q)$. Let $\hat{\mathcal{C}}_{\mathcal{H}}$ be the $\mathbb{F}_q$-linear code of length $|\mathcal{H}|$ spanned by the hyperplane sections of $\mathcal{H}$. Then*

$$\dim(\hat{\mathcal{C}}_{\mathcal{H}}) = \left[\binom{p+n-1}{n}^2 - \binom{p+n-2}{n}^2\right]^{r/2} + 1.$$

Bounds for ovoids in unitary polar spaces, similar to those of Theorem 4 and Corollary 1, are obtained [25] using Theorem 5.

Theorem 6. *Let $\mathcal{Q}$ be a nondegenerate (parabolic) quadric in $P^4(\mathbb{F}_q)$, and consider the incidence system of points of $P^4(\mathbb{F}_q)$ versus lines of $\mathcal{Q}$. The p-rank of this incidence system is*

(a) *(for $q = 2^r$)* $1 + \left(\frac{1+\sqrt{17}}{2}\right)^{2r} + \left(\frac{1-\sqrt{17}}{2}\right)^{2r}$;

(b) *(for $q = p$)* $1 + \frac{p(p+1)^2}{2}$;

(c) *(for $q = p^r$)* $1 + \alpha_+^r + \alpha_-^r$ *where* $\alpha_\pm = \frac{p(p+1)^2}{4} \pm \frac{p(p^2-1)}{12}\sqrt{17}$.

The incidence system of Theorem 6 is a classical generalized quadrangle of order (q, q); and it immediately follows that the dual generalized quadrangle also has p-rank as given by Theorem 6. This dual generalized quadrangle is the symplectic polar space in $P^3(\mathbb{F}_q)$ formed by a nondegenerate alternating form. Proofs of (a) and (c), using representation theory, appear in [29] and [11]; and in the prime case (b) a proof appears in [8] using methods from Sect. 3.

In the following, we embed the collection of all projective s-subspaces of $P^m(\mathbb{F}_q)$ in $P^n(\mathbb{F}_q)$ via the Plücker embedding, where $n = \binom{m+1}{s+1}-1$. The image of this embedding is the set $\mathcal{G}_s^n(\mathbb{F}_q)$ of $\mathbb{F}_q$-rational points of the Grassmann variety $\mathcal{G}_s^n$. Recall that $\mathcal{G}_s^n = \mathcal{V}(I)$ where the ideal $I \subset R$ is generated by certain homogeneous polynomials of degree 2 (the van der Waerden syzygies). The Hilbert function for this variety is known:

$$h_I(d) = \prod_{0\le j\le s} \frac{(m+d-s+j)!\,j!}{(m-s+j)!\,(d+j)!}.$$

Theorem 7. *Let $\mathcal{G} = \mathcal{G}_s^n(\mathbb{F}_q) = \mathcal{V}(I+J)$ be the set of $\mathbb{F}_q$-rational points of the Grassmann variety, with n, I, h_I as above, and let $\hat{C}_\mathcal{G}$ be the $\mathbb{F}_q$-linear code of length $|\mathcal{G}| = \left[\begin{smallmatrix} n+1 \\ s+1 \end{smallmatrix}\right]_q$ spanned by the hyperplane sections of $\mathcal{G}$. Then*

$$\dim(\hat{C}_\mathcal{G}) = h_I(p-1)^r + 1$$

with $h_I(d)$ as above.

Note that the Grassmann variety $\mathcal{G}_1^3(\mathbb{F}_q)$ is in fact the Klein quadric, i.e. the hyperbolic quadric in $P^5(\mathbb{F}_q)$; in this case the dimension of the code $\hat{C}_\mathcal{G}$ is

$$\left[\frac{1}{12}p(p+1)^2(p+2)\right]^r + 1,$$

as given by either Theorem 4 or 7.

Acknowledgments

The author is pleased to acknowledge the support and encouragement of Professor Bruno Buchberger and the coordinators of the Special Semester on Gröbner Bases (February 1 – July 31, 2006) organized by RICAM, Austrian Academy of Sciences, and RISC, Johannes Kepler University, Linz, Austria, for partial support of this project.

References

1. M. Bardoe and P. Sin, The permutation modules for $\mathrm{GL}(n+1, \mathbb{F}_q)$ acting on $P^n(\mathbb{F}_q)$ and $\mathbb{F}_q^{n+1}$, *J. London Math. Soc.*, **61** (2000), 58–80.
2. A. Barlotti, Un'estensione del teorema di Segre-Kustaanheimo, *Bull. Un. Mat. Ital.*, **10** (1955), 498–506.
3. S. C. Black and R. J. List, On certain abelian groups associated with finite projective geometries, *Geom. Dedicata*, **33** (1990), 13–19.
4. A. Blokhuis and G. E. Moorhouse, Some p-ranks related to orthogonal spaces, *J. Algebr. Comb.*, **4** (1995), 295–316.
5. A. E. Brouwer and H. E. Wilbrink, Block designs, in F. Buekenhout (ed.) *Handbook of Incidence Geometry*, pp. 349–382, Elsevier, Amsterdam, 1995.
6. M. R. Brown, Ovoids of $\mathrm{PG}(3, q)$, q even, with a conic section, *J. London Math. Soc.*, **62** (2000), 569–582.

7. M. R. Brown, The determination of ovoids of PG$(3, q)$ containing a pointed conic, *J. Geom.*, **67** (2000), 61–72.
8. D. de Caen and G. E. Moorhouse, The p-Rank of the Sp$(4, p)$ Generalized Quadrangle, Preprint, 2000, available at http://www.uwyo.edu/moorhouse/pub/sp4p.pdf.
9. P. J. Cameron, Finite geometries, in R. Graham et al. (eds.) *Handbook of Combinatorics*, pp. 647–691, Elsevier, Amsterdam, 1995.
10. D. B. Chandler, P. Sin, and Q. Xiang, The invariant factors of the incidence matrices of points and subspaces in $PG(n, q)$ and $AG(n, q)$, *Trans. Amer. Math. Soc.*, **358** (2006), 4935–4957.
11. D. B. Chandler, P. Sin, and Q. Xiang, *The Permutation Action of Finite Symplectic Groups of Odd Characteristic on Their Standard Modules*, Preprint, 2006.
12. J. H. Conway, P. B. Kleidman, and R. A. Wilson, New families of ovoids in O_8^+, *Geom. Dedicata*, **26** (1988), 157–170.
13. J. M. Goethals and P. Delsarte, On a class of majority-logic decodable cyclic codes, *IEEE Trans. Inform. Theory*, **14** (1968), 182–188.
14. D. R. Grayson and M. E. Stillman, *Macaulay 2, A Software System for Research in Algebraic Geometry*, available at http://www.math.uiuc.edu/Macaulay2/.
15. A. Gunawardena and G. E. Moorhouse, The non-existence of ovoids in $O_9(q)$, *Eur. J. Comb.*, **18** (1997), 171–173.
16. N. Hamada, The rank of the incidence matrix of points and d-flats in finite geometries, *J. Sci. Hiroshima Univ. Ser. A—I Math.*, **32** (1968), 381–396.
17. J. W. P. Hirschfeld, *Projective Geometries over Finite Fields*, 2nd ed., Oxford University Press, Oxford, 1998.
18. J. W. P. Hirschfeld and J. A. Thas, *General Galois Geometries*, Oxford University Press, Oxford, 1991.
19. W. M. Kantor, Ovoids and translation planes, *Canad. J. Math.*, **24** (1982), 1195–1207.
20. F. J. MacWilliams and H. B. Mann, On the p-rank of the design matrix of a difference set, *Inf. Control*, **12** (1968), 474–489.
21. M. B. Monagan et al., *Maple 10 Programming Guide*, Maplesoft, Waterloo, 2005.
22. G. E. Moorhouse, Bruck nets, codes, and characters of loops, *Designs Codes Cryptogr.*, **1** (1991), 7–29.
23. G. E. Moorhouse, Ovoids from the E_8 root lattice, *Geom. Dedicata*, **46** (1993), 287–297.
24. G. E. Moorhouse, Root lattice constructions of ovoids, in F. De Clerck et al. (eds.) *Finite Geometry and Combinatorics*, pp. 269–275, Cambridge University Press, Cambridge, 1993.
25. G. E. Moorhouse, Some p-ranks related to Hermitian varieties, *J. Stat. Plan. Inf.*, **56** (1996), 229–241.

26. G. E. Moorhouse, Ovoids and translation planes from lattices, in N. L. Johnson (ed.) *Mostly Finite Geometries*, pp. 123–134, Dekker, New York, 1997.
27. G. E. Moorhouse, Some p-ranks related to finite geometric structures, in N. L. Johnson (ed.) *Mostly Finite Geometries*, pp. 353–364, Dekker, New York, 1997.
28. G. Panella, Caratterizzazione delle quadriche di uno spazio (tridimensionale) lineare sopra un corpo finito, *Bull. Un. Mat. Ital.*, **10** (1955), 507–513.
29. N. Sastry and P. Sin, The codes of generalized quadrangles of even order, in *Proceedings of the AMS Summer Research Institute on Group Actions and Cohomology*, Seattle, 1996.
30. J. P. Serre, *A Course in Arithmetic*, Springer, New York, 1973.
31. K. J. C. Smith, On the p-rank of the incidence matrix of points and hyperplanes in a finite projective geometry, *J. Combin. Theory*, **1** (1969), 122–129.
32. J. A. Thas, Projective geometry over a finite field, in F. Buekenhout (ed.) *Handbook of Incidence Geometry*, pp. 295–347, Elsevier, Amsterdam, 1995.
33. J. A. Thas, Ovoids and spreads of finite classical polar spaces, *Geom. Dedicata*, **10** (1981), 135–144.

Computer Aided Investigation of Total Graph Coherent Configurations for Two Infinite Families of Classical Strongly Regular Graphs

Matan Ziv-Av

Department of Mathematics, Ben-Gurion University of the Negev, Beer Sheva, Israel. matan@svgalib.org

Summary. In this chapter we introduce the notion of total graph coherent configuration, and use computer tools to investigate it for two classes of strongly regular graphs – the triangular graphs $T(n)$ and the lattice square graphs $L_2(n)$. For $T(n)$, we show that its total graph coherent configuration has exceptional mergings only in the cases $n = 5$ and $n = 7$.

Key words: Triangular graph, Lattice square graph, Total graph coherent closure, Coherent subalgebra

1 Introduction

The notion of total graph coherent configuration was introduced and used in [14], where an imprimitive rank 5 association scheme on 40 points was constructed as a merging of relations in the total graph configuration of the triangular graph $T(5)$.

In this paper we investigate systematically the total graph coherent configurations of two infinite series of classical strongly regular graphs.

Section 2 contains all preliminaries which make this presentation mostly self-contained. In Sect. 3 we consider the triangular graphs $T(n)$ (for $n \geq 6$) and show that in the corresponding total graph coherent configuration $\mathcal{T}(n)$ there are two merging association schemes with two and three classes. Besides this there are sporadic mergings for the cases $n = 5, 7$. Using a computer we prove that these mergings expire all possible merging association schemes in $\mathcal{T}(n)$. We also show that $\mathcal{T}(n)$ coincides for all $n > 4$ with the Schurian coherent configuration defined by the automorphism group of the total graph of $T(n)$ (this group is actually S_n, in action on the edges and paths of length 2 of K_n). In Sect. 4, similar results are presented for the total graph coherent configuration defined by the lattice square graphs $L_2(n)$.

M. Klin et al. (eds.), Algorithmic Algebraic Combinatorics and Gröbner Bases, DOI 10.1007/978-3-642-01960-9_12,

Our results provide an example of successful amalgamation of essential computer algebra experimentation with subsequent theoretical analysis and generalization. An important feature of presented approach is that the structure constants for both series of coherent configurations appear as polynomials in variable n. In Sect. 5 we present a detailed outline of the algorithm that we used for the search of mergings, and its implementation in GAP.

Finally, in Sect. 6 we discuss further possibilities for investigation of the introduced class of coherent configurations. Those include relations to the famous graph isomorphism problem and potential applications of Gröbner bases.

2 Preliminaries

2.1 Coherent Configurations and Association Schemes

2.1.1 Axioms

Let $X = [1, n]$, and let $\mathfrak{R} = \{R_1, \ldots, R_r\}$ be a collection of binary relations on X (subsets of X^2) such that:

CC1 $R_i \cap R_j = \emptyset$ for $1 \leq i \neq j \leq r$;
CC2 $\bigcup_{i=1}^{r} R_i = X^2$;
CC3 $\forall i \in [1, r]\ \exists i' \in [1, r] R_i^t = R_{i'}$, where $R_i^t = \{(y, x)|(x, y) \in R_i\}$;
CC4 $\exists I' \subseteq [1, r] \bigcup_{i \in I'} R_i = \Delta$, where $\Delta = \{(x, x)|x \in X\}$;
CC5 $\forall i, j, k \in [1, r]\ \forall (x, y) \in R_k |\{z \in X|(x, z) \in R_i \wedge (z, y) \in R_j\}| = p_{ij}^k$,

then $\mathfrak{M} = (X, R)$ is called a *coherent configuration*. The relations in R are the *basis relations* of $\mathfrak{M}$. The parameters p_{ij}^k are the *structure constants* of the configuration. The graphs $\Gamma_i = (X, R_i)$ are called the *basis graphs* of the coherent configuration.

See [11] for the original definitions.

Let (G, Ω) be a permutation group. G acts naturally on Ω^2 by $(x, y)^g = (x^g, y^g)$. Following H. Wielandt in [18], the orbits of this action, (G, Ω^2) are called the *2-orbits of* (G, Ω), denoted by 2-$orb(G, \Omega)$.

For every permutation group (G, Ω), $(\Omega$, 2-$orb(G, \Omega))$ is a coherent configuration. Conversely, if the set of relations of a coherent configuration, $\mathfrak{M}$, coincides with 2-$orb(G, \Omega)$ for a suitable group G, then $\mathfrak{M}$ is called a *Schurian* coherent configuration.

A coherent configuration which has $\Delta = \{(x, x)|x \in X\}$ as one of its basis relations is called an *association scheme*. In this case, all basis relations except for Δ are called *classes*.

A *fusion* configuration (or a *merging*) of a coherent configuration $\mathfrak{M} = (X, R)$ is a coherent configuration $\mathfrak{M}' = (X, S)$ on the same set such that each basis relation S_i of $\mathfrak{M}'$ is a union of basis relations of $\mathfrak{M}$.

Coherent configurations can be alternatively described in matrix language. The adjacency matrix $A(R)$ of a relation R on X is a $(0, 1)$-matrix $A(R) =$

(a_{ij}) of dimension $|X| \times |X|$ such that $a_{ij} = 1$ iff $(i, j) \in R$. If $R = \{R_1, \ldots, R_r\}$ are the basis relations of a coherent configuration then their adjacency matrices $\{A_i = A(R_i)\}_{i=1}^{r}$ form a basis of a matrix algebra which is closed under Schur-Hadamard (element wise) product.

This leads to equivalent formulation of the axioms of coherent configuration:

Let $W \subseteq \mathbb{C}^{n \times n}$ be a matrix algebra of square matrices of order n over the complex field, such that

CA1 Was a linear space over $\mathbb{C}$ has some basis, $A_1, A_2, \ldots, A_r$, consisting of $(0,1)$-matrices;
CA2 $\sum_{i=1}^{r} A_i = J_n$, where J_n is the square matrix of order n all entries of which are equal to 1;
CA3 $\forall i \in [1, r] \exists i' \in [1, r] A_i^t = A_{i'}$;
CA4 $I \in W$ (I denotes the identity matrix),

then W is called a *coherent algebra* of rank r and order n with the *standard basis* $\mathcal{C} = \{A_1, A_2, \ldots, A_r\}$. We write $W = \langle A_1, \ldots, A_r \rangle$.

The notion corresponding to a fusion scheme in this notation is a *coherent subalgebra*, that is a subalgebra which is also a coherent algebra.

2.1.2 Weisfeiler-Leman Closure

Using matrix notation, it is easy to see that the intersection of coherent algebras is a coherent algebra, and that each square matrix is contained in some coherent algebra (since $M_n(\mathbb{C})$ is coherent). Therefore, we can define the *coherent closure* of a matrix A, denoted $\langle\langle A \rangle\rangle$ as the smallest coherent algebra containing this matrix (or in other words, the intersection of all coherent algebras containing it).

An efficient algorithm for computing $\langle\langle A \rangle\rangle$ was suggested by Weisfeiler and Leman [17] and is frequently called the WL-stabilization of the matrix A.

2.1.3 Wreath Product

If $\mathfrak{M}_1 = (X_1, \{R_0, R_1, \ldots, R_{r-1}\})$ and $\mathfrak{M}_2 = (X_2, \{S_0, S_1, \ldots, S_{l-1}\})$ are association schemes (R_0 and S_0 are the reflexive relations), then the wreath product of $\mathfrak{M}_1$ with $\mathfrak{M}_2$ is defined as $\mathfrak{M}_1 \wr \mathfrak{M}_2 = (Y = X_1 \times X_2, \{T_0, T_1, \ldots, T_{r+l-2}\})$ where T_0 is the identity relation on Y, $T_i = \{((a,b),(c,d)) | (a,c) \in R_i\}$ for all $1 \leq i \leq r-1$, and $T_{r-1+i} = \{((a,b),(a,c)) | (b,c) \in S_i\}$ for all $1 \leq i \leq l-1$.

The wreath product of association schemes of ranks r and l is an association scheme of rank $r + l - 1$.

2.2 Total Configuration

Let $\Gamma = (V, E)$ be a graph. The *total graph* $T(\Gamma)$ is the graph with the vertex set $V \cup E$, two such vertices in $T(\Gamma)$ are adjacent if and only if they are adjacent or incident in Γ (here edges of Γ are incident if they have a joint vertex).

The coherent closure of $T(\Gamma)$ will be called the *total coherent configuration* of Γ.

The *Schurian total coherent configuration* of a graph Γ is $(X, 2\text{-}Orb(Aut(T(\Gamma))))$ where X is the set of vertices of $T(\Gamma)$.

The total configuration is a fusion of the Schurian total configuration. Indeed, since an automorphism of $T(\Gamma)$ maps edges of $T(\Gamma)$ to edges, any 2-orbit of $Aut(T(\Gamma))$ either contains only edges, or does not contain edges at all.

2.3 Computational Tools

2.3.1 COCO

COCO is a set of programs for dealing with coherent configurations. The current version was developed during the years 1990–1992 in Moscow, USSR, mainly by Faradžev and Klin [8, 7].

COCO can be used to construct a coherent configuration $(\Omega, 2\text{-}orb(G, \Omega))$ from a prescribed permutation group (G, Ω), as well as to calculate the structure constants of the constructed coherent configuration, and find all association schemes which are mergings of the coherent configuration, together with their automorphism groups.

COCO was originally written for DOS, and the version currently in use is the UNIX port by A.E. Brouwer, available from Brouwer's home page [4].

2.3.2 WL-stabilization

Two implementations of the Weisfeiler-Leman stabilization [17] are available, under the name stabil [1] and stabcol [2]. The two implementations differ slightly in memory usage and run time, but both are adequate for the coherent configurations used in this article.

2.3.3 GAP

GAP [9, 16], an acronym for "Groups, Algorithms and Programming", is a system for computation in discrete abstract algebra. The system is extensible in the sense that it supports easy addition of extensions (packages, in GAP nomenclature), that are written in the GAP programming language which can extend the abilities of the GAP system.

Within GAP framework, COCO-II (a reimplementation of COCO functionality as a GAP package, currently in development by S. Reichard et al.) will be used. COCO-II improves on the original COCO by adding functionality such as WL-stabilization, as well as using algorithms developed since the release of COCO.

The author modified some COCO-II functions to handle polynomial structure constants, instead of the usual numeric constants, and those functions are used to handle the general case in this paper.

3 Total Configuration of Triangular Graph

3.1 Definition and Basic Properties

Let $\mathbb{T}(n)$ denote total graph of the triangular graph $T(n)$ (recall that $T(n)$ is the line graph $L(K_n)$ of the complete graph K_n), and let $\mathcal{T}(n)$ denote the coherent closure of $\mathbb{T}(n)$.

In more detail, the total graph $\mathbb{T}(n)$ is the total graph of the triangular graph $T(n)$. The vertices of $\mathbb{T}(n)$ are the edges and the paths of length 2 of K_n. The edges of $\mathbb{T}(n)$ are partitioned to three types:

$\{e, f\}$, where e and f are edges of K_n which share a common point.
$\{e, P\}$, where e is an edge of K_n, P is a path of length 2 in K_n, and P includes e.
$\{P, Q\}$, where P and Q are paths of length 2 in K_n which share a common edge. $\mathbb{T}(n)$ has $\frac{n(n-1)^2}{2}$ vertices.

To investigate $\mathcal{T}(n)$ we will first consider the Schurian total coherent configuration $\mathcal{S}(n)$, and later show that $\mathcal{T}(n) = \mathcal{S}(n)$.

Let Ω be the set of vertices of $\mathbb{T}(n)$. In other words,

$$\Omega = \{\{a, b\} | a \neq b \in [1, n]\} \cup \{\{\{a, b\}, \{a, c\}\} | a, b, c \in [1, n], a \neq b, a \neq c, b \neq c\}.$$

Let G be the automorphism group of $\mathbb{T}(n)$; it is well known that G is the permutation group (S_n, Ω) (with the natural action of S_n on Ω) for $n > 4$. Then, $\mathcal{S}(n) = (\Omega, 2 - orb(G))$ is the Schurian total coherent configuration of the triangular graph.

For $n \geq 6$, $\mathcal{S}(n)$ has 2 fibres and 25 relations as follows (for the sake of brevity, we will list a standard compact description for each relation), see also Table 1.

Two reflexive relations:
$R_0 = (\{a, b\}, \{a, b\})$,

Table 1. Sizes of basis relations of $\mathcal{S}(n)$

Relation	Size	Relation	Size
0	$\binom{n}{2}$	1	$3\binom{n}{3}$
2	$n(n-1)(n-2)$	3	$\binom{n}{2}\binom{n-2}{2}$
4	$\binom{n}{2}(n-2)\binom{n-3}{2}$	5	$\binom{n}{2}2\binom{n-2}{2}$
6	$\binom{n}{2}2(n-2)(n-3)$	7	$\binom{n}{2}2(n-2)$
8	$\binom{n}{2}(n-2)$	14	$n\binom{n-1}{2}(n-3)\binom{n-4}{2}$
15	$n\binom{n-1}{2}\binom{n-3}{2}$	16	$n\binom{n-1}{2}(n-3)(n-4)$
17	$n\binom{n-1}{2}2\binom{n-3}{2}$	18	$n\binom{n-1}{2}2(n-3)(n-4)$
19	$n\binom{n-1}{2}2(n-3)$	20	$n\binom{n-1}{2}2(n-3)$
21	$n\binom{n-1}{2}2(n-3)$	22	$n\binom{n-1}{2}(n-3)$
23	$n\binom{n-1}{2}2(n-3)$	24	$n\binom{n-1}{2}2$

$R_1 = (\{\{a,b\},\{a,c\}\},\{\{a,b\},\{a,c\}\})$.
Two relations within first fibre:
$R_2 = (\{a,b\},\{a,c\})$ (arcs of the triangular graph),
$R_3 = (\{a,b\},\{c,d\})$ (arcs of its complement),
five relations between first and second fibre:
$R_4 = (\{a,b\},\{\{c,d\},\{c,e\}\})$,
$R_5 = (\{a,b\},\{\{a,c\},\{a,d\}\})$,
$R_6 = (\{a,b\},\{\{c,a\},\{c,d\}\})$,
$R_7 = (\{a,b\},\{\{a,b\},\{a,c\}\})$,
$R_8 = (\{a,b\},\{\{c,a\},\{c,b\}\})$,
and the five inverses $R_9,\dots,R_{13}$ respectively,
eleven relations within second fibre:
$R_{14} = (\{\{a,b\},\{a,c\}\},\{\{d,e\},\{d,f\}\})$,
$R_{15} = (\{\{a,b\},\{a,c\}\},\{\{a,d\},\{a,e\}\})$,
$R_{16} = (\{\{a,b\},\{a,c\}\},\{\{d,a\},\{d,e\}\})$,
$R_{17} = (\{\{a,b\},\{a,c\}\},\{\{b,d\},\{b,e\}\})$,
$R_{18} = (\{\{a,b\},\{a,c\}\},\{\{d,b\},\{d,e\}\})$,
$R_{19} = (\{\{a,b\},\{a,c\}\},\{\{a,b\},\{a,d\}\})$,
$R_{20} = (\{\{a,b\},\{a,c\}\},\{\{b,a\},\{b,d\}\})$,
$R_{21} = (\{\{a,b\},\{a,c\}\},\{\{d,a\},\{d,b\}\})$,
$R_{22} = (\{\{a,b\},\{a,c\}\},\{\{d,b\},\{d,c\}\})$,
$R_{23} = (\{\{a,b\},\{a,c\}\},\{\{b,c\},\{b,d\}\})$,
$R_{24} = (\{\{a,b\},\{a,c\}\},\{\{b,a\},\{b,c\}\})$.
(Note, that of the last 11 relations, 14, 15, 18, 19, 20, 22 and 24 are symmetric, $(16,17)$ $(21,23)$ are the anti-symmetric pairs.)

The set of edges of the total graph of the triangular graph is the union

$$R_2 \cup R_7 \cup R_{12} \cup R_{19} \cup R_{20} \cup R_{24}.$$

3.2 Structure Constants of $\mathcal{S}(n)$

Proposition 1. *The structure constants of $\mathcal{S}(n)$ are functions of n. For $n \geq 9$, each such function is a polynomial function in n of degree at most 3.*

Proof. To calculate a structure constant $p_{i,j}^k$, we take the general representative pair, (X,Y), of R_k, and try to find a general vertex Z such that $(X,Z) \in R_i$ and $(Z,Y) \in R_j$. The selection of (X,Y) partitions the points $[1,n]$ into at most 6 parts of constant size (not dependent on n, but dependent on X,Y), and one part of size $n-k$ where k is again dependent on X,Y but not on n. So, the number of ways of selecting Z is a product of two or three of the sizes of parts, or sizes of parts minus one, or sizes of parts minus 2 (or maybe half this product, in case a set of two points needs to be selected), and therefore is a polynomial of degree at most 3 in n. □

Examples:

If we want to calculate $p_{10,5}^{15}$:

$(\{\{a,b\},\{a,c\}\},\{\{a,d\},\{a,e\}\})$ partitions the set $[1,n]$ to parts $\{a\}$, $\{b,c\}$, $\{d,e\}$, and the rest, of size $n-5$. A vertex $\{x,y\}$ such that $(\{\{a,b\},\{a,c\}\}, \{x,y\})$ is in R_{10} and $(\{x,y\},\{\{a,d\},\{a,e\}\})$ is in R_5 must satisfy $x \in \{a\}$, $y \notin \{a,b,c\}$, $x \in \{a\}$, $y \notin \{a,d,e\}$. We have one way of selecting x and $n-5$ ways of selecting y, so $p_{10,5}^{15} = n-5$.

If we want to calculate $p_{14,14}^{14}$:

$(\{\{a,b\},\{a,c\}\},\{\{d,e\},\{d,f\}\})$ partitions $[1,n]$ into $\{a\}$, $\{b,c\}$, $\{d\}$, $\{e,f\}$, and we need to find $\{\{x,y\},\{x,z\}\}$ such that $x,y,z \notin \{a,b,c,d,e,f\}$, so we have $n-6$ ways to select x and $\frac{(n-7)(n-8)}{2}$ ways to select y and z, so $p_{14,14}^{14} = \frac{(n-6)(n-7)(n-8)}{2}$.

This last example shows why we need to assume $n \geq 9$ for the general argument. This is the case that requires the maximal number of points from the original graph K_n.

Now we can use a computer to calculate the actual polynomials: Using COCO, we find numerical values for all structure constants in the cases $n = 9, 10, 11, 12$. Then for each triplet i, j, k we use Lagrange interpolation in GAP to find the polynomial $p_{i,j}^k(n)$.

3.3 $\mathcal{S}(n)$ and $\mathcal{T}(n)$

Proposition 2. $\mathcal{S}(n) = \mathcal{T}(n)$.

Proof. First we recall that the total graph $\mathbb{T}(n)$ is the union of the relations $R_2 \cup R_7 \cup R_{12} \cup R_{19} \cup R_{20} \cup R_{24}$, so $\mathcal{T}(n)$ is a fusion configuration of $\mathcal{S}(n)$. It is enough to show that no proper fusion configuration of $\mathcal{S}(n)$ contains $\mathbb{T}(n)$.

For $4 < n < 9$ we check by a computer implementation of the Weisfeiler-Leman closure algorithm that the closure of the total graph is indeed the configuration $\mathcal{S}(n)$ (without the empty relation R_{14} in the case $n = 5$). By using COCO, we find that for $n \geq 6$ there are two fusion association schemes of ranks 3 and 4, except for $n = 7$ where there is another scheme of rank 3.

For $n \geq 9$, we confirm by computer search that those two mergings always appear, no other mergings appear for all n, and that there are no sporadic mergings other then those described in 3.4.

This proves that $\mathcal{S}(n) = \mathcal{T}(n)$, since $\mathcal{T}(n)$ is a merging of $\mathcal{S}(n)$, but the mergings we found do not admit $\mathbb{T}(n)$ as a union of relations. □

3.4 Mergings of $\mathcal{S}(n)$

We are looking for mergings of $\mathcal{S}(n)$ (for $n \geq 6$) resulting in association schemes. The result of the search as described in Sect. 5 is that in the general case (that is, for $n \geq 6$) there are only two mergings:

A strongly regular graph, Γ, with parameters $(\frac{n(n-1)^2}{2}, n-2, n-3, 0)$. This SRG is the union of the relations $R_8 \cup R_{13} \cup R_{22}$. This graph can be defined on the vertices of $\mathbb{T}(n)$, denoted by $\{a,b\}$ and $(\{a,b\},c)$ (the latter standing for $\{\{a,c\},\{b,c\}\}$), as follows: two vertices are adjacent if they share the same two-set. Since $\mu=0$, this graph is isomorphic to $\frac{n(n-1)}{2}$ copies of K_{n-1}.

A rank 4 association scheme, whose classes are the unions of the relations:
$R_0 \cup R_1$;
$R_8 \cup R_{13} \cup R_{22}$;
$R_2 \cup R_6 \cup R_7 \cup R_{11} \cup R_{12} \cup R_{18} \cup R_{19} \cup R_{21} \cup R_{23} \cup R_{24}$;
$R_3 \cup R_4 \cup R_5 \cup R_9 \cup R_{10} \cup R_{14} \cup R_{15} \cup R_{16} \cup R_{17} \cup R_{20}$.

When using the above notation ($\{a,b\}$ and $(\{a,b\},c)$) for the set of vertices of $\mathbb{T}(n)$, the relations of this merging association scheme can be defined by the number of points their two-sets share.

With this observation we recognize that the rank 4 scheme is the wreath product $\mathfrak{M} \wr K_{n-1}$, where $\mathfrak{M}$ is the rank 3 association scheme with basis graphs Δ, $T(n)$, and $\overline{T(n)}$.

The only exception is for $n=7$ which has another SRG as a merging of the relations: $R_3 \cup R_4 \cup R_8 \cup R_9 \cup R_{13} \cup R_{15} \cup R_{18} \cup R_{20}$. This SRG has parameters $(126,45,12,18)$, and will be discussed in subsequent publication [13].

For $n=5$, relation R_{14} is actually empty, since it requires 6 different points of K_n. So $\mathcal{S}(5)$ is a rank 24 coherent configuration. This configuration has 9 merging association schemes listed in Table 2. (All mergings also merge the reflexive relations R_0 and R_1.) Some of the mergings are discussed in [14, 13].

4 Total Configuration of $L_2(n)$

4.1 Definition and Basic Properties

The lattice square graph, $L_2(n)$, is a graph with n^2 vertices, usually the vertex set is denoted by $[1,n]^2$, with (a,b) adjacent to (c,d) if $a=c$ or $b=d$. It is useful to regard the vertices as the points of the $n \times n$-grid. Two vertices are adjacent if and only if they are in the same row or the same column. For our purposes it is also useful to see this graph as the line graph of the complete bipartite graph $K_{n,n}$.

$L_2(n)$ is a regular graph of valency $2(n-1)$, so it has $n^2(n-1)$ edges. The total graph $T(L_2(n))$ has $n^2+n^2(n-1)=n^3$ vertices.

Let us denote the vertices of $L_2(n)$ by (a,x), where $a,x \in [1,n]$. The automorphism group $G=Aut(L_2(n))$ of order $2(n!)^2$ is generated by S_n acting on first coordinate, S_n acting on second coordinate, and involution mapping (a,x) to (x,a), denoted by t. In other words, G is the exponentiation $S_n \uparrow S_2$, as in [8].

In this notation, edges of $L_2(n)$ are of the form $\{(a,x),(a,y)\}$ (a pair of vertices in the same row) or $\{(x,a),(y,a)\}$ (a pair of vertices in the same column), here $x \neq y$.

Table 2. Mergings of $\mathcal{S}(5)$

Rank	Mergings	$\|aut\|$	SRG parameters
5	$(3,4,5,9,10,15,16,17,20)$ $(8,13,22)$ $(2,7,12,18,19,24)$ $(6,11,21,23)$	1920	
5	$(3,4,9,15,20)$ $(8,13,22)$ $(5,10,16,17)$ $(2,6,7,11,12,18,19,21,23,24)$	7680	
5	$(3,7,8,12,13,15,16,17,18)$ $(4,9,24)$ $(2,5,10,19,20,22)$ $(6,11,21,23)$	1920	
5	$(3,8,13,15,18)$ $(4,9,24)$ $(2,5,6,10,11,19,20,21,22,23)$ $(7,12,16,17)$	7680	
4	$(3,4,5,9,10,15,16,17,20)$ $(8,13,22)$ $(2,6,7,11,12,18,19,21,23,24)$	$2^{33}3^{11}5$	
4	$(3,7,8,12,13,15,16,17,18)$ $(4,9,24)$ $(2,5,6,10,11,19,20,21,22,23)$	$2^{33}3^{11}5$	
3	$(2,3,4,5,6,7,9,10,11,12,15,16,17,18,19,20,21,23,24)$ $(8,13,22)$	$2^{38}3^{14}5^27$	$(40,3,2,0)$
3	$(2,3,4,5,7,8,9,10,12,13,15,16,17,18,19,20,22,24)$ $(6,11,21,23)$	51,840	$(40,12,2,4)$
3	$(2,3,5,6,7,8,10,11,12,13,15,16,17,18,19,20,21,22,23)$ $(4,9,24)$	$2^{38}3^{14}5^27$	$(40,3,2,0)$

We shall denote the total coherent configuration of $L_2(n)$ by $\mathfrak{T}(n)$ and the Schurian total configuration of $L_2(n)$ by $\mathfrak{S}(n)$.

In the following listing of representatives of relations of $\mathfrak{S}(n)$, a, b, c, d stand for distinct elements of $[1, n]$, and x, y, z, w stand for distinct elements of $[1, n]$. The sets $\{a, b, c, d\}$ and $\{x, y, z, w\}$ are not necessarily disjoint.

In relations $R_4, \dots, R_{11}$ the representative of edges appear all as a pair of vertices in the same row. Since the involution t is in the automorphism group, and maps a pair of vertices in the same row to a pair of vertices in the same column, those edges are also represented. For example, $((a, x), \{(a, x), (b, x)\})$ is in R_4, since it is the result of action of t on $((x, a), \{(x, a), (x, b)\})$ which is clearly in R_4.

In the same manner, when looking at relations $R_{12}, \dots, R_{20}$ (pairs of edges of $L_2(n)$), it does not matter if the first edge is a pair of vertices in the same row or a pair of vertices in the same column, but it does matter whether both edges are of the same kind (row or column) or of different kinds. Relations $R_{12}, \dots, R_{16}$ are of the former type, while relations $R_{17}, \dots, R_{20}$ are of the latter type.

$\mathfrak{S}(n)$ has the following 21 relations (for $n \geq 3$), see also Table 3.

Reflexive relations:

$R_0 = ((a, x), (a, x))$, $R_1 = (\{(a, x), (a, y)\}, \{(a, x), (a, y)\})$.

Relations within first fibre:

$R_2 = ((a, x), (a, y))$, $R_3 = ((a, x), (b, y))$.

Relations between first and second fibre:

$R_4 = ((a, x), \{(a, x), (a, y)\})$, $R_5 = ((a, x), \{(b, x), (b, y)\})$,
$R_6 = ((a, x), \{(a, y), (a, z)\})$, $R_7 = ((a, x), \{(b, y), (b, z)\})$.

Their inverses:

$R_8 = (\{(a, x), (a, y)\}, (a, x))$, $R_9 = (\{(b, x), (b, y)\}, (a, x))$,
$R_{10} = (\{(a, y), (a, z)\}, (a, x))$, $R_{11} = (\{(b, y), (b, z)\}, (a, x))$.

And relations within second fibre:

$R_{12} = (\{(a, x), (a, y)\}, \{(a, x), (a, z)\})$, $R_{13} = (\{(a, x), (a, y)\}, \{(a, z), (a, w)\})$,
$R_{14} = (\{(a, x), (a, y)\}, \{(b, x), (b, y)\})$, $R_{15} = (\{(a, x), (a, y)\}, \{(b, x), (b, z)\})$,
$R_{16} = (\{(a, x), (a, y)\}, \{(b, z), (b, w)\})$, $R_{17} = (\{(a, x), (a, y)\}, \{(a, x), (b, x)\})$,
$R_{18} = (\{(a, x), (a, y)\}, \{(a, z), (b, z)\})$, $R_{19} = (\{(a, x), (a, y)\}, \{(b, x), (c, x)\})$,
$R_{20} = (\{(a, x), (a, y)\}, \{(b, z), (c, z)\})$.

R_{18} and R_{19} form an anti-symmetric pair. All other relations within second fibre are symmetric.

The total graph $T(L_2(n))$ is the union of the relations: $R_2 \cup R_4 \cup R_8 \cup R_{12} \cup R_{17}$.

4.2 Structure Constants of $\mathfrak{S}(n)$

As in the case of the triangular graph in the previous section, when we calculate p_{ij}^k for a given triplet (i, j, k), we actually take an element (M, N) of relation R_k, and count the amount of elements P such that $(M, P) \in R_i$ and $(P, N) \in R_j$. Here P is either a vertex or an edge of $L_2(n)$, so we need to

Table 3. Sizes of basis relations of $\mathfrak{S}(n)$

Relation	Size	Relation	Size
0	n^2	1	$n^2(n-1)$
2	$2n^2(n-1)$	3	$n^2(n-1)^2$
4	$2n^2(n-1)$	5	$2n^2(n-1)^2$
6	$n^2(n-1)(n-2)$	7	$n^2(n-1)^2(n-2)$
12	$2n^2(n-1)(n-2)$	13	$\frac{1}{2}n^2(n-1)(n-2)(n-3)$
14	$n^2(n-1)^2$	15	$2n^2(n-1)^2(n-2)$
16	$\frac{1}{2}n^2(n-1)^2(n-2)(n-3)$	17	$2n^2(n-1)^2$
18	$n^2(n-1)^2(n-2)$	19	$n^2(n-1)^2(n-2)$
20	$\frac{1}{2}n^2(n-1)^2(n-2)^2$		

select two or three elements of $[1, n]$. For each element of $[1, n]$ that we need to select, it either needs to be one already used in M or N, in which case the number of options is a constant independent of n, or not used, in which case the number of options is $n - r$, where r is dependent on i, j, k, but not on n. After all selections, we might need to multiply by 2 (if the representative element of R_i is not invariant under the involution t), and similarly for R_j an R_k. We also need to divide by 2, if P is and edge of $L_2(n)$, since we selected two elements the order of which is irrelevant. Finally, we conclude that p_{ij}^k is a polynomial function in n of degree at most 3.

For finding the minimal n for which this argument will work, we note that the worst case is in the calculation of $p_{16,16}^{16}$, where we have a pair $(\{(a, x), (a, y)\}, \{(b, z), (b, w)\})$, and need to find an edge of $L_2(n)$ $\{(c, u), (c, v)\}$, such that u is different from x, y, w, z, and so is v. In conclusion:

Proposition 3. *The structure constants of* $\mathfrak{S}(n)$ *are functions of* n. *For* $n \geq 6$, *each such function is a polynomial function in* n *of degree at most* 3.

4.3 Mergings of $\mathfrak{S}(n)$

$\mathfrak{S}(n)$ admits no association schemes as mergings (for $n \geq 3$).

4.4 $\mathfrak{S}(n)$ and $\mathfrak{T}(n)$

Proposition 4. $\mathfrak{S}(n) = \mathfrak{T}(n)$

Proof. First we recall that the total graph $T(L_2(n))$ is the union of the relations $R_2 \cup R_4 \cup R_8 \cup R_{12} \cup R_{17}$. So, $\mathfrak{T}(n)$ is a fusion configuration of $\mathfrak{S}(n)$. It is enough to show that no proper fusion configuration of $\mathfrak{S}(n)$ contains $T(L_2(n))$. This is done by computer search. □

5 Details of Computer Search

The computer search mentioned in Sects. 3.3 and 4.4 is based on the notion of good sets, which goes back to [8]:

If $W = \langle A_0, \dots, A_r \rangle$ is a coherent algebra, then we define a good set to be a subset $B \subseteq [0, r]$ such that:

GS1 $M = \sum_{i \in B} A_i$ is an adjacency matrix of a symmetric or an anti-symmetric relation;
GS2 if $M^2 = \sum_{i=0}^{r} b_i A_i$ then for every $i, j \in B$, $b_i = b_j$;
GS3 $I \circ M = M$ or $I \circ M = 0$ ($\circ$ is Schur-Hadamard product).

With this definition of a good set, we see that for a partition $P = \{P_1, \dots, P_k\}$ of $[0, r]$ to induce a coherent subalgebra, each P_i must be a good set. This reduces the computational search for subalgebras from a search through all partitions of $[0, r]$, to a search through partitions consisting of good sets only.

This method, originally developed in [8] for use with a numerical tensor of structure constants, also works for a polynomial tensor. A set that is good by its polynomial parameters, that is good for all $n \geq 9$, is called *polynomially good set.*

I The graphs $\mathbb{T}(n)$ are constructed for $n = 9, 10, 11, 12$, and for each such n, the WL-closure, $\mathcal{T}(n)$ is calculated. We then check that it coincides with $\mathcal{S}(n)$, which is calculated by COCO.
II The structure constants of the four coherent configurations are used to generate the polynomial tensor of structure constants (using Lagrange interpolation).
III Instead of searching for all mergings, we limit our search to specific kinds of mergings:
   i Mergings resulting in association schemes: Since the basis graphs of an association scheme are regular, we add another requirement for a good set: the graph with edge set $\bigcup_{i \in B} R_i$ must be regular. The number of good sets with this additional condition is 5, and a quick search shows that only two mergings (those described in 3.4) appear.
   ii Mergings that admit $\mathcal{T}(n)$ as a merging. This means that $Q = \{2, 7, 12, 19, 20, 24\}$ is a union of sets from the partition, or in other words an additional condition for a good set B is that either $B \subseteq Q$ or $B \cap Q = \emptyset$. Since none of the previous two mergings fulfill this condition, we know that such mergings do not result in association schemes.
   Since we only have two fibres, the mergings we are looking for also have two fibres. This allows us to partition the relations into cells according to the fibres:

$$\{\{0\}, \{1\}, \{2, 3\}, \{4, 5, 6, 7, 8\}, \{9, 10, 11, 12, 13\}, \{14, \dots, 24\}\}$$

   and require a good set to comply with partition, that is, to be a subset of one of the sets in the partition.

Those two conditions leave 48 good sets and a simple search shows that none of the partitions result in a coherent configuration.

IV The previous step is enough to show that there is no merging (except for the two described in 3.4) that appear for every $n \geq 9$. To confirm that no other mergings appear for particular $n > 7$ we use the following principle: When we check if the set $B = \{a_1, \ldots, a_l\}$ is good, we actually calculate the sums $Q_k = \sum_{i,j \in B} p_{i,j}^k$ for each $k \in B$. Clearly, if all these sums are equal, then the set is good. If it is not good, then we have (at least) two elements $i, j \in B$ such that the polynomials Q_i and Q_j differ. If a natural number n_0 is a root of the polynomial $Q_i - Q_j$, it means that while the set B is not a good set for all n, it might be a good set for n_0. In this case we add this n_0 to the list of n for which an exceptional merging might appear.
In the case of $\mathcal{S}(n)$, the list of possible exceptions included all the integers in the range $[1, 22]$, and a computerized brute force search shows that for no n in the range $7 < n \leq 22$ does an exceptional merging appear.

A similar search is performed for the total configurations of the lattice square graphs. In this case there were no mergings resulting in association schemes which appear for all n, and the list of possible exceptions is $\{3, 4, 5, 6, 7, 8, 9, 10, 11, 17\}$. A computerized search shows that there is no exceptional merging resulting in an association scheme for any of those values.

6 Conclusions

It is easy to check that the total graph of the complete graph K_n with n vertices is isomorphic to the triangular graph $T(n+1)$. This observation by [3], see also [10] was one of the earliest stimuli of interest in this concept.

We again refer to [14] for a detailed discussion of exceptional Schurian association scheme with the automorphism group of order 1920 which appears as a merging of classes in $T(5)$.

The triangular graph $T(5)$ is a complement of the Petersen graph. Petersen graph may be considered as the smallest Moore graph (see, e.g. [5]). This is why we are also investigating the total graph configurations of the complements of the unique Moore graph of valency 7 and of a potential Moore graph of valency 57, see [13].

It seems as a very attractive task to search for other examples of strongly regular graphs, which have total coherent configuration admitting exceptional mergings. Our results for the graphs $L_2(n)$ provide a small evidence for believing that such examples are quite rare.

We hope that the use of Gröbner bases (cf. [15]) may help in computerized investigation of other infinite series of classical strongly regular graphs.

The problem of the description of the 2-orbits of the automorphism group of an arbitrary graph Γ is closely related to the graph isomorphism problem, see e.g. [12]. In case when for a graph Γ its total graph coherent configuration and the Schurian total graph coherent configuration coincide, using WL-closure, we get as a by-product a polynomial-time procedure for the description of $2\text{-}orb(Aut(\Gamma))$. At this moment we are not aware of an example where those two coherent configurations are distinct for strongly regular graphs. Certain counterexamples for arbitrary graphs will be discussed in [13], in particular, in view of the results from [6] and related publications.

Acknowledgments

This project is a part of the graduate thesis which the author is writing under the supervision of M. Klin.

The author is pleased to acknowledge Prof. Bruno Buchberger and the coordinators of the Special Semester on Gröbner Bases (February 1 – July 31, 2006), organized by RICAM, Austrian Academy of Sciences, and RISC, Johannes Kepler University, Linz Austria.

The conceptualization of this text goes back to the time of workshop D1 at Linz, May 2006, which was in the scope of the Special Semester.

References

1. L. Babel, I. V. Chuvaeva, M. Klin, and D. V. Pasechnik, *Algebraic Combinatorics in Mathematical Chemistry. Methods and Algorithms. II. Program Implementation of the Weisfeiler-Leman Algorithm*, Preprint, Report TUM-M9701, 1997, Fakultät für Mathematik, TU Münc. http://www-lit.ma.tum.de/veroeff/html/960.68019.html.
2. L. Babel, S. Baumann, and M. Luedecke, *STABCOL: An Efficient Implementation of the Weisfeiler-Leman Algorithm*, Technical report, Technical University Munich, TUM-M9611, 1996.
3. M. Behzad, G. Chartrand, and E. A. Nordhaus, Triangles in line-graphs and total graphs, *Indian J. Math.*, **10** (1968), 109–120.
4. http://www.win.tue.nl/~aeb/.
5. P. J. Cameron and J. H. van Lint, *Designs, Graphs, Codes and Their Links*, London Math. Soc. Student Texts, Vol. 22, Cambridge University Press, Cambridge, 1991.
6. S. Evdokimov and I. Ponomarenko, On highly closed cellular algebras and highly closed isomorphisms, *Electr. J. Comb.*, **6** (1999), #R18.
7. I. A. Faradžev and M. H. Klin, Computer package for computations with coherent configurations, in *Proc. ISSAC-91*, pp. 219–223, ACM, Bonn, 1991.
8. I. A. Faradžev, M. H. Klin, and M. E. Muzichuk, Cellular rings and groups of automorphisms of graphs, in I. A. Faradžev et al. (eds.) *Investigations in Algebraic Theory of Combinatorial Objects*, pp. 1–152, Kluwer Academic, Dordrecht, 1994.

9. http://www.gap-system.org.
10. F. Harary, *Graph Theory*, Addison-Wesley, Reading, 1969.
11. D. G. Higman, Coherent configurations, *I. Rend. Sem. Mat. Univ. Padova*, **44** (1970), 1–25.
12. M. Klin, C. Rücker, G. Rücker, and G. Tinhofer, Algebraic combinatorics in mathematical chemistry. Methods and algorithms. I. Permutation groups and coherent (cellular) algebras, *MATCH*, **40** (1999), 7–138.
13. M. Klin, M. Ziv-Av, and L. Jorgensen, *Small rank 5 Higmanian association schemes and total graph coherent configurations*, in preparation.
14. M. Klin and M. Ziv-Av, *A family of Higmanian association schemes on 40 points: A computer algebra approach*, in Proceedings of an International Conference in Honor of Eiichi Bannai's 60th Birthday, June 26–30, 2006, Algebraic Combinatorics, Sendai International Center, Sendai, Japan, pp. 190–203.
15. D. A. Leonard, Using Gröbner bases to investigate flag algebras and association scheme fusion, in M. Klin, G. A. Jones, A. Jurišić, M. Muzychuk, and I. Ponomarenko (eds.) *Algorithmic Algebraic Combinatorics and Gröbner Bases*, pp. 113–135, Springer, Berlin, 2009 (this volume).
16. M. Schönert, et al., *GAP – Groups, Algorithms, and Programming*, Lehrstuhl D für Mathematik, fifth edition, Rheinisch-Westfälische Technische Hochschule, Aachen, Germany, 1995.
17. B. Weisfeiler, *On Construction and Identification of Graphs*, Lecture Notes in Math., Vol. 558, Springer, Berlin, 1976.
18. H. W. Wielandt, *Permutation Groups Through Invariant Relations and Invariant Functions*, Lecture Notes, Ohio State University Press, Columbus, 1969.

MIX
Papier aus verantwortungsvollen Quellen
Paper from responsible sources
FSC® C105338

If you have any concerns about our products,
you can contact us on
ProductSafety@springernature.com

In case Publisher is established outside the EU,
the EU authorized representative is:
Springer Nature Customer Service Center GmbH
Europaplatz 3, 69115 Heidelberg, Germany

Printed by Libri Plureos GmbH
in Hamburg, Germany